Handbook of Pollution Prevention Practices

Environmental Science and Pollution Control Series

1. Toxic Metal Chemistry in Marine Environments, *Muhammad Sadiq*
2. Handbook of Polymer Degradation, *edited by S. Halim Hamid, Mohamed B. Amin, and Ali G. Maadhah*
3. Unit Processes in Drinking Water Treatment, *Willy J. Masschelein*
4. Groundwater Contamination and Analysis at Hazardous Waste Sites, *edited by Suzanne Lesage and Richard E. Jackson*
5. Plastics Waste Management: Disposal, Recycling, and Reuse, *edited by Nabil Mustafa*
6. Hazardous Waste Site Soil Remediation: Theory and Application of Innovative Technologies, *edited by David J. Wilson and Ann N. Clarke*
7. Process Engineering for Pollution Control and Waste Minimization, *edited by Donald L. Wise and Debra J. Trantolo*
8. Remediation of Hazardous Waste Contaminated Soils, *edited by Donald L. Wise and Debra J. Trantolo*
9. Water Contamination and Health: Integration of Exposure Assessment, Toxicology, and Risk Assessment, *edited by Rhoda G. M. Wang*
10. Pollution Control in Fertilizer Production, *edited by Charles A. Hodge and Neculai N. Popovici*
11. Groundwater Contamination and Control, *edited by Uri Zoller*
12. Toxic Properties of Pesticides, *Nicholas P. Cheremisinoff and John A. King*
13. Combustion and Incineration Processes: Applications in Environmental Engineering, Second Edition, Revised and Expanded, *Walter R. Niessen*
14. Hazardous Chemicals in the Polymer Industry, *Nicholas P. Cheremisinoff*
15. Handbook of Highly Toxic Materials Handling and Management, *edited by Stanley S. Grossel and Daniel A. Crowl*
16. Separation Processes in Waste Minimization, *Robert B. Long*
17. Handbook of Pollution and Hazardous Materials Compliance: A Sourcebook for Environmental Managers, *Nicholas P. Cheremisinoff and Madelyn Graffia*

18. Biosolids Treatment and Management: Processes for Beneficial Use, *edited by Mark J. Girovich*
19. Biological Wastewater Treatment: Second Edition, Revised and Expanded, *C. P. Leslie Grady, Jr., Glen T. Daigger, and Henry C. Lim*
20. Separation Methods for Waste and Environmental Applications, *Jack S. Watson*
21. Handbook of Polymer Degradation, Second Edition, Revised and Expanded, *S. Halim Hamid*
22. Bioremediation of Contaminated Soils, *edited by Donald L. Wise, Debra J. Trantolo, Edward J. Cichon, Hilary I. Inyang, and Ulrich Stottmeister*
23. Remediation Engineering of Contaminated Soils, *edited by Donald L. Wise, Debra J. Trantolo, Edward J. Cichon, Hilary I. Inyang, and Ulrich Stottmeister*
24. Handbook of Pollution Prevention Practices, *Nicholas P. Cheremisinoff*

Additional Volumes in Preparation

Handbook of Pollution Prevention Practices

Nicholas P. Cheremisinoff

N&P Consultants, Limited
Harpers Ferry, West Virginia

Marcel Dekker, Inc. New York • Basel

ISBN: 0-8247-0542-4

This book is printed on acid-free paper.

Headquarters
Marcel Dekker, Inc.
270 Madison Avenue, New York, NY 10016
tel: 212-696-9000; fax: 212-685-4540

Eastern Hemisphere Distribution
Marcel Dekker AG
Hutgasse 4, Postfach 812, CH-4001 Basel, Switzerland
tel: 41-61-261-8482; fax: 41-61-261-8896

World Wide Web
http://www.dekker.com

The publisher offers discounts on this book when ordered in bulk quantities. For more information, write to Special Sales/Professional Marketing at the headquarters address above.

Current printing (last digit):
10 9 8 7 6 5 4 3 2 1

PRINTED IN THE UNITED STATES OF AMERICA

PREFACE

Pollution Prevention, rather than concentrating on the treatment and disposal of wastes, focuses on the elimination or reduction of undesired byproducts within the production process itself. In the long run, pollution prevention through waste minimization and cleaner production is more cost-effective and environmentally sound than traditional pollution control methods. Pollution prevention techniques apply to any manufacturing process and range from relatively easy operational changes and good housekeeping practices to more extensive changes such as making substitutions for toxic substances, the implementation of clean technology, and the installation of state-of-the-art recovery equipment. Pollution prevention can improve plant efficiency, enhance the quality and quantity of natural resources for production, and make it possible to invest more financial resources in economic development.

In the United States, pollution prevention (P2) practices have resulted in many millions of dollars in operational cost savings. These savings have largely been achieved through incremental savings associated with both direct and indirect benefits. Some of the benefits that translate into direct savings include reduced O&M costs associated with elimination of end-of-pipe pollution controls, cost savings from elimination of transport and off-site disposal of wastes, elimination of large capital investments into pollution control technologies, reduced permit fees for pollution controls and discharges, and raw materials and energy savings. Some of the indirect savings associated with P2 programs include improved productivity of workforces and operations, improvement in product quality, reduction in health risks associated with manufacturing operations and exposure to toxic and hazardous materials, reduced manpower requirements for recordkeeping and permitting, reduced laboratory, analytical, and monitoring services for emissions, and increased consumer confidence in the company, its products, and its services. But perhaps among the greatest savings category associated with P2 programs is that related to future liabilities. By eliminating or reducing pollution at the source, a company avoids potential liabilities such as off-site disposal to future Superfund sites where liability is always tied to the waste generator.

This handbook has been compiled and written as a concise reference source of ideas and approaches to P2 practices. The reader will find useful tables and matrixes that provide industry-specific suggestions, ideas, and proven technologies and practices. Step-wise procedures are provided for determining the economic viability of P2 projects and for implementing in-plant assessments or P2 audits that can identify cost savings as well as pollution reduction measures. Many industry examples and case studies are based on first-hand experiences and audits I have conducted in various international consulting assignments, as well as on a review of the technical, government, and trade literature.

The volume is organized into nine chapters. Chapter 1 provides an overview of the principles of waste minimization and pollution prevention and introduces terminology used throughout the book. Chapter 2 provides an overview of the properties and environmental fate of priority pollutants (i.e., those pollutants recognized as contributors to global pollution problems), as well as descriptions of standard pollution control equipment. Although the goal of pollution prevention is ideally to eliminate pollution, and hence costly control, treatment, and disposal methods, the reality is that end-of-pipe treatment technologies still comprise the majority of techniques needed to manage industry discharges and emissions. In addition, many P2 projects will make use of or incorporate these equipment choices and technologies into better solutions for pollution management. Thus, it is important to understand the equipment options that are available to us.

Chapter 3 deals with project financing. Sometimes investment projects in P2 are among the first to be postponed in times of budget shortfalls in companies. This has been due in large part to the inadequate support and defensive posture taken by corporations toward environmental projects on an economic basis. Typically, when a production division requests money, all the necessary documentation, facts, and figures are ready for presentation. The production project is justified by showing how the project will increase revenues and how the added revenue will not only recover costs, but substantially increase earnings for the company or company's operating division as well. A P2 project justification requires this same emphasis. To be competitive and to get "management buy-in," an understanding of the financial system or project financing and "bankability" of a project is essential. Financial tools demonstrate the importance of the P2 investment on a life cycle or total cost basis, in terms of revenues, expenses, and profits. In this chapter, the principles and practices of cost accounting are discussed and applied to industry examples.

Chapter 4 covers the audit, or the *pollution prevention in-plant assessment*. This chapter provides a stepwise approach to conducting a P2 audit aimed at identifying waste and pollution reduction, and value-added cost savings that can be captured through energy efficiency, raw materials savings, productivity gains, product quality improvements, and other incentives. The chapter emphasizes a simple spreadsheet approach to performing material balances that are needed to assess benefits derived from a P2 opportunity, as identified through a team approach.

Chapters 5 through 8 focus attention on specific industry practices. The reader will find concise industry profiles that describe the manufacturing technologies, the sources of pollution from within unit processes, and the environmental fates associated with the major pollutants from each process. Using these industry profiles, discussions focus on current and recommended P2 practices for each industry covered. In many cases, industry-specific P2 action lists are provided.

Sectors covered include the chemical process industry, the petroleum refining industry, the iron and steel and allied metals refining industries, and a variety of other industry sectors, including pulp and paper, tanneries, food processing, and electronics. Chapter 8 provides in-depth case studies that will give the reader ideas on practical approaches to industry-specific problems.

Chapter 9 is a summary - pulling together many of the important concepts laid out in the volume and providing final guidance on developing and implementing P2 practices within your own company. General rules and practical tips are provided in this final chapter.

Specific references are cited for further readings throughout the book. In addition, an extensive list of abbreviations has been provided at the beginning of the volume. The reader will need to refer to this section often during the reading of chapters. Many of the industry profiles and general P2 practices as applied to specific industries were obtained by reviewing publications of the World Bank Organization, the USEPA, the World Health Organization, interviews conducted at industrial sites, and some of my own consulting assignments. It is hoped that many readers will find this volume useful, and I welcome direct comments or suggestions for improving the next edition.

A special thanks is extended to Marcel Dekker, Inc., for the fine production of this volume.

Nicholas P. Cheremisinoff, Ph.D.

CONTENTS

ABBREVIATIONS

A

AATCC	American Association of Textile Chemists and Colorists
ABS	acrylonitrile butadiene styrene
acfm	actual cubic feet per minute
ACM	asbestos-containing materials
ACRS	accelerated cost recovery system
ACS	American Chemical Society
ADP	air-dried pulp
AHE	acute hazards event
AIRS	aerometric information retrieval system
AMD	acid mine drainage
AN	ammonium nitrate
API	American Petroleum Institute
ASN	ammonium sulfate nitrate
B	
BAT	best available technology
BATNEEC	best available technology not entailing excessive cost
BB	butane/butylene
B/C	benefit-to-cost ratio
BIFs	boilers and industrial furnaces
BOD	biochemical oxygen demand
BOF	basic oxygen furnace
BPT	best practicable technology
BS	black smoke or British smokeshade method
BS	British Standard
C	
CAA	Clean Air Act
CAAA	Clean Air Act Amendments
CAN	calcium ammonium nitrate
CCD	continuous countercurrent decanting
CERCLA	Comprehensive Environmental Response, Compensation, and Liability Act
CFCs	chlorofluorocarbons
cfm	cubic feet per minute
CFR	Code of Federal Regulations
CI	color index
COD	chemical oxygen demand
COG	coke oven gas
CP	cleaner production

CPI	chemical process industry
cpm	cycles per minute
CSM	continuous stack monitoring
CTC	carbon tetrachloride
CTMP	chemithermomechanical pulping
CTSA	Cleaner Technology Substitute Assessment
CWA	Clean Water Act
D	
DAF	dissolved air flotation
DAP	diammonium phosphate
DDT	dichlorodiphenyltrichloroethane
DEA	diethanolamine
DIPA	di-isopropanolamine
DMT	dimethyl terephthalate
DO	dissolved oxygen
E	
EA	environmental assessment
EAF	electric arc furnace
ECF	elemental chlorine-free (bleaching)
EIA	environmental impact assessment
EIS	environmental impact statement
EMAS	European Eco-Management and Audit Scheme
EMS	environmental management system
emf	electromotive force
EMS	environmental management systems
EPA	Environmental Protection Agency
EPCRA	Emergency Planning and Community Right-to-Know Act
EPT	Environmental Policy and Technology
ERNS	Emergency Response Notification System
ESP	electrostatic precipitator
EU	European Union
F	
FBC	fluidized bed combustion
FCC	fluidized catalytic cracking
FCCUs	fluidized-bed catalytic cracking units
FGD	flue gas desulfurization
FGR	flue gas recirculation
FGT	flue gas treatment
G	
GHG	greenhouse gas

GJ	gigajoule
GLPTS	Great Lakes Persistent Toxic Substances
GMP	Good Management Practices
gpm	gallons per minute
gr	grain
GW	gigawatt
GWP	global warming potential
H	
HAPs	hazardous air pollutants
HCFCs	hydrochlorofluorocarbons
HCN	hydrogen cyanide
HCs	hydrocarbons
HDPE	high density polyethylene
HEPA	high efficiency particulate air filter
HFC	hydrofluorocarbon
HSDB	Hazardous Substances Data Bank
HSWA	Hazardous and Solid Waste Amendments
I	
IAF	International Accreditation Forum
IARC	International Agency for Cancer Research
ID	identification
IER	Initial Environmental Review
IFC	International Finance Corporation
IPCC	Intergovernmental Panel on Climate Change
IQ	intelligence quotient
IRIS	integrated risk information system
ISO	International Organization for Standardization
L	
LAB	linear alkyl benzene
LCA	life cycle analysis
LCC	life cycle costing or life cycle checklist
LDAR	leak detection and repair
LDPE	low density polyethylene
LDRs	land disposal restrictions
LEA	low excess air
LEPCs	local emergency planning committees
LLDPE	linear low density polyethylene
LPG	liquefied petroleum gas
M	
MACT	maximum allowable control technology

MAP monoammonium phosphate
MCLGs maximum contaminant level goals
MCLs maximum contaminant levels
MEK methyl ethyl ketone
MIBK methyl isobutyl ketone
MLAs multilateral agreements
MMT methylcyclopentadienyl magnesium tricarbonyl
MOS metal oxide semiconductor
MSDSs Material Safety Data Sheets
MTBE methyltertbutylether
N
NAAQSs National Ambient Air Quality Standards
NCP National Contingency Plan
NESHAPs National Emission Standards for Hazardous Air Pollutants
NGVs natural gas vehicles
NIS Newly Independent States of the former Soviet Union
NPDES National Pollutant Discharge Elimination System
NPK nitrogen-phosphorus-potassium
NPL National Priority List
NRC National Response Center
NSCR nonselective catalytic reduction
NSPSs New Source Performance Standards
O
ODP ozone-depleting potential
ODSs ozone-depleting substances
OFA overfire air
OSHA Occupational Safety and Health Act
OTA Office of Technology Assessment
OTC over-the-counter (medicines)
P
PAH polynuclear aromatic hydrocarbons
P2M pollution prevention matrix
P3 pollution prevention practices
P3AW Pollution Prevention Project Analysis Worksheet
PAHs polynuclear aromatic hydrocarbons
PBR polybutadiene rubber
PCB polychlorinated biphenyls
PCE perchloroethylene
PFA pulverized fly ash
PICs products of incomplete combustion

PM	particulate matter
PM10	particulate matter smaller than 10 microns in size
PMN	premanufacture notice
POM	prescription only medicines
POTW	publicly owned treatment works
PP	propane/polypropylene
ppm	parts per million
ppb	parts per billion
ppmv	parts per million by volume
PVC	polyvinyl chloride
PVNB	present value of net benefits
PWB	printed wiring board
Q	
QA	quality assurance
QC	quality control
R	
R&D	research and development
RCRA	Resource Conservation and Recovery Act
ROI	return on investment
rpm	revolutions per minute
S	
SARA	Superfund Amendments and Reauthorization Act
SBR	styrene butadiene rubber
scfm	standard cubic feet per minute
SCF	supercritical cleaning fluid
SCR	selective catalytic reduction
SCW	supercritical fluid
SDWA	Safe Drinking Water Act
SERCs	State Emergency Response Commissions
SIC	Standard Industrial Code
SIP	state implementation plan
SMT	surface mount technology
SNCR	selective noncatalytic reduction
SS	suspended solids
SSP	single phosphate
T	
TAME	tertiary amyl methyl ether
TCA	total cost accounting or total cost analysis
TCE	1,1,1-trichloroethane
TCF	total chlorine-free

TCLP	toxic characteristic leachate procedure
TDI	toluenediisocyanate
TEWI	total equivalent waming impact
tpy	tons per year
TQEM	Total Quality Environmental Management
TQM	Total Quality Management
TRI	Toxic Release Inventory
TRS	total reduced sulfur
TSCA	Toxic Substances Control Act
TSD	transport, storage, and disposal
TSP	total suspended particulates
TSS	total suspended solids
U	
UIC	underground injection control
UNEP	United Nations Environmental Programme
USEPA	United States Environmental Protection Agency
USTs	underground storage tanks
UV	ultraviolet
V	
VAT	value-added taxes
VCM	vinyl chloride monomer
VOCs	volatile organic compounds
W	
WBO	World Bank Organization
WEC	World Environmental Center
WHO	World Health Organization

Handbook of Pollution Prevention Practices

Chapter 1
Principles of Pollution Prevention and Waste Minimization

INTRODUCTION

The terms *Pollution Prevention*, *Waste Minimization*, *Clean Production, Recycling*, *Waste Utilization*, and a number of other loosely related terminology that the reader will come across in this Handbook, all sound like they have about the same objective - namely, to eliminate or minimize environmental problems associated with a manufacturing operation. For the most part that is true, although there are both subtle and sometimes stark contrasts among the objectives of the practices associated with each of these terms. They do all have similar overall objectives in that their intent is to displace, but in some instances, enhance end-of-pipe treatment technologies, and to eliminate disposal practices. There are a number of reasons why these actions are taken, but the principal one should be economics.

It is well recognized in Western corporations that end-of-pipe treatment of pollution is cost ineffective, and only in extreme cases or those situations where the costs of installing a piece of equipment and ultimate disposal of wastes represent incrementally small costs to manufacturing, that the conventional approach of pollution engineering practices be used. What has created the economic driving force, at least in the United States, has been strict enforcement of environmental legislation. Compliance with environmental statutes has created the need to find more cost effective approaches to managing environmental affairs. Non-compliance means fines, penalties, possible imprisonment of the CEOs and other corporate officers responsible for willful violations, but in addition, loss of business can result from public outcry, loss of confidence in products and companies that do not comply.

In the early stages of implementing pollution prevention programs, many companies in certain industry sectors saw this as an opportunity to improve their image - creating a pro-active policy of environmental management that created an image of being a *Good Corporate Citizen*. But as pollution prevention practices were more commonly implemented and monitored, it also became clear that as a general approach, pollution prevention can have a positive impact on the *bottom line* economic performance of a business operation. There are now many examples where pollution prevention practices have dramatically improved the operating performances of manufacturing operations, have identified waste streams that can be recycled - thereby providing savings in raw materials, have eliminated or reduced the need to utilize third parties in waste disposal – thereby not only eliminating direct costs, but potential future liabilities. These are but a few examples of the cost savings associated with P2 practices, which in some companies have had dramatic impacts on improving their profitability.

Countries in transitioning economies don't quite see these advantages for their industries, and quite often, pollution prevention practices performed on very small scales, are implemented inconsistently, and often pollution prevention projects are given the same low priority as any other environmental project. This situation occurs more in countries that have weak environmental enforcement policies.

Despite the lack of understanding on the importance of cost-effective environmental management practices in some countries, globalization of industry practices is creating the need for adoption of more sound environmental management practices. Pollution prevention is an integral part of environmental management systems (EMS) which are becoming more universally accepted in world business practices. Pollution prevention is an integral component of the ISO 14000 standards that are now so widely recognized and in the process of being adopted by many companies.

Despite than many P2 successes, there are still many U.S. corporations that put too little emphasis on pollution prevention. Part of the problem lies with a lack of training with many older professionals, and all too often, senior management touting strong environmental policy statements but with poor environmental management systems in place to enforce their policies. As industry is forced to move more toward a systems-based approach to managing their operations, including their environmental issues (which are integral to many types of business), a greater standardization in management practices will take place.

One way to view pollution prevention is as an incentive to developing and investing in an environmental management systems (EMS). Adoption/certification of an EMS (like ISO 14001) can be a very expensive investment for a company, and indeed the long-term economic benefits for this investment are sometimes hard to justify or recognize. Therefore, pollution prevention, when implemented and monitored, and rolled-out in stages, can help demonstrate the need for an EMS, and even pay for a good part of it. Some may disagree with this interpretation and believe that pollution prevention practices are simply a part of an EMS, and may not necessarily the true economic driving force for implementing formal EMSs. Whatever the proper viewpoint, few can argue against the financial incentives for implementing dedicated pollution prevention programs.

This chapter is offered for orientation purposes. A range of ideas, concepts and terminology are introduced, as well as the overall philosophy of the book. The Abbreviations section at the beginning should be consulted for unfamiliar terms used throughout the volume.

WASTE AVOIDANCE AND UTILIZATION

From society's standpoint, the minimization of wastes requiring disposal is increasingly important as available disposal options become more and more constrained, and particularly as more substances enter into everyday use that are not readily decomposed in the natural environment and that can present long-term hazards. The term *minimization* is taken to include avoidance of the generation of wastes, when practical, and the productive utilization of any wastes that are generated (i.e., recycling).

There are many parts of the world where we can find rural or nonindustrialized areas. In these regions wastes are typically organic or inert and do not pose major disposal problems, particularly since they are often utilized for animal food or other purposes. However, as the level of industrialization increases, and certainly in industrialized countries like the United States and European Union members, waste disposal becomes an increasing problem. The problems are typically associated with non-biodegradable or bioaccumulative substances such as waste pesticides, solvents, heavy metals, and chemical sludges. These are often production wastes, but they can also arise from inappropriate application (e.g., pesticides) or poor consumer behavior (e.g., motor oil waste).

The development and widespread use of new substances such as specialty elastomers and plastics and the products that they have made possible have improved the standard of living for tens of millions, but they have also introduced new threats to the environment, as typified by the histories of DDT and polychlorinated biphenyls. The long-term solution to the problem of persistent or hazardous wastes must lie in efforts to find alternatives to the hazardous substances. In the meantime, high priority should be given to minimizing the use of resources and reducing the discharge of wastes. From industry's standpoint, the minimization of wastes also has a direct impact on operating costs and profits.

The need to avoid or minimize the release of complex organic and inorganic substances into the environment is all the greater because of uncertainty about their effects on human health and the natural environment and the very high costs of retrofitting or cleanup. At the same time a realistic attitude must be maintained regarding developing countries. Much industrial and product design is based on industrial country practice, and almost all of the fundamental science on which regulation is based has been carried out in the more advanced economies. Although there will be some opportunities to leapfrog to more sophisticated systems, the priority in developing countries should be to ensure that policymakers and regulators are up-to-date and informed, and avoid repeating fundamental mistakes made in the course of industrialization elsewhere.

Waste minimization is simply one of a number of related terms and concepts that, despite having similar overall goals and often being used interchangeably, may differ significantly in basic principles and in emphasis from pollution prevention. However, in this handbook, *waste* is used to refer to any material from a manufacturing process that has no perceived value to the manufacturer and that has to be disposed of in some manner. There is also *waste of energy*, which indeed may have a more defined value to the manufacturer. In conjunction with waste minimization are the terms *avoidance* and *utilization*.

Avoidance refers to actions by the producer to avoid generating the waste. *Utilization* includes the range of actions that make the waste a useful input to other processes, eliminating the need for disposal. Waste minimization thus comprises both avoidance and utilization. Processes that reduce the toxicity or potentially harmful impacts of a waste can in some cases be regarded as minimization, although in other circumstances such changes represent treatment before ultimate disposal.

Although the terminology used may vary, a number of important activities can be distinguished. *Reuse* refers to the repeated use of a waste material in a process (often after some treatment or makeup). *Recycling* refers to the use by one producer of a waste generated by another, or reuse as a raw material component within an existing manufacturing process. *Recovery is* the extraction from a waste of some components that have value in other uses.

With few exceptions, most companies in the private sector are concerned to one degree or another with sustainable development. Waste avoidance and utilization can be viewed as part of a broader hierarchy of approaches to achieving sustainable development. At the highest level are approaches that seek to satisfy human needs and requirements in ways that do not waste resources or generate harmful by-products or residuals. These approaches include changing consumer behavior and reexamining the range and character of the products and services produced. At a slightly lower level are efforts to redesign products and services and to raise consumers' awareness about the impacts of their decisions.

Application of techniques such as life cycle analysis (LCA) is part of the difficult analysis of the overall impacts of products and services on the environment. Such approaches are at present adopted mainly by industries in more advanced countries. In developing countries, much focus is on improvements in production processes. These approaches include cleaner production, pollution prevention, and waste minimization, all of which are related, to a greater or lesser degree, to better management, improvements in production processes, substitution of hazardous inputs, reuse and recycling of wastes, and so on. But the same holds true for many companies in the West, but perhaps to a lesser extent in some industry sectors, simply because many heavy polluting industries made the investments to green technologies more than a decade ago.

Looking at these concepts in terms of a hierarchy, the next step, which should be minimized but is not to be neglected, is *treatment* and proper *disposal* of wastes. The lowest level in the hierarchy, and the one that all the other levels strive to eliminate, is *remediation* of the impacts of wastes discharged to the environment. Cleanup is simply costlier than prevention.

A clear and effective governmental framework for waste management is necessary. Such a framework includes the delegation of relevant powers to the lower levels of government that are typically responsible for implementation. It is based on a clear and broadly accepted long-term policy and includes a predictable and flexible regulatory regime and targeted economic incentives. At the same time, programs should be put in place to increase awareness and education, with the long-term objective of changing the behavior of manufacturers and consumers in the direction of minimizing waste generation. This indeed has been the situation in the United States, and simply stated, strict enforcement of environmental legislation has created significant driving forces - both from the standpoints of legal and economics, to minimize wastes in manufacturing operations.

Manufacturers have learned that they can improve their environmental and economic performances through both management changes and technological improvement. This is where LCA, although still an evolving tool, is making an impact. The LCA approach focuses attention on the overall impact of the production, use, and disposal of products.

There are differences in philosophies between countries on what really constitutes waste minimization and pollution prevention. Consumers in some of the wealthier countries are moving toward a greater awareness about the need for waste reduction, as shown by participation in recycling schemes and some demand for environmentally friendly products. However, progress is often slow, and there is a need for ongoing education and awareness, as well as careful analysis of options and incentives. In developing countries, the demand for resources often leads to significant recycling of materials such as glass, metals, and plastics. These recycling systems have important social and economic consequences at the local level, and their improvement must be approached with care.

As has been noted by the World Bank Organization (refer to references 1 though 4), waste management efforts are linked closely with income levels. There is a broad progression from recycling of most materials in the poorest societies, through increasing consumerism - often with little concern for waste problems in low- and middle-income countries, to the environmental activism of some wealthier countries. The appropriate waste avoidance and utilization strategy for any situation must take into account the level of the economy, the capabilities of government at different levels, and the environmental circumstances.

As with any other national environmental strategy, there is a need for public involvement and political support in the identification of priorities and the implementation of the necessary enabling measures. Discussions of waste minimization as a matter of national policy are beyond the objectives of this handbook, however, where appropriate, comments and brief discussions are interjected.

ENERGY EFFICIENCY

Efficient use of energy is one of the main strategic measures not only for the conservation of fossil energy resources but also for abatement of air pollution and the slowing down of anthropogenic climate change. Accordingly, economic and technical measures to reduce specific energy demand should be priorities across all sectors of an economy. Many opportunities exist for improving efficiency, but progress has been disappointingly slow in many cases.

The phrase *efficient use of energy* includes all the technical and economical measures aimed at reducing the specific energy demand of a production system or economic sector. Although implementation of energy-saving techniques may require initial investments, short-term financial returns can often be achieved through lower fuel costs due to the reduced energy demand.

On a global basis, energy is vital to economic development in developing countries. Poverty will not be reduced without greater use of modern forms of energy. Assuming that energy demand in developing countries grows by 2.6 % per year, their total consumption of energy will be double the level of total consumption in industrial countries by 2050 (5). Even then, each person in the developing countries will be using, on average, a mere quarter of the energy consumed by each inhabitant of the industrial world. As these societies seek to improve their standards of living, developing countries have the opportunity to do things differently from what has happened in the past. The challenge is to break the link between economic growth and energy consumption by pursuing efficient production processes and reducing waste and, at the same time, to break the link between energy consumption and pollution by relying more on renewables and by using fossil fuels more efficiently.

According to World Energy Council projections (6), fossil fuels will still account for almost two thirds of primary energy even decades from now. Some long-term scenarios (for example, by Shell International and the Intergovernmental

Panel on Climate Change) IPCC postulate a rapidly increasing share of renewable technologies - solar, wind, geothermal, and biomass, as well as the more traditional hydroelectric (7). Under these scenarios, with appropriate policies and new technological developments, renewables could reach up to 50% of the total by the middle of the twenty-first century. However, even in optimistic scenarios, carbon emissions from burning fossil fuels (in the form of carbon dioxide) are predicted to increase dramatically.

Industrial countries are responsible for the bulk of the buildup of heat-trapping gases currently in the atmosphere, and only they have made firm commitments to cut their emissions at the Conference of the Parties to the United Nations Framework Convention on Climate Change in Kyoto in December 1997 (8). Yet emissions from developing countries are already growing rapidly, and by early in this century they are expected to exceed those of industrial countries. The fundamental issue is how to reconcile economic growth, primarily fueled by coal, oil, and gas, with protection of the environment.

New energy technologies are being developed, such as integrated gasification combined-cycle power plants, pressurized pulverized-coal-firing technology, humid air turbines, and fuel cells. Some of these technologies, although they are capable of efficiencies well in advance of current technology and show greatly reduced emissions, are yet not in a mature state of development. Currently, several large integrated gasification combined-cycle demonstration projects are being assessed, but it is too early to rely on these approaches as technically and economically viable alternatives to more conventional plants.

Industrial production processes often show a high specific energy demand. Industry is estimated to account for between 25% and 35% of total final energy consumption. Although great progress has been made in the rational use of energy in the industrial sector during the last two decades, improvements in cost-effective energy utilization have not nearly been exhausted. This holds true for new plants as well as for existing plants.

Improvement in energy end-use efficiency offers the largest opportunity of all alternatives for meeting the energy requirements of a growing world economy. It is impossible to list all the measures that have been implemented or that show promise for further improvements in special industrial branches. Many of the technical options for energy saving require only small investments and are easy to implement. In a number of cases, even simple organizational changes bring about considerable energy savings, yielding not only environmental benefits but also financial returns.

Energy-saving measures often show very short payback times, especially in industrial applications. However, as in the case of cleaner production approaches, it is often difficult to generate management interest in and support for the identification and implementation of energy-saving measures. Without such support, success is almost always limited.

The first step in identifying the energy-savings potential within an industrial plant is to conduct an *energy audit*, taking into account the specific conditions at the plant and the local conditions at the production site. An energy audit is needed to determine the scope of the energy efficiency project, to achieve a broad view of all the equipment installed at the production site, and to establish a consistent methodology of evaluation.

Preparation of an improved energy utilization scheme starts with an inventory of the equipment, its energy demand, and the flow of energy through the plant. Electrical energy and heat should be recorded separately, and the time dependence of the energy demand should be taken into account.

A few key areas can be identified on which to focus conservation efforts. These are (1) electricity production typically requires three times as much primary energy as direct heat use. Therefore, electricity should only be used if it cannot be replaced by other, more direct energy sources. (2) The chemical energy contained in fuels should be utilized as efficiently as possible. When combustion processes are used to meet the energy demand of a process or an industrial plant, high combustion efficiencies should be achieved by utilizing as much as possible the thermal energy contained in the flue gases, by minimizing heat losses (through use of insulation), and by recovering the thermal energy contained in combustion by-products such as ashes and slag. (3) Special attention should be given to separation processes for recovering and purifying products, which account for up to 40% of the total energy demand of chemical processes.

Energy savings of 10 to 40% are achievable through heat integration of the reboiler and the condenser of distillation columns, by using heat pumps or water compression systems. In several applications, it may also be possible to replace common but very energy-intensive distillation process with advanced separation processes, such as membrane techniques, that show a significantly reduced energy demand.

The first step in breaking the energy-environment link is to capture the opportunities for reaping environmental benefits through economically attractive solutions at no additional cost (i.e., no-cost measures). These opportunities include, at the very least, improvements in energy efficiency on the supply and demand sides, and a switch to less polluting energy sources. These can be referred to as "win-win" measures and can go a long way toward reducing local environmental degradation. Alone, however, they will not be sufficient. The objective must be to integrate local environmental and social externality costs into energy pricing and investment decisions so that the polluter pays for the additional costs of environmental protection and pollution abatement.

It is now well recognized that emissions of greenhouse gases from human activities are affecting the global climate. The consequences of climate change will disproportionately affect both poor people and poor countries. Under the 1997 Kyoto Protocol, some countries with economies in transition have obligations to reduce emissions of greenhouse gases. Developing nations have obligations to measure and monitor GHG (Green House Gas) emissions within their countries, but do not yet have to reduce emissions.

Use of cogeneration plants, which produce both electricity and heat, can reduce overall energy consumption by 10 to 30%, in comparison with separate generation of electricity and heat. Cogeneration plants are based on currently available standard technologies, and thus no technical risks are involved. However, reasonable and cost-effective utilization of this technology is only feasible if the heat can be supplied to a district heating network or to a nearby industrial plant where it can be used for process heating purposes.

THE REGULATORY DRIVING FORCE

Pollution prevention practices likely would have evolved in industry even without strict environmental enforcement. As an example, the author remembers a story by his father from the 1960s, where dust was collected in a baghouse at a precious metal refining operation. The dust traditionally was sent to a landfill, until an analytical test showed that the waste stream contained a large concentration of platinum. Beyond regulatory driving forces, industry has always had incentives for minimizing wastes, product recovery, recycling, and pollution or more accurately - waste prevention. Nonetheless, environmental standards, and their strict enforcement have created strong economic incentives for identifying and implementing projects that minimize end-of-pipe equipment investments that do not enhance products, as well as disposal practices that add to manufacturing costs. This section provides an overview of the U.S. regulatory standards and legislation. The key environmental standards that industry must comply with are briefly summarized.

Ambient Standards

Ambient standards set maximum allowable levels of a pollutant in the receiving medium (air, water, or soil). Ambient standards can offer a simple method of establishing priorities, since areas (or stream lengths) that comply with the relevant ambient standards are considered to require no further intervention, while other areas may be ranked by the extent to which concentrations exceed the ambient standards. Setting ambient standards requires an explicit agreement on the environmental quality objectives that are desired and the costs that society is willing to accept to meet those objectives. However, because ambient standards can be set at different levels for different locations, it is possible to use them to protect valuable ecosystems in a way that would not be possible by using emissions standards. It has been usual to establish an ambient standard for a pollutant by referring to the health effects of different levels of exposure, although some countries are moving toward ambient standards aiming to protect natural ecosystems. Historically, ambient standards in the industrial market economies have been continually tightened in the light of medical evidence on the impact of certain pollutants and in response to increased demand for better environmental quality. In particular, as reductions are achieved in the levels of simple pollutants such as biochemical oxygen demand (BOD), the focus has moved to the control of less obvious but more persistent pollutants such as heavy metals, polychlorinated biphenyls (PCBs), and the like, which are accumulative and essentially not biodegradable.

Emissions Standards

Emissions standards set maximum amounts of a pollutant that may be given off by a plant or other source. They have typically been expressed as concentrations, although there is increasing use of load-based standards, which reflect more directly the overall objective of reducing the total load on the environment. Emissions standards may be established in terms of what can be achieved with available technology or in terms of the impacts of the emissions on the ambient environment. Technology-based standards are based on knowledge of what can be achieved with current equipment and practices.

A wide range of principles has been used, including "best available technology" (BAT), "best practicable technology" (BPT), and "best available technology not entailing excessive cost" (BATNEEC). All these approaches are open to interpretation and are related to establishing what are the highest levels of equipment and performance that can reasonably be demanded from industrial operations. Alternatively, emissions standards can be established by estimating the discharges that are compatible with ensuring that receiving areas around the plant meet the ambient standards defined for the pollutant. This, however, requires considerable information on both the sources and the ambient environment and varies from area to area.

New source performance standards (NSPSs) are specific emissions standards in which the standard is applied only to new plants. They represent a special form of grandfathering, since emissions from existing plants are treated differently from those from new plants. Where NSPSs are significantly stricter than standards imposed on existing plants and are therefore costly, they may have the effect of prolonging the economic life of existing plants - subject, of course, to the influence of other economic and technological factors. On the other hand, it is easier for new plants to adopt cleaner processes and to incorporate treatment requirements in the initial design. Therefore, the costs of well-designed NSPSs need not be excessive.

Federal Statutes and Regulations

The following descriptions are intended solely for general information. Depending upon the nature or scope of the activities at a particular facility, these summaries may or may not necessarily describe all applicable environmental requirements. Moreover, they do not constitute formal interpretations or clarifications of the statutes and regulations. For further information, readers should consult the *Code of Federal Regulations* (CFR), the USEPA and state or local regulatory agencies.

Resource Conservation and Recovery Act (RCRA)

The Resource Conservation and Recovery Act (RCRA) of 1976, which amended the Solid Waste Disposal Act, addresses solid (Subtitle D) and hazardous (Subtitle C) waste management activities. The Hazardous and Solid Waste Amendments (HSWA) of 1984 strengthened RCRA's hazardous waste management provisions and added Subtitle I, which governs underground storage tanks (USTs). Regulations promulgated pursuant to Subtitle C of RCRA (40 CFR Parts 260-299) establish a "cradle-to-grave" system governing hazardous waste from the point of generation to disposal. RCRA hazardous wastes include the specific materials listed in the regulations (commercial chemical products, designated with the code "P" or "U"; hazardous wastes from specific industries/sources, designated with the code "K"; or hazardous wastes from non-specific sources, designated with the code "F") and materials which exhibit a hazardous waste characteristic (ignitability, corrosivity, reactivity, or toxicity and designated with the code "D"). Regulated entities that generate hazardous waste are subject to waste accumulation, manifesting, and record keeping standards. Facilities that treat, store, or dispose of hazardous waste must obtain a permit, either from EPA or from a state agency which EPA has authorized to implement the permitting program. Subtitle C permits contain general facility standards such as contingency plans, emergency procedures, record keeping and reporting requirements, financial assurance mechanisms, and unit-specific standards. RCRA also contains provisions (40 CFR Part 264, Subpart S and §264.10) for conducting corrective actions which govern the cleanup of releases of hazardous waste or constituents from solid waste management units at RCRA-regulated facilities. Although RCRA is a Federal statute, many states implement the RCRA program. Most RCRA requirements are not industry specific but apply to any company that transports, treats, stores, or disposes of hazardous waste. The following are some important RCRA regulatory requirements:

Identification of Hazardous Wastes (40 CFR Part 26 1) outlines the procedure every generator should follow to determine whether the material created is considered a hazardous waste, solid waste, or is exempted from regulation.

Standards for Generators of Hazardous Waste (40 CFR Part 262) establishes the responsibilities of hazardous waste generators including obtaining an ID number, preparing a manifest, ensuring proper packaging and labeling, meeting standards for waste accumulation units, and record keeping and reporting requirements. Generators can accumulate hazardous waste for up to 90 days (or 180 days depending on the amount of waste generated) without obtaining a permit.

Land Disposal Restrictions (LDRs) are regulations prohibiting the disposal of hazardous waste on land without prior treatment. Under the LDRs (40 CFR Part 268), materials must meet land disposal restriction (LDR) treatment standards prior to placement in a RCRA land disposal unit (landfill, land treatment unit, waste pile, or surface impoundment). Wastes subject to the LDRs include solvents, electroplating wastes, heavy metals, and acids. Generators of waste subject to the LDRs must provide notification of such to the designated TSD facility to ensure proper treatment prior to disposal.

Used Oil Management Standards (40 CFR Part 279) impose management requirements affecting the storage, transportation, burning, processing, and re-refining of the used oil. For parties that merely generate used oil, regulations establish storage standards. For a party considered a used oil marketer (one who generates and sells off-specification used oil directly to a used oil burner), additional tracking and paperwork requirements must be satisfied.

Tanks and Containers used to store hazardous waste with a high volatile organic concentration must meet emission standards under RCRA. Regulations (40 CFR Part 264-265, Subpart CC) require generators to test the waste to determine the concentration of the waste, to satisfy tank and container emissions standards, and to inspect and monitor regulated units. These regulations apply to all facilities who store such waste, including generators operating under the 90-day accumulation rule.

Underground Storage Tanks (USTs) containing petroleum and CERCLA hazardous substance are regulated under Subtitle I of RCRA. Subtitle I regulations (40 CFR Part 280) contain tank design and release detection requirements, as well as financial responsibility and corrective action standards for USTs. The UST program also establishes increasingly stringent standards, including upgrade requirements for existing tanks.

Boilers and Industrial Furnaces (BIFs) that use or burn fuel containing hazardous waste must comply with strict design and operating standards. BIF regulations (40 CFR Part 266, Subpart H) address unit design, provide performance standards, require emissions monitoring, and restrict the type of waste that may be burned.

Comprehensive Environmental Response, Compensation, and Liability Act (CERCLA)

CERCLA, a 1980 law commonly known as Superfund, authorizes EPA to respond to releases, or threatened releases, of hazardous substances that may present an imminent and substantial endangerment to public health, welfare, or the environment. CERCLA also enables EPA to force parties responsible for environmental contamination to clean it up or to reimburse the Superfund for response costs incurred by EPA. The Superfund Amendments and Reauthorization Act (SARA) of 1986 revised various sections of CERCLA, extended the taxing authority for the Superfund, and created a free-standing law, SARA Title III also known as the Emergency Planning and Community Right-to-Know Act (EPCRA). The CERCLA hazardous substance release reporting regulations (40 CFR Part 302) direct the person in charge of a facility to report to the National Response Center (NRC) any environmental release of a hazardous substance which exceeds a reportable quantity.

Reportable quantities are defined and listed in 40 CFR §302.4. A release report may trigger a response by EPA or by one or more Federal or State emergency response authorities. EPA implements *hazardous substance responses* according to procedures outlined in the National Oil and Hazardous Substances Pollution Contingency Plan (NCP) (40 CFR Part 300). The NCP includes provisions for permanent cleanups, known as remedial actions, and other cleanups referred to as "removals." EPA generally takes remedial actions only at sites on the National Priorities List (NPL), which currently includes approximately 1,300 sites. Both EPA and states can act at other sites; however, EPA provides responsible parties the opportunity to conduct removal and remedial actions and encourages community involvement throughout the Superfund response process.

Emergency Planning and Community Right-To-Know Act (EPCRA)

The Superfund Amendments and Reauthorization Act (SARA) of 1986 created EPCRA, also known as SARA Title III, a statute designed to improve community access to information about chemical hazards and to facilitate the development of chemical emergency response plans by State and local governments. EPCRA required the establishment of State emergency response commissions (SERCs), responsible for coordinating certain emergency response activities and for appointing local emergency planning committees (LEPCs). EPCRA and the EPCRA regulations (40 CFR Parts 350-372) establish four types of reporting obligations for facilities which store or manage specified chemicals: **EPCRA §302** requires facilities to notify the SERC and LEPC of the presence of any "extremely hazardous substance" (the list of such substances is in 40 CFR Part 355, Appendices A and B) if it has such substance in excess of the substance's threshold planning quantity, and directs the facility to appoint an emergency response coordinator. **EPCRA §304** requires the facility to notify the SERC and the LEPC in the event of a non-exempt release exceeding the reportable quantity of a CERCLA hazardous substance or an EPCRA extremely hazardous substance. **EPCRA §311 and §312** require a facility at which a hazardous chemical, as defined by the Occupational Safety and Health Act, is present in an amount exceeding a specified threshold of chemical use to submit to the SERC, LEPC and local fire department material safety data sheets (MSDSs) or lists of MSDS's and hazardous chemical inventory forms (also known as Tier I and II forms). This information helps the local government respond in the event of a spill or release of the chemical. **EPCRA §313** requires manufacturing facilities included in SIC codes 20 through 39, which have ten or more employees, and which manufacture, process, or use specified chemicals in amounts greater than threshold quantities, to submit an annual toxic chemical release report. This report, commonly known as the Form R, covers releases and transfers of toxic chemicals to various facilities and environmental media, and allows EPA to compile the national Toxic Release Inventory (TRI) database. All information submitted pursuant to EPCRA regulations is publicly accessible, unless protected by a trade secret claim.

Clean Water Act (CWA)

The primary objective of the Federal Water Pollution Control Act, commonly referred to as the CWA, is to restore and maintain the chemical, physical, and biological integrity of the nation's surface waters. Pollutants regulated under the CWA include "priority" pollutants, including various toxic pollutants; "conventional" pollutants, such as biochemical oxygen demand (BOD), total suspended solids (TSS), fecal coliform, oil and grease, and pH; and "nonconventional" pollutants, including any pollutant not identified as either conventional or priority. The CWA regulates both direct and indirect discharges.

The **National Pollutant Discharge Elimination System (NPDES)** program (CWA §402) controls direct discharges into navigable waters. Direct discharges or "point source" discharges are from sources such as pipes and sewers. NPDES permits, issued by either EPA or an authorized state (EPA has presently authorized forty States to administer the NPDES program), contain industry specific, technology-based and/or water quality-based limits, and establish pollutant monitoring reporting requirements. A facility that intends to discharge into the nation's waters must obtain a permit prior to initiating a discharge. A permit applicant must provide quantitative analytical data identifying the types of pollutants present in the facility's effluent. The permit will then set forth the conditions and effluent limitations under which a facility may make a discharge.

A NPDES permit may also include discharge limits based on Federal or state water quality criteria or standards, that were designed to protect designated uses of surface waters, such as supporting aquatic life or recreation. These standards, unlike the technological standards, generally do not take into account technological feasibility or costs. Water quality criteria and standards vary from State to State, and site to site, depending on the use classification of the receiving body of water. Most states follow EPA guidelines which propose aquatic life and human health criteria for many of the 126 priority pollutants.

Storm Water Discharges

In 1987 the CWA was amended to require EPA to establish a program to address storm water discharges. In response, EPA promulgated the NPDES storm water permit application regulations. Stormwater discharge associated with industrial activity means the discharge from any conveyance which is used for collecting and conveying stormwater and which is directly related to manufacturing, processing or raw material storage areas at an industrial plant (40 CFR 122.26(b)(14)). These regulations require that facilities with the following storm water discharges apply for an NPDES permit: (1) a discharge associated with industrial activity; (2) a discharge from a large or medium municipal storm sewer system; or (3) a discharge which EPA or the State determines to contribute to a violation of a water quality standard or is a significant contributor of pollutants to waters of the United States. The term "storm water discharge associated with industrial activity" means a storm water discharge from one of 11 categories of industrial activity defined at 40 CFR 122.26. Six of the categories are defined by SIC codes while the other five are identified through narrative descriptions of the regulated industrial activity. If the primary SIC code of the facility is one of those identified in the regulations, the facility is subject to the storm water permit application requirements. If any activity at a facility is covered by one of the five narrative categories, storm water discharges from those areas where the activities occur are subject to storm water discharge permit application requirements. Those facilities/activities that are subject to storm water discharge permit application requirements are identified below. To determine whether a particular facility falls within one of these categories, the regulation should be consulted. Category i: Facilities subject to storm water effluent guidelines, new source performance standards, or toxic pollutant effluent standards. Category ii: Facilities classified as SIC 24-lumber and wood products (except wood kitchen cabinets); SIC 26-paper and allied products (except paperboard containers and products); SIC 28-chemicals and allied products (except drugs and paints); SIC 291 -petroleum refining; and SIC 311-leather tanning and finishing. Category iii: Facilities classified as SIC 10-metal mining; SIC 12-coal mining; SIC 13-oil and gas extraction; and SIC 14-nonmetallic mineral mining. Category iv: Hazardous waste treatment, storage, or disposal facilities. Category v: Landfills, land application sites, and open dumps that receive or have received industrial wastes. Category vi: Facilities classified as SIC 5015-used motor vehicle parts; and SIC 5093-automotive scrap and waste material recycling facilities. Category vii: Steam electric power generating facilities. Category viii: Facilities classified as SIC 40-railroad transportation; SIC 41-local passenger transportation; SIC 42-trucking and warehousing (except public warehousing and storage); SIC 43-U.S. Postal Service; SIC 44-water transportation; SIC 45-transportation by air; and SIC 5171-petroleum bulk storage stations and terminals. Category ix: Sewage treatment works, Category x: Construction activities except operations that result in the disturbance of less than five acres of total land area. Category xi: Facilities classified as SIC 20-food and kindred products; SIC 21-tobacco products; SIC 22-textile mill products; SIC 23-apparel related products; SIC 2434-wood kitchen cabinets manufacturing; SIC 25-furniture and fixtures; SIC 265-paperboard containers and boxes; SIC 267-converted paper and paperboard products; SIC 27-printing, publishing, and allied industries; SIC 283-drugs; SIC 285-paints, varnishes, lacquer, enamels, and allied products; SIC 30-rubber and plastics; SIC 31-leather and leather products (except leather and tanning and finishing); SIC 323-glass products; SIC 34-fabricated metal products (except fabricated structural metal); SIC 35-industrial and commercial machinery and computer equipment; SIC 36-electronic and other electrical equipment and components; SIC 37-transportation equipment (except ship and boat building and repairing); SIC 38-measuring, analyzing, and controlling instruments; SIC 39-miscellaneous manufacturing industries; and SIC 4221-4225-public warehousing and storage.

Pretreatment Program

Another type of discharge that is regulated by the CWA is one that goes to a publicly-owned treatment works (POTWs). The national pretreatment program (CWA §307(b)) controls the indirect discharge of pollutants to POTWs by industrial users. Facilities regulated under §307(b) must meet certain pretreatment standards. The goal of the pretreatment program is to protect municipal wastewater treatment plants from damage that may occur when hazardous, toxic, or other wastes are discharged into a sewer system and to protect the toxicity characteristics of sludge generated by these plants. Discharges to a POTW are regulated primarily by the POTW itself, rather than the state or EPA. EPA has developed general pretreatment standards and technology-based standards for industrial users of POTWs in many industrial categories. Different standards may apply to existing and new sources within each category. "Categorical" pretreatment standards applicable to an industry on a nationwide basis are developed by EPA. In addition, another kind of pretreatment standard, "local limits," are developed by the POTW in order to assist the POTW in achieving the effluent limitations in its NPDES permit. Regardless of whether a state is authorized to implement either the NPDES or the pretreatment program, if it develops its own program, it may enforce requirements more stringent than Federal standards.

Safe Drinking Water Act (SDWA)

The SDWA mandates that EPA establish regulations to protect human health from contaminants in drinking water. The law authorizes EPA to develop national drinking water standards and to create a joint Federal-State system to ensure compliance with these standards. The SDWA also directs EPA to protect underground sources of drinking water through the control of underground injection of liquid wastes. EPA has developed primary and secondary drinking water standards under its SDWA authority. EPA and authorized States enforce the primary drinking water standards, which are, contaminant-specific concentration limits that apply to certain public drinking water supplies. Primary drinking water standards consist of maximum contaminant level goals (MCLGs), which are non-enforceable health-based goals, and maximum contaminant levels (MCLs), which are enforceable limits set as close to MCLGs as possible, considering cost and feasibility of attainment. The SDWA **Underground**

Injection Control (UIC) program (40 CFR Parts 144-148) is a permit program which protects underground sources of drinking water by regulating five classes of injection wells. UIC permits include design, operating, inspection, and monitoring requirements. Wells used to inject hazardous wastes must also comply with RCRA corrective action standards in order to be granted a RCRA permit, and must meet applicable RCRA land disposal restrictions standards. The UIC permit program is primarily state-enforced, since EPA has authorized all but a few states to administer the program. The SDWA also provides for a Federally-implemented Sole Source Aquifer program, which prohibits Federal funds from being expended on projects that may contaminate the sole or principal source of drinking water for a given area, and for a state-implemented Wellhead Protection program, designed to protect drinking water wells and drinking water recharge areas.

Toxic Substances Control Act (TSCA)

TSCA granted EPA authority to create a regulatory framework to collect data on chemicals in order to evaluate, assess, mitigate, and control risks which may be posed by their manufacture, processing, and use. TSCA provides a variety of control methods to prevent chemicals from posing unreasonable risk. TSCA standards may apply at any point during a chemical's life cycle. Under TSCA §5, EPA has established an inventory of chemical substances. If a chemical is not already on the inventory, and has not been excluded by TSCA, a premanufacture notice (PMN) must be submitted to EPA prior to manufacture or import. The PMN must identify the chemical and provide available information on health and environmental effects. If available data are not sufficient to evaluate the chemicals effects, EPA can impose restrictions pending the development of information on its health and environmental effects. EPA can also restrict significant new uses of chemicals based upon factors such as the projected volume and use of the chemical. Under TSCA §6, EPA can ban the manufacture or distribution in commerce, limit the use, require labeling, or place other restrictions on chemicals that pose unreasonable risks. Among the chemicals EPA regulates under §6 authority are asbestos, chlorofluorocarbons (CFCs), and polychlorinated biphenyls (PCBs).

Clean Air Act (CAA)

The CAA and its amendments, including the Clean Air Act Amendments (CAAA) of 1990, are designed to "protect and enhance the nation's air resources so as to promote the public health and welfare and the productive capacity of the population." The CAA consists of six sections, known as Titles, which direct EPA to establish national standards for ambient air quality and for EPA and the states to implement, maintain, and enforce these standards through a variety of mechanisms. Under the CAAA, many facilities were required to obtain permits for the first time. State and local governments oversee, manage, and enforce many of the requirements of the CAAA. CAA regulations appear at 40 CFR Parts 50-99. Pursuant to Title I of the CAA, EPA has established national ambient air quality standards (NAAQSs) to limit levels of "criteria pollutants," including carbon monoxide, lead, nitrogen dioxide, particulate matter, ozone, and sulfur dioxide. Geographic areas that meet NAAQSs for a given pollutant are classified as attainment areas; those that do not meet NAAQSs are classified as non-attainment areas. Under § 110 of the CAA, each state must develop a State Implementation Plan (SIP) to identify sources of air pollution and to determine what reductions are required to meet Federal air quality standards.

Title I also authorizes EPA to establish New Source Performance Standards (NSPSs), which are nationally uniform emission standards for new stationary sources falling within particular industrial categories. NSPSs are based on the pollution control technology available to that category of industrial source but allow the affected industries the flexibility to devise a cost-effective means of reducing emissions. Under Title I, EPA establishes and enforces National Emission Standards for Hazardous Air Pollutants (NESHAPs), nationally uniform standards oriented towards controlling particular hazardous air pollutants (HAPs). Title III of the CAAA further directed EPA to develop a list of sources that emit any of 189 HAPs, and to develop regulations for these categories of sources. The emission standards are being developed for both new and existing sources based on "maximum achievable control technology (MACT)." The MACT is defined as the control technology achieving the maximum degree of reduction in the emission of the HAPs, taking into account cost and other factors.

Title II of the CAA pertains to mobile sources, such as cars, trucks, buses, and planes. Reformulated gasoline, automobile pollution control devices, and vapor recovery nozzles on gas pumps are a few of the mechanisms EPA uses to regulate mobile air emission sources.

Title IV establishes a sulfur dioxide emissions program designed to reduce the formation of acid rain (see Chapter 2 for a discussion on acid rain). Reduction of sulfur dioxide releases are obtained by granting to certain sources limited emissions allowances, which are set below previous levels of sulfur dioxide releases.

Title V of the CAAA of 1990 created an operating permit program for all "major sources" (and certain other sources) regulated under the CAA. One purpose of the operating permit is to include in a single document all air emissions requirements that apply to a given facility. States are developing the permit programs in accordance with guidance and regulations from EPA. Once a State program is approved by EPA, permits are issued and monitored by that state.

Title VI is intended to protect stratospheric ozone by phasing out the manufacture of ozone-depleting chemicals and restricting their use and distribution. Production of Class I substances, including 15 kinds of chlorofluorocarbons (CFCs), will be phased out entirely by the year 2000, while certain hydrochlorofluorocarbons (HCFCs) will be phased out by 2030.

EMS AND ISO 14000

EMS is the abbreviation for *Environmental Management Systems*, of which *ISO 14000* is one of several EMSs. ISO 14000 is not argued to be the best EMS, but it does embody the elements of other systems, and clearly is becoming universally recognized among industry and the public.

EMSs such as ISO 14000 are seen as mechanisms for achieving improvements in environmental performance and for supporting the trade prospects of "clean" firms. The advantages of EMSs are becoming more clear. An environmental management system is a structured program of continuous environmental improvement that follows procedures drawn from established business management practices.

The concept is straightforward, and the principles can be easily applied, given the necessary support. The first steps in the control of industrial pollution have been the creation of the necessary regulatory framework and the specification and design of control equipment to reduce emissions. These efforts have been broadly successful in improving the performance of many polluting companies, but at high costs.

In addition, companies in transitioning economies, for example, Russia and Newly Independent States of the former Soviet Union, investments in pollution equipment are often wasted, because the equipment is not operated properly. The potential benefits of ecoefficiency are unequivocal: good operational practices, supported by committed management, can achieve considerable improvements in environmental performance at low cost and can get the maximum benefits from investments in hardware. Without management and worker support, the best equipment can be useless. The challenge is to achieve long-lasting improvements in performance, and EMS is seen as one of the key tools in achieving this.

An important related issue, in a context of increasingly free trade, is the concern that environmental performance may become an important commercial factor, either as a positive attribute or as a potential trade barrier. The implementation of an EMS, and particularly of the ISO 14000 system, is seen as a way to demonstrate an acceptable level of environmental commitment.

A good EMS allows an enterprise to understand and track its environmental performance. It provides a framework for implementing improvements that may be desirable for financial or other corporate reasons or that may be required to meet regulatory requirements. Ideally, it is built on an existing quality management system.

If an EMS were adopted purely as an internal management tool, the details of the system and its structure would not be important. However, the EMS is becoming more and more a matter of interest outside the management of the enterprise - to workers, regulators, local residents, commercial partners, bankers and insurers, and the general public. In this context, the EMS is no longer an internal system and becomes a mechanism for communicating the enterprise's performance to outside parties, and a level of standardization and common understanding are required.

As noted, the ISO 14000 series of standards has become the best-known common framework for EMS. This series is based on the overall approach and broad success of the quality management standards prepared and issued as the ISO 9000 series. ISO 14000 consists of a series of standards covering ecolabeling and life cycle assessment (LCA), as well as EMS. There are two other major EMS standards: the British BS 7750, which was one of the first broadly accepted systems and has been adopted by a number of other countries, and EMAS, the European Eco-Management and Audit Scheme. A process of harmonization has been under way to ensure reciprocal acceptability of these systems with ISO 14001. BS 7750 and EMAS are, however, broader in their requirements than ISO 14000. In particular, EMAS includes requirements for continued improvement of performance and for communication with the public, which are not part of ISO 14001.

Within the ISO system, ISO 14001 sets out the basic structure for an EMS, while ISO 14004 provides guidance. The crucial feature of the ISO 14001 standard is that it identifies the elements of a system which can be independently audited and certified. The issue of certification underlies much of the discussion about environmental management systems. The presentation in these standards is clear and concise and provides a framework that can be used as the starting point for a simple system for a small company or a highly detailed one for a multinational enterprise.

Compliance with ISO 14001 does not by itself automatically ensure that an enterprise will actually achieve improved environmental performance. The standard requires that there be an environmental policy that "**includes a commitment to continual improvement and pollution prevention**" and "**a commitment to comply with relevant environmental legislation and**

regulations." It also requires that the company establish procedures for taking corrective and preventive action in cases of nonconformance. It may seem to be splitting hairs to say that these requirements for a policy and procedures would not result in improved performance, but in essence the issue is one of following the spirit and not just the letter of the standard.

Basically, an EMS can be described as a program of continuous environmental improvement that follows a defined sequence of steps drawn from established project management practice and routinely applied in business management. These steps are as follows:

- Review the environmental consequences of the operations.
- Define a set of policies and objectives for environmental performance.
- Establish an action plan to achieve the objectives.
- Monitor performance against these objectives.
- Report the results appropriately.
- Review the system and the outcomes and strive for continuous improvement.

Not every system will present these steps in exactly the same manner, but the basic principles are clear and easily understandable. The ISO 14000 series is a series of standards for different aspects of environmental management. A number of these standards relating to environmental management systems have been adopted formally by the members of the ISO, while others are in different stages of preparation.

The desirable approach is for management to make a commitment to specific environment performance improvements within a defined period and then use ISO 14000 as the mechanism for demonstrating that it is complying with that commitment. As a manager for a multinational firm observed, "Having a certificate doesn't mean that you have a clean company. The bad guys who pollute today will still do it, and they'll have a certificate." It should be noted that ISO 14000 standards are voluntary. "Adoption" by a company normally means that the standards organization has said that the ISO version is the EMS standard that is recognized.

The direct benefits to an enterprise of implementing an EMS usually come from savings through cleaner production and waste minimization approaches. An order of magnitude estimate is that about 50% of the pollution generated in a typical "uncontrolled" plant can be prevented, with minimal investment, by adopting simple and inexpensive process improvements. In industrial countries, increased discharge fees and waste disposal charges provide much of the incentives for cost-effective pollution reduction which, further demonstrates the importance of an appropriate framework of regulations and incentives to drive the performance improvements. The major impact of the introduction of an EMS can be the identification of waste minimization and cleaner production possibilities. Management and worker commitment to improving performance is a key element. The process of introducing the EMS can be a catalyst for generating support for environmental performance improvements, including the simple changes that make up "good housekeeping," and also for making the best use of existing pollution control equipment. Just as important, the development of good management systems is one of the best hopes for sustaining the improvements that can be achieved when attention is focused on environmental performance. Perhaps the greatest concern about the ISO 14001 system is the lack of a clear commitment to improvements in actual environmental performance. The EMS approach is designed to improve performance, but critics of the rush to implement ISO 14001 argue that the standard can be misused. It is not yet clear how valid this point is, and its resolution will depend on how the overall approach is used in trade and regulatory areas. However, there is a legitimate concern that some may view ISO 14000 as an end rather than a means. Given the current stage of development of auditing and certification systems, it is possible in some parts of the world to obtain (or claim) certification with a minimum level of real environmental improvement.

As noted, ISO 14001 sets out a system that can be audited and certified. In many cases, it is the issue of certification that is critical and is at the heart of the discussion about the trade implications. Certification means that a qualified body (an "accredited certifier") has inspected the EMS system that has been put in place and has made a formal declaration that the system is consistent with the requirements of ISO 14001. The standard does allow for "self-certification," a declaration by an enterprise that it conforms to ISO 14001. There is skepticism as to whether the self declaration approach is capable of wide acceptance, especially when certification is viewed by many as having legal or commercial consequences.

Perhaps the most significant problem with ISO 14000 certification is that it can entail significant costs. There are also issues relating to the international acceptance of national certification that may make it particularly difficult for companies in some countries to achieve credible certification at reasonable cost. For firms concerned about certification that carries real credibility, the cost of bringing in international auditors can be quite high, partly because the number of internationally recognized firms of certifiers at present is limited; and Western consultant daily rates simply are excessive to many countries like those in Central and Eastern Europe.

The issue of accreditation of certifiers is increasingly important as the demand increases. Countries that have adopted ISO 14001 as a national standard can accredit qualified parties as certifiers, and this will satisfy national legal or contractual requirements. But it must be remembered that the fundamental purpose of ISO is to achieve consistency internationally. If certificates from countries or agencies are not fully accredited or are regarded as "second class," the goal will not have been achieved. It is likely that the international marketplace will evolve to a state of real commercial value on high-quality certificates, but this level of sophistication and discrimination has not yet been achieved. Ultimately, there must be a mechanism to ensure that certification in any one country has credibility and acceptability elsewhere. The ISO has outlined procedures for accreditation and certification (Guides 61 and 62), and a formal body, which has been established to operationalize the process. At the same time, a number of established national accreditation bodies involved in ISO have set up the informal International Accreditation Forum (IAF) to examine mechanisms for achieving international reciprocity through multilateral agreements (MLAs). However, these systems are in the early stages, and many enterprises continue to use the established international certifiers, even at additional cost, because of lack of confidence in the acceptability of local certifiers. Given the variability in the design of individual EMS and the substantial costs of the ISO 14000 certification process, there is a growing tendency for large companies that are implementing EMS approaches to pause before taking this last step. After implementing an EMS and confirming that the enterprise is broadly in conformance with ISO 14001, it is becoming routine to carry out a *gap analysis* to determine exactly what further actions would be required to achieve certification and to examine the benefits and costs of bringing in third-party certifiers. The gap analysis is most often accompanied by an *Initial Environmental Review* (IER), which established a baseline on the environmental performance of a company's operations.

A very practical question is to what extent an EMS can reduce the costs of compliance, in terms of both the overall government enforcement effort and the costs by the individual enterprise. The use of ISO 14001 certification to replace some statutory reporting requirements is a topic of considerable discussion in a number of countries, particularly those where regulatory requirements are extensive enough to be a serious burden on industry. It is now clear that an EMS is not a substitute for a regulatory framework, but the monitoring and reporting systems of a well-managed enterprise might substitute for some of the statutory inspections, audits, and reports normally required under government regulations. The issue is when and how the government can trust the capabilities and commitment of a company to self-monitor its environmental performance and whether some formal EMS and certification system, such as ISO 14000, would provide the mechanism to convince regulators that limited government resources would be better used elsewhere in pursuing less cooperative organizations. This approach seems attractive, but there are a number of obstacles to clear before it can be put into place on a widespread basis. Reaching agreement on such matters is proving to be a more difficult and complex task than might at first be assumed. Some of the difficulties are legal (lack of flexibility in regulations or the need to ensure that voluntary reports are not unreasonably used to prosecute enterprises that are making good-faith efforts to improve), but often they relate to the necessary level of confidence on both sides that the other parties are genuine in their efforts.

It should not be overlooked that an informed public has a strong influence on the environmental performance of industrial enterprises, through a variety of mechanisms that include market forces, social pressures, and support for improved regulatory controls. ISO 14000 does not include specific requirements for the disclosure or publication of environmental performance measures or audit results, but other EMS models do have some such requirements. There is also a growing interest on the part of commercial banks and insurance companies in environmental risk (in a purely business sense). Such organizations are considering whether EMS certification (typically EMAS, in Europe) demonstrates that a firm has real control over its environmental risk and potential liability. It is possible that certification may lead to commercial benefits, such as lower insurance rates, in certain high-risk sectors. Public release of the main environmental information from an EMS can also be used as a central component of a community relations program.

The author has been exposed to a number of projects in Central and Eastern Europe where Western consultants have made strong statements to the effect that before long, ISO 14000 certification will be an essential passport for developing countries wishing to trade with the industrial nations. Such statements, in the author's opinion, are speculative and likely incorrect. It is, however, unclear to what extent ISO 14001 might become a barrier to trade, in direct contradiction to the basic objectives of the ISO, or, alternatively, might provide a competitive edge for certified firms. The trade implications are of concern to many countries. The World Trade Organization is beginning to consider some of the issues under its mandate on technical barriers to trade. In this context, a distinction needs to be made between product standards, such as the ecolabeling and LCA standards under ISO 14000, and production process standards such as ISO 14001; the impacts are likely to be different.

In many cases in developing countries and countries in transitioning economies, the environmental pressures come through the supplier chain - i.e., the ongoing relationship between a major company (often a multinational) and its smaller national suppliers. The sensitivity of multinationals to pressures regarding their performance on environmental and other issues is causing them to look for better performance from the suppliers. This relationship is typically a cooperative one in which large companies work with smaller ones to achieve better performance in such areas as quality and price. In a number of cases multinationals are indeed asking their suppliers to achieve and demonstrate environmental performance improvements, but there is no evidence that unreasonable targets or time scales are being applied. Where ISO 14001 certification is an ultimate aim, certification by many is seen as a long-term objective rather than a short-term requirement. Even if ISO 14001 is not likely to be a contractual constraint in the foreseeable future, environmental performance is increasingly becoming a factor in commercial transactions, and companies looking to establish a presence in the international marketplace are considering whether a so-called "green badge" would be an advantage to them. Unfortunately, in practice, it is often marketing rather than environmental concerns that drive the ISO certification process. By far, the majority of the application of EMS has taken place in large companies. The use of such systems in small and medium-size enterprises has been limited - although it is in this segment of industry that some of the largest benefits might be anticipated, because of the difficulty of regulating large numbers of small firms and the potential efficiency improvements that are likely to exist. The characteristics of the typical small to medium size enterprise make the adoption of EMS difficult: many small companies for example do not have a formal management structure, they lack technically trained personnel, and they are subject to severe short-term pressures on cash flow.

It is likely that an EMS cannot be used to drive improved performance in a poorly organized small firm. Targeted training in management and quality control can improve overall performance, including its environmental aspects, and can provide a basis for more specific EMS development. Many firms can reap significant benefits from introducing quality management concepts, even where they are not aiming at formal certification. Any steps in this direction should be encouraged.

An EMS really is intended to build on existing production and quality management systems. Where such systems are weak or ineffective, as is often the case in enterprises that have poor environmental performance, a better management framework has to be established before focusing on the details of the EMS. The costs of establishing an EMS will therefore obviously depend on the starting point in terms of both management systems and environmental performance. The ecoefficiency savings can, in some cases, pay for the costs of establishing the EMS, particularly if most of the planning and organizational work is carried out in-house. However, a poor performer will very likely have to invest in production upgrading or pollution control in order to meet environmental requirements. Depending upon the size of the company operations, these costs can be significant. It is also important to recognize that a full EMS can be complex and can require an appreciable commitment of operational resources. However, the final system can be reached reasonably through a series of discrete steps, starting from a basic, simple procedure and becoming more comprehensive and sophisticated as capabilities and resources allow. In this way, even a small enterprise can begin to put in place the basic elements of an ISO 14001 system and can develop them at an appropriate pace. Once the basic EMS is in place, it is possible to carry out a gap analysis and to make a balanced judgment on the costs and benefits of seeking formal certification.

Although the focus of this Handbook is on industry, we would be remiss by not making note that there is also a special role for government in the EMS. ISO 14000 is a set of voluntary standards that individual companies may or may not choose to adopt, but governments can clearly have a role in providing information, establishing the necessary framework and infrastructure, and, in some cases, helping companies to develop the basic capabilities to adopt ISO 14000. There are two areas in which government action is useful: (a) providing information on the sectors and markets where ISO 14001 certification is a significant issue and assisting sector organizations to develop appropriate responses, and (b) helping to establish a certification framework, based on strengthening national standards organizations and encouraging competitive private sector provision of auditing and certification services. Governments should see EMS approaches as part of a broad environmental strategy that includes regulatory systems, appropriate financial incentives, and encouragement of improved industrial performance. Such encouragement can really only be effective where there is cooperation at the government level between the relevant departments, including industry and trade, as well as environment.

The evolution of ISO 14000 is leading to increased understanding of the benefits of better environmental management and greater awareness of environmental performance as a factor in succeeding in increasingly competitive markets. At the same time, the standards alone do not enable environmental improvements in those countries where regulation and enforcement are ineffective, nor can the standards open markets where competition is strong. The standards only provide a framework on which to build better performance, greater efficiency, and a competitive image. With serious commitment and effort from the organization, implementing a system such as ISO 14001 can yield significant benefits. And, the reader should recognize that one of the principles in this EMS focuses on pollution prevention practices. For details on ISO standards,

the reader can contact national standards organizations or the International Organization for Standardization:
ISO Central Secretariat 1, rue de Varambe Case postale 56
CH-1211 Geneva 20 Switzerland
Tel: +4122 749 0111, fax: +4122 734 1079

CLOSING REMARKS

This introductory chapter has provided the reader with a broad overview of terminology and issues as they relate to environmental management. Part of a sound environmental management system is pollution prevention. With few exceptions in industry, an EMS is simply not complete or properly working without applying the principles of pollution prevention. In fact, pollution prevention is a cornerstone issues in ISO 14000. It is well recognized in Western corporations that significant reductions in pollution loads can often be achieved at little cost. The efficient use of resources and the reduction in wastage in manufacturing are obviously preferable to reliance on costly end-of-pipe treatment. In the U.S. and European Union countries, strong enforcement of environmental regulations was the initial driving force behind industry adoption of pollution prevention. But there are many parts of the world today where regulatory enforcement is weak. In these situations, the true driving force for pollution prevention programs is the economic incentives - i.e., the savings in raw materials and energy, which many Western corporations understand only now as the major incentive for implementing pollution prevention practices. In the following chapters, we will explore the basic elements that go into establishing and maintaining a pollution prevention program. Where possible, typical examples from industrial settings are included into discussions.

REFERENCES

1. *Greening Industry: New Roles for Communities, Markets and Governments*, A World Bank Policy Research Report, Oxford University Press, New York, 2000.
2. Lovei, M., *Financing Pollution Abatement,: Theory and Practice*, The World Bank, Washington, D.C., October, 1995.
3. Hansen, S., *Towards a Sustainable Economy: Can We Survive*, The World Bank, Washington, D.C., June 1996.
4. *Pollution Prevention and Abatement Handbook*, The World Bank, Washington, D.C., 1998.

5. Ahmed, K., *Renewable Energy Technologies*, The World Bank, Washington, D.C., 1994.
6. *Fuel for Thought: A New Environmental Strategy for the Energy Sector*, The World Bank, Washington, D.C., 1998.
7. Author interviews with IPPS - Environment, Infrastructure, and Agriculture Division, Policy Research Department, World Bank, Washington, D.C., 2000.
8. *The World Bank and UN Framework Convention on Climate Change*, Environment Department Papers, Climate Change Series 8, Environment Department, The World Bank, Washington, D.C., 1995.

Chapter 2
Managing Hazardous Chemicals

INTRODUCTION

The management of hazardous chemicals can be a complex process depending on the magnitude of an industrial operation. Not only are there extensive environmental statutes that must be met, but long term health risk issues, as well as occupational exposure issues must be dealt with. An important objective in pollution prevention is to reduce pollution through the application of cleaner technologies. This can be accomplished by process changes and modifications, re-engineering, and using chemical substitutes as raw materials that are less polluting or harmful. The USEPA applies the terminology *cleaner technology substitute assessment*, which is another term for applying a pollution prevention audit, assessing the financial benefits of the proposed changes, and implementing cleaner production. The methodology and protocol for applying the audit and conducting a financial analysis are discussed in later chapters. The current chapter focuses on the hazardous properties and most recent pollution prevention practices for the major pollutants that industry generates. The chapter also provides a description of the more common pollution control equipment used throughout industry. The reader should refer tp references 1 through 3 for information of the hazards and safe handling of common chemicals and risks associated with exposures to pollutants.

AIRBORNE PARTICULATE MATTER

Airborne particulate matter, which includes dust, dirt, soot, smoke, and liquid droplets emitted into the air, is small enough to be suspended in the atmosphere. Airborne particulate matter may be a complex mixture of organic and inorganic substances. They can be characterized by their physical attributes, which influence their transport and deposition, and their chemical composition, which influences their effect on health. The physical attributes of airborne particulates include mass concentration and size distribution. Ambient levels of mass concentration are measured in micrograms per cubic meter (mg/m^3); size attributes are usually measured in aerodynamic diameter. Particulate matter (PM) exceeding 2.5 μ (μ) in aerodynamic diameter is generally defined as coarse particles, while particles smaller than 2.5 mm ($PM_{2.5}$) are called fine particles.

The acid component of particulate matter, and most of its mutagenic activity, are generally contained in fine particles, although some coarse acid droplets are also present in fog. Samples taken in the United States showed that about 30% of particulate matter was in the fine fraction range. Particles interact with various substances in the air to form organic or inorganic chemical compounds. The most common combinations of fine particles are those with sulfates. In the United States, sulfate ions account for about 40% of fine particulate matter and may also be present in concentrations exceeding about 10 micrograms per normal cubic meter (mg/Nm^3). The smaller particles contain the secondarily formed aerosols, combustion particles, and recondensed organic and metal vapors. The carbonaceous component of fine particles - products of incomplete combustion - contains both elemental carbon (graphite and soot) and nonvolatile organic carbon (hydrocarbons emitted in combustion exhaust, and secondary organic compounds formed by photochemistry). These species may be the most abundant fine particles after sulfates. Additionally, atmospheric reactions of nitrogen oxides produce nitric acid vapor (HNO_3) that may accumulate as nitrate particles in both fine and coarse forms. The most common combination of coarse particles consists of oxides of silicon, aluminum, calcium, and iron.

There are several terms that are used to describe particulate matter. Generally, these terms are associated with the sampling method, and are briefly described as follows:

Total suspended particulates (TSP) includes particles of various sizes. Some proportion of TSP consists of particles too large to enter the human respiratory tract; therefore, TSP is not a good indicator of health-related exposure. TSP is measured by a high-volume gravimetric sampler that collects suspended particles on a glass-fiber filter. The upper limit for TSP is 45 mm in diameter in the United States and up to 160 μm in Europe. TSP sampling and TSP-based standards were used in the United States until 1987. Several countries in Central and Eastern Europe, Latin America, and Asia still monitor and establish standards based on measurements of TSP. As monitoring methods and data analysis have become more sophisticated, the focus of attention has gradually shifted to fine particulates. Recent evidence shows that fine particulates, which can reach the thoracic regions of the respiratory tract, or lower, are responsible for most of the excess mortality and morbidity associated with high levels of exposure to particulates. Most sophisticated studies suggest that fine particulates are the sole factor accounting for this health damage, while exposure to coarse particulates has little or no independent effect. The particles most likely to cause adverse health effects are the *fine particulates*, in particular, particles smaller than 10 μ and 2.5 mm in aerodynamic diameter, respectively. They are sampled using (a) a high-volume sampler with a size-selective inlet using a quartz filter or (b) a dichotomous sampler that operates at a slower flow rate, separating on a Teflon filter particles smaller than 2.5 mm and sizes between 2.5 mm and 10 mm. No generally accepted conversion method exists between TSP and PM_{10}, which may constitute between 40% and 70% of TSP. In 1987, the USEPA switched its air quality standards from TSP to PM_{10}. PM_{10} standards have also been adopted in, for example, Brazil, Japan, and the Philippines. In light of the emerging evidence on the health impacts of fine particulates, the USEPA has proposed that U.S. ambient standards for airborne particulates be defined in terms of fine particulate matter.

Black smoke (BS) is a particulate measure that typically contains at least 50% respirable particulates smaller than 4.5 mm in aerodynamic diameter, sampled by the British smokeshade (BS) method. The reflectance of light is measured by the darkness of the stain caused by particulates on a white filter paper. The result of BS sampling depends on the density of the stain and the optical properties of the particulates. Because the method is based on reflectance from elemental carbon, its use is recommended in areas where coal smoke from domestic fires is the dominant component of ambient particulates (WHO and UNEP). Most investigators conclude that BS is roughly equivalent to PM_{10}. However, there is no precise equivalence of the black smoke measurements with other methods. The BS measure is most widely used in Great Britain and other parts of Europe.

Some particulates come from natural sources such as evaporated sea spray, windborne pollen, dust, and volcanic or other geothermal eruptions. Particulates from natural sources tend to be coarse. Almost all fine particulates are generated as a result of combustion processes, including the burning of fossil fuels for steam generation, heating and household cooking, agricultural field burning, diesel-fueled engine combustion, and various industrial processes. Emissions from these anthropogenic sources tend to be in fine fractions. However, some industrial and other processes that produce large amounts of dust, such as cement manufacturing, mining, stone crushing, and flour milling, tend to generate particles larger than 1 mm and mostly larger than 2.5 mm. In cold and temperate parts of the world, domestic coal burning has been a major contributor to the particulate content of urban air. Traffic-related emissions may make a substantial contribution to the concentration of suspended particulates in areas close to traffic. Some agroindustrial processes and road traffic represent additional anthropogenic sources of mostly coarse particulate emissions. The largest stationary sources of particulate emissions include fossil-fuel-based thermal power plants, metallurgical processes, and cement manufacturing. The physical and chemical composition of particulate emissions is determined by the nature of pollution sources. Most particles emitted by anthropogenic sources are less than 2.5 mm in diameter and include a larger variety of toxic elements than particles emitted by natural sources. Fossil fuel combustion generates metal and sulfur particulate emissions, depending on the chemical composition of the fuel used. The USEPA estimates that more than 90% of fine particulates emitted from stationary combustion sources are combined with sulfur dioxide (SO_2). Sulfates, however, do not necessarily form the largest fraction of fine particulates. In locations such as Bangkok, Chongqing (China), and Sao Paulo (Brazil), organic carbon compounds account for a larger fraction of fine particulates, reflecting the role of emissions from diesel and two-stroke vehicles or of smoke from burning coal and charcoal. Although sulfates represent a significant share (30 to 40%) of fine particulates in these cases, caution is required before making general assertions about the relationship between sulfates and fine particulates, since the sources and species characteristics of fine particulates may vary significantly across locations. Combustion devices may emit particulates comprised of products of incomplete combustion and toxic metals, which are present in the fuel and in some cases may also be carcinogenic. Particulates emitted by thermal power generation may contain lead, mercury, and other heavy metals. The melting, pouring, and torch-cutting procedures of metallurgy emit metal particulates containing lead, cadmium, and nickel. Particles emitted by the cement industry are largely stone or clay-based particulate matter that may contain toxic metals such as lead.

The respiratory system is the major route of entry for airborne particulates. The deposition of particulates in different parts of the human respiratory system depends on particle size, shape, density, and individual breathing patterns (mouth or nose breathing). The effect on the human organism is also influenced by the chemical composition of the particles, the duration of exposure, and individual susceptibility. While all particles smaller than 10 mm in diameter can reach the human lungs, the retention rate is largest for the finer particles. Products of incomplete combustion, which form a significant portion of the fine particulates, may enter deep into the lungs. Clinical, epidemiologic, and toxicological sources are used to estimate the mortality and morbidity effects of short- and long-term exposure to various particulate concentration levels. Several studies have found statistically significant relationships between high short-term ambient particulate concentrations and excess mortality in London and elsewhere. The estimated 4,000 excess deaths in the London metropolitan area in December 1952 were associated with BS measurements equivalent to a 4,000 mg/m^3 maximum daily average ambient concentration of particulates. Studies have also found a significant association between daily average PM_{10} concentrations and mortality at concentrations below the current U.S. standard of 150 mg/m^3 for short-term PM10 concentrations.

Vegetation exposed to wet and dry deposition of particulates may be injured when particulates are combined with other pollutants. Coarse particles, such as dust, directly deposited on leaf surfaces can reduce gas exchange and photosynthesis, leading to reduced plant growth. Heavy metals that may be present in particulates, when deposited on soil, inhibit the process in soil that makes nutrients available to plants. This, combined with the effects of particulates on leaves, may contribute to reduction of plant growth and yields. In addition, particulates contribute to the soiling and erosion of buildings, materials, and paint, leading to increased cleaning and maintenance costs and to loss of utility.

Particulate emissions have their greatest impact on terrestrial ecosystems in the vicinity of emissions sources. Ecological alterations may be the result of particulate emissions that include toxic elements. Furthermore, the presence of fine particulates may cause light scattering, known as atmospheric haze, reducing visibility and adversely affecting transport safety, property values, and aesthetics.

The most frequently used reference guidelines for ambient particulate concentration are those of WHO, the EU, and the USEPA. These guidelines are based on clinical, toxicological, and epidemiologic evidence and were established by determining the concentrations with the lowest observed adverse effect (implicitly accepting the notion that a lower threshold exists under which no adverse human health effects can be detected), adjusted by an arbitrary margin of safety factor to allow for uncertainties in extrapolation from animals to humans and from small groups of humans to larger populations. The WHO guidelines are based on health considerations alone; the EU and USEPA standards also reflect the technological feasibility of meeting the standards. In the EU, a prolonged consultation and legislative decision making process took into account the environmental conditions and the economic and social development of the various regions and countries and acknowledged a phased approach to compliance. A potential tradeoff was also recognized in the guidelines for the combined effects of sulfur dioxide and particulate matter.

The main objective of air quality guidelines and standards is the protection of human health. Since fine particulates (PM_{10}) are more likely to cause adverse health effects than coarse particulates, guidelines and standards referring to fine particulate concentrations are preferred to those referring to TSP, which includes coarse particulate concentrations. Scientific studies provide ample evidence of the relationship between exposure to short-term and long-term ambient particulate concentrations and human mortality and morbidity effects. However, the dose response mechanism is not yet fully understood. Furthermore, according to WHO, there is no safe threshold level below which health damage does not occur. A difficulty that should not be overlooked is that airborne particulates are rarely homogeneous. They vary greatly in size and shape, and their chemical composition is determined by factors specific to the source and location of the emissions. The combined effects and interactions of various substances mixed with particulates have not yet been established (except for sulfur dioxide), but they are believed to be significant, especially where long-term exposure occurs. Measurement techniques and their reliability may vary across regions and countries, and so may other factors, such as diet, lifestyle, and physical fitness, that influence the human health effects of exposure to particulates.

Pollution Prevention Practices

Airborne particulate matter (PM) emissions can be minimized by pollution prevention and emission control measures. Prevention is frequently more cost-effective than control, and therefore, should be emphasized. Special attention should be given to pollution abatement measures in areas where taxies and buses associated with particulate emissions may pose a significant environmental risk.

Measures such as improved process design, operation, maintenance, housekeeping, and other management practices can reduce emissions. By improving combustion efficiency, the amount of products of incomplete combustion (PICs) a component of particulate matter, can be significantly reduced. Proper fuel-firing practices and combustion zone configuration, along with an adequate amount of excess air, can achieve lower PICs.

Atmospheric particulate emissions can be reduced by choosing cleaner fuels. Natural gas used as fuel emits negligible amounts of particulate matter. Oil-based processes also emit significantly fewer particulates than coal-fired combustion processes. Low-ash fossil fuels contain less noncombustible, ash-forming mineral matter and thus generate lower levels of particulate emissions. Lighter distillate oil-based combustion results in lower levels of particulate emissions than heavier residual oils. However, the choice of fuel is usually influenced by economic as well as environmental considerations.

Reduction of ash by fuel cleaning reduces the generation of PM emissions. Physical cleaning of coal through washing and beneficiation can reduce its ash and sulfur content, provided that care is taken in handling the large quantities of solid and liquid wastes that are generated by the cleaning process. An alternative to coal cleaning is the co-firing of coal with higher and lower ash content. In addition to reduced particulate emissions, low-ash coal also contributes to better boiler performance and reduced boiler maintenance costs and downtime, thereby recovering some of the coal cleaning costs.

The use of more efficient technologies or process changes can reduce PIC emissions. Advanced coal combustion technologies such as coal gasification and fluidized-bed combustion are examples of cleaner processes that may lower PICs by approximately 10%. Enclosed coal crushers and grinders emit lower PM.

A variety of particulate removal technologies, with different physical and economic characteristics, are available. Some of these are as follows:

Inertial or impingement separators rely on the inertial properties of the particles to separate them from the carrier gas stream. Inertial separators are primarily used for the collection of medium-size and coarse particles. They include settling chambers and centrifugal cyclones (straight-through, or the more frequently used reverse-flow cyclones). Cyclones are low-cost, low-maintenance centrifugal collectors that are typically used to remove particulates in the size range of 10-100 μ. The fine-dust-removal efficiency of cyclones is typically below 70%, whereas electrostatic precipitators (ESPs) and baghouses can have removal efficiencies of 99.9% or more. Cyclones are therefore often used as a primary stage before other PM removal mechanisms. They typically cost about US$35 per cubic meter/minute flow rate (m^3/min), or US$1 per cubic foot/ minute (cu. ft/min).

Electrostatic precipitators (ESPs) remove particles by using an electrostatic field to attract the particles onto the electrodes. Collection efficiencies for well-designed, well-operated, and well maintained systems are typically in the order of 99.9% or more of the inlet dust loading. ESPs are especially efficient in collecting fine particulates and can also capture trace emissions of some toxic metals with an efficiency of 99%. They are less sensitive to maximum temperatures than are fabric filters, and they operate with a very low pressure drop. Their consumption of electricity is similar to that of fabric filters. ESP performance is affected by fly-ash loading, the resistance of fly ash, and the sulfur content of the fuel. Lower sulfur concentrations in the flue gas can lead to a decrease in collection efficiency. ESPs have been used for the recovery of process materials such as cement, as well as for pollution control. They typically add 1-2% to the capital cost of a new industrial plant.

Filters and dust collectors (baghouses) collect dust by passing flue gases through a fabric that acts as a filter. The most commonly used is the bag filter, or baghouse. The various types of filter media include woven fabric, needled felt, plastic, ceramic, and metal. The operating temperature of the baghouse gas influences the choice of fabric. Accumulated particles are removed by mechanical shaking, reversal of the gas flow, or a stream of high-pressure air. Fabric filters are efficient (99.9% removal) for both high and low concentrations of particles but are suitable only for dry and free-flowing particles. Their efficiency in removing toxic metals such as arsenic, cadmium, chromium, lead, and nickel is greater than 99%. They also have the potential to enhance the capture of sulfur dioxide (SO_2) in installations downstream of sorbent injection and dry-scrubbing systems. They typically add 1-2% to the capital cost of new power plants.

Wet scrubbers rely on a liquid spray to remove dust particles from a gas stream. They are primarily used to remove gaseous emissions, with particulate control a secondary function. The major types are venturi scrubbers, jet (fume) scrubbers, and spray towers or chambers. Venturi scrubbers consume large quantities of scrubbing liquid (such as water) and electric power and incur high pressure drops. Jet or fume scrubbers rely on the kinetic energy of the liquid stream. The typical removal efficiency of a jet or fume scrubber (for particles 10 μ or less) is lower than that of a venturi scrubber. Spray towers can handle larger gas flows with minimal pressure drop and are therefore often used as precoolers. Because wet scrubbers may contribute to corrosion, removal of water from the effluent gas of the scrubbers may be necessary. Another consideration is that wet scrubbing results in a liquid effluent. Wet-scrubbing technology is used where the contaminant cannot be removed easily in a dry form, soluble gases and wettable particles are present, and the contaminant will undergo some subsequent wet process (such as recovery, wet separation or settling, or neutralization). Gas flow rates range from 20 to 3,000 (m^3/min), Gas flow rates of approximately 2,000 (m^3/min), may have a corresponding pressure drop of 25 cm water column.

The selection of PM emissions control equipment is influenced by environmental, economic, and engineering factors.

- *Environmental factors* include (a) the impact of control technology on ambient air quality; (b) the contribution of the pollution control system to the volume and characteristics of wastewater and solid waste generation; and (c) maximum allowable emissions requirements.
- *Economic factors* include (a) the capital cost of the control technology; (b) the operating and maintenance costs of the technology; and (c) the expected lifetime and salvage value of the equipment.
- *Engineering factors* include (a) contaminant characteristics such as physical and chemical properties--concentration, particulate shape, size distribution, chemical reactivity, corrosivity, abrasiveness, and toxicity; (b) gas stream characteristics such as volume flow rate, dust loading, temperature, pressure, humidity, composition, viscosity, density, reactivity, combustibility, corrosivity, and toxicity; and (c) design and performance characteristics of the control system such as pressure drop, reliability, dependability, compliance with utility and maintenance requirements, and temperature limitations, as well as size, weight, and fractional efficiency curves for particulates and mass transfer or contaminant destruction capability for gases or vapors.

ESPs can handle very large volumetric flow rates at low pressure drops and can achieve very high efficiencies (99.9%). They are roughly equivalent in costs to fabric filters and are relatively inflexible to changes in process operating conditions. Wet scrubbers can also achieve high efficiencies and have the major advantage that some gaseous pollutants can be removed simultaneously with the particulates.

However, they can only handle smaller gas flows (up to 3,000 m^3/min), can be very costly to operate (owing to a high pressure drop), and produce a wet sludge that can present disposal problems. For a higher flue gas flow rate and greater than 99% removal of PM, ESPs and fabric filters are the equipment of choice, with very little difference in costs. A discussion of the above air pollution control equipment, as well as other devices are given later in this chapter.

ARSENIC

Arsenic is a metalloid that is distributed widely in the earth's crust. Pure arsenic is rarely found in the environment. More commonly, it bonds with various elements such as oxygen, sulfur, and chlorine to form inorganic arsenic compounds and with carbon and hydrogen to form organic arsenic compounds. The water-soluble trivalent and pentavalent oxidation states of inorganic arsenic are the most toxic arsenic compounds. Atmospheric arsenic exists primarily in inorganic form and is absorbed by particulate matter, while soluble arsenate and arsenite salts are the most typical forms in water. Atmospheric arsenic deposits to the soil, and is then absorbed by plants, leached to groundwater and surface water, and taken up by plants and animals. Airborne concentrations of arsenic range from a few nanograms per cubic meter (ng/m^3) to a few tenths of a microgram per cubic meter, but concentrations may exceed 1 $\mu g/m^3$ near stationary sources of emissions. A few micrograms per liter ($\mu g/\ell$) of arsenic are normally found in drinking water. In some locations, however, concentrations may exceed 1 milligram per liter (mg/ℓ). Uncontaminated soil typically contains about 7 micrograms per gram ($\mu g/g$) arsenic, on average, but levels in the range of 100 to 2,500 $\mu g/g$ have been detected near stationary sources, and up to 700 $\mu g/g$ in agricultural soils treated with arsenic-containing pesticides. High concentrations of arsenic, mainly fat-soluble or water-soluble organoarsenic compounds, have been observed in seafood.

The highest mineral concentrations can be found as arsenides of copper, lead, silver, and gold, but high levels may also be found in some coal. The principal natural sources of arsenic in the atmosphere are volcanic activity and, to a lesser degree, low-temperature volatilization.

White arsenic (arsenic trioxide), a by-product of roasting sulfide ores, is the basis for manufacturing all arsenicals. The main uses of arsenicals, as components of pesticides and herbicides, have been banned in many countries. Arsenicals are also used in leather pigments. Chromated copper, sodium, and zinc arsenates are used in antifungal wood preservatives, and in some places, arsanilic acid is added to farm animal feed as a growth stimulant. Metallic arsenic is used in electronics and as a metal alloy, and sodium arsenite has been included in drugs for treating leukemia and other diseases. Arsenic is also used in lead crystal glass manufacturing, contributing to atmospheric emissions and the generation of highly toxic wastes. The greatest part of anthropogenic arsenic emissions originates from stationary sources, including copper smelting (about 50%), combustion of coal, especially low-grade brown coal (about 20%), and other nonferrous metal industries (around 10%). The drying of concentrates in mining operations also contributes to atmospheric emissions of arsenic. The contribution of agriculture to anthropogenic arsenic releases, through the use of arsenicals as pesticides and herbicides and through the burning of vegetation and of wood treated with arsenic-containing preservatives, is estimated at around 20%. The largest contributors of arsenic in terrestrial waters are landfills, mines, pit heaps, wastewater from smelters, and arsenic-containing wood preservatives. Some iron and steel plants that use iron pyrites from metal mines, as well as other industries, such as sulfuric acid plants, that use pyrite as a source of sulfur for production, could be substantial sources of arsenic pollution of both air and water.

Ingestion is the main route of exposure to arsenic for the public. Arsenic can have both acute and chronic toxic effects on humans. It affects many organ systems including the respiratory, gastrointestinal, cardiovascular, nervous, and hematopoietic systems. When ingested in dissolved form, both inorganic and organic soluble arsenic compounds are readily absorbed from the gastrointestinal tract; less soluble forms have lower absorption rates. Short-term acute poisoning cases involving the daily ingestion of 1.3-3.6 g arsenic by children in Japan resulted in acute renal damage, disturbed heart function, and death. Chronic exposure leads to accumulation of arsenic in the bone, muscle, and skin and, to a smaller degree, in the liver and the kidneys. Mild chronic poisoning causes fatigue and loss of energy. More severe symptoms include peripheral vascular disorders ("blackfoot disease"), gastrointestinal problems, kidney degeneration, liver dysfunction, bone marrow injury, and severe neuralgic pain. Such symptoms have been reported in populations consuming water with 500-1,000 mg/ℓ arsenic content. Chronic exposure also results in dermatological disorders such as palm and sole hyperkeratosis, allergic contact dermatitis, and cancerous lesions. Long term consumption of drinking water with arsenic concentrations exceeding 200 mg/ℓ has been connected with the prevalence of skin cancer.

Inhalation is a less significant pathway for arsenic exposure for the general population, although smokers are constantly exposed to some arsenic due to the natural arsenic content of tobacco leaf and the effect of arsenate insecticide treatment used by tobacco plantations. There are some indications that smoking may exacerbate the effects of exposure to airborne arsenic. About 30% of inhaled arsenic is absorbed by the human body. Acute inhalation of inorganic arsenic compounds can result in local damage to the respiratory system, including perforation of the nasal septum. Increased mortality from cardiovascular diseases and lung cancer was associated with exposure of smelter workers to high levels of airborne arsenic. The carcinogenic potential of inorganic arsenic is considered the key criterion in assessing the hazard from both environmental and occupational exposures. Ingested organic arsenic compounds have no proven health effects even at relatively high concentrations.

CADMIUM

Cadmium is a relatively rare soft metal that occurs in the natural environment typically in association with zinc ores and, to a lesser extent, with lead and copper ores. Some inorganic cadmium compounds are soluble in water, while cadmium oxide and cadmium sulfide are almost insoluble. In the air, cadmium vapor is rapidly oxidized. Wet and dry deposition transfers cadmium from the ambient air to soil, where it is absorbed by plants and enters the food chain. This process may be influenced by acidification that increases the availability of cadmium in soil. Atmospheric levels of cadmium range up to 5 ng/m^3 in rural areas, from 0.005 to 0.015 $\mu g/m^3$ in urban areas, and up to 0.06 $\mu g/m^3$ in industrial areas. Concentrations may reach 0.3 $\mu g/m^3$ weekly mean values near metal smelters. Atmospheric cadmium is generally associated with particulate matter of respirable size. Fresh water typically contain levels of cadmium below 1 $\mu g/l$, but concentrations up to 10 $\mu g/l$ may occur on rare occasions due to environmental disturbances such as acid rain. Concentration in nonpolluted agricultural soils varies between 0.01 and 0.7 micrograms per gram ($\mu g/g$). Food contains cadmium as a result of uptake from the soil by plants and bioaccumulation in terrestrial and aquatic animals. The highest concentrations of cadmium are found in shellfish (over 1 $\mu g/g$) and in the liver and kidneys of farm animals (0.1-1 $\mu g/g$).

Cadmium is emitted into the atmosphere from natural sources, mainly volcanic activities, and from anthropogenic sources. Metal production (drying of zinc concentrates and roasting, smelting, and refining of ores) is the largest source of anthropogenic atmospheric cadmium emissions, followed by waste incineration and by other sources, including the production of nickel-cadmium batteries, fossil fuel combustion, and generation of dust by industrial processes such as cement manufacturing.

The largest contributors to the contamination of water are mines (mine water, concentrate processing water, and leakages from mine tailings); process water from smelters; phosphate mining and related fertilizer production; and electroplating wastes. The largest sources of cadmium in landfills are smelters, iron and steel plants, electroplating wastes, and battery production. Mine tailings generated as the result of zinc mining also have the potential to transfer cadmium to the ambient environment.

Cadmium is mainly used as an anticorrosion coating in electroplating, as an alloying metal in solders, as a stabilizer in plastics (organic cadmium), as a pigment, and as a component of nickel-cadmium batteries. Cadmium production may use by-products and wastes from the primary production of zinc.

Ingestion via food, especially plant-based foodstuffs, is the major route by which cadmium enters the human body from the environment. Average human daily intake of cadmium from food has been estimated at around 10-50 μg. This may increase to several hundred micrograms per day in polluted areas. The intake of cadmium through inhalation is generally less than half that via ingestion, while daily intake from drinking water ranges from below 1 μg to over 10 μg. The kidney, especially the renal tract, is the critical organ of intoxication after exposure to cadmium. Excretion is slow, and renal accumulation of cadmium may result in irreversible impairment in the reabsorption capacity of renal tubules. Only a small proportion (5-10%) of ingested cadmium is absorbed by humans, and large variations exist among individuals. Severe renal dysfunction and damage to the bone structure, a syndrome termed itaiitai disease, have been associated with long-term exposure to cadmium in food (mainly rice) and water in Japan. The WHO estimated that long-term daily ingestion of 200 μg of cadmium via food can be connected with 10% prevalence of adverse health effects. Deficiencies of iron, zinc, and calcium in the human body generally facilitate cadmium absorption. Since most crops, with the exception of rice, contain zinc that inhibits the uptake of cadmium by animals and humans, there is no scientific proof that populations in general are at risk of cadmium exposure via the food chain.

Less than 50% of inhaled cadmium is absorbed from the lungs. Acute and chronic exposure to cadmium dust and fumes, occurring mainly under working conditions, can result in cadmium poisoning. Acute respiratory effects can be expected at cadmium fume concentrations above 1 mg/m^3. Chronic effects occur at exposures to 20 μg/m^3 cadmium concentrations after about 20 years. Because of the cadmium content of tobacco, heavy smokers have elevated absorption of airborne cadmium. Cigarettes containing 0.5 mg cadmium per gram of tobacco can result in up to 3 mg daily cadmium absorption via the lungs, assuming a 25% absorption factor. Considering various sources of exposure and applying a safety factor, WHO estimated that 0.2 μg/m^3 was a safe level of atmospheric cadmium concentrations with regard to renal effects through inhalation. Animal studies have yielded sufficient evidence of the carcinogenicity of cadmium in animals. Limited evidence of human carcinogenicity is also available in studies linking long-term occupational exposure to cadmium to increased occurrence of prostate and lung cancer cases. The USEPA estimated the incremental cancer risk from continuous lifetime exposure to 1 μg/ m^3 concentrations to be 0.0018.

Because of the indirect route of exposure to cadmium through the food chain, the accumulation of cadmium without natural degradation, and incomplete understanding of the relationship between emissions into the different media and long-term environmental and biological impacts, ambient environmental standards may not be the best tools for protecting human health from the effects of exposure to environmental cadmium.

LEAD

Lead is a gray-white, soft metal with a low melting point, a high resistance to corrosion, and poor electrical conducting capabilities. It is highly toxic. In addition to its highly concentrated ores, lead is naturally available in all environmental media in small concentrations. From the atmosphere lead is transferred to soil, water, and vegetation by dry and wet deposition. A significant part of lead particles from emissions sources is of submicron size and can be transported over large distances. Larger lead particles settle more rapidly and closer to the source. Lead in soil binds hard, with a half-life of several hundred years. New depositions, primarily atmospheric, therefore contribute to increased concentrations. Atmospheric deposition is the largest source of lead in surface water, as well. Only limited amounts are transported to water from soil. Terrestrial and aquatic plants show a strong capability to bioaccumulate lead from water and soil in industrially contaminated environments. Lead can also be taken up by grazing animals, thus entering the terrestrial food chain. Natural atmospheric lead concentrations are estimated to be in the range of 0.00005 $\mu g/m^3$. Urban concentrations are around 0.5 $\mu g/m^3$ and annual average concentrations may reach 3 $\mu g/m^3$ or more in cities with heavy traffic.

Mining, smelting, and processing of lead and lead-containing metal ores generate the greatest part of lead emissions from stationary sources. In addition, the combustion of lead-containing wastes and fossil fuels in incinerators, power plants, industries, and households releases lead into the atmosphere. Airborne ambient lead concentrations reaching over 100 $\mu g/m^3$ have occasionally been reported in the vicinity of uncontrolled stationary sources, decreasing considerably with distance from the source due to the deposition of larger lead particles. As a result of the extensive use of alkyl-lead compounds as fuel additives, vehicular traffic is the largest source of atmospheric lead in many urban areas, accounting for as much as 90% of all lead emissions into the atmosphere. High concentrations of lead in urban air have been attributed to vehicular emissions in various countries. Traffic-generated lead aerosols are mostly of the submicron size; they can penetrate deeply into the lungs after inhalation, and they are transported and dispersed over large distances. With the worldwide phaseout of leaded gasoline, the relative contribution of traffic to environmental lead concentrations is changing. Due to its special physical characteristics, lead has been used in a variety of products. Water distribution systems frequently contain lead pipes or lead solder, contaminating drinking water. Lead carbonate ("white lead") was highly popular as a base for oil paints before its use was banned in most countries in the first half of the twentieth century. Lead-based paint and dust contaminated by such paint still represent significant sources of human exposure in several countries. Lead-acid batteries contribute to the contamination of all environmental media during their production, disposal, and incineration. Lead compounds may be also used as stabilizers in plastics. Other lead-based products include food-can solder, ceramic glazes, crystal glassware, lead-jacketed cables, ammunition, and cosmetics.

The main pathways of lead to humans are ingestion and inhalation. Children up to about six years of age constitute the population group at the highest risk from lead exposure through ingestion: their developing nervous systems are susceptible to lead-induced disruptions; their intake of food is relatively high for their body weight; they are exposed to high intake from dust, dirt, soil, and lead-containing paint due to their hand-to-mouth behavior; and their absorption through the gut is very efficient. According to the WHO, the proportion of lead absorbed from the gastrointestinal tract is four to five times higher in children than in adults. The main sources of lead exposure of children are dust and dirt; the role of dissolved lead in water supply systems, lead-based paint, and other sources varies across locations. The contribution of drinking water to exposure is highest in infants under one year of age and children under five years of age. Lack of essential trace elements such as iron, calcium, and zinc and poor nourishment may increase the absorption of lead by the human body. Inhalation poses the highest risk of exposure to environmental lead in adults. Inhaled airborne lead represents a relatively small part of the body burden in children, but in adults it ranges from 15 to 70%. About 30-50% of lead inhaled with particles is retained in the respiratory system and absorbed into the body. In addition to environmental exposure, alcohol consumption and tobacco smoking have been shown to contribute to human exposure to lead. Lead affects several organs of the human body, including the nervous system, the blood-forming system, the kidneys, and the cardiovascular and reproductive systems. Of most concern are the adverse effects of lead on the nervous system of young children: reducing intelligence and causing attention deficit, hyperactivity, and behavioral abnormalities. These effects occur at relatively low blood lead levels without a known lower threshold. Many of these symptoms can be captured by standardized intelligence tests. Various studies have found a highly significant association between lead exposure and the measured intelligence quotient (IQ) of school-age children.

Prenatal exposure to lead was demonstrated to produce toxic effects in the human fetus, including reduced birth weight, disturbed mental development, spontaneous abortion, and premature birth. Such risks were significantly greater at blood lead levels of 15 $\mu g/d\ell$ and more.

High lead concentrations, generally due to occupational exposure or accidents, result in encephalopathy a life-threatening condition at blood lead levels of 100 to 120 $\mu g/d\ell$ in adults and 80 to 100 $\mu g/d\ell$ in children. An acute form of damage to the gastrointestinal tract known as "lead colic" is also associated with high lead levels. The hematological effects of lead exposure are attributed to the interruption of biosynthesis of heme by lead, severely inhibiting the metabolic pathway and resulting in reduced output of hemoglobin. Reduced heme synthesis has been associated with blood levels over 20 $\mu g/d\ell$ in adults and starting from below 10 $\mu g/dl$ in children.

Pollution Prevention Practices

People are exposed to lead from a variety of sources and in a variety of ways, and ambient guidelines and standards for individual media may not provide sufficient protection. A comprehensive approach and strategy is therefore necessary to protect human health. Ambient environmental quality guidelines and standards should be only the starting point for such a strategy. Environmental monitoring of ambient concentrations in soil, air, and drinking water should help to identify highly polluted areas and high risk population groups. This step should be followed by targeted biological screening and policy intervention. Such an approach should be the core of a comprehensive policy intervention that deals with lead exposure from all sources.

MERCURY

Mercury is a toxic heavy metal that can be found in cinnabar (red sulfide) and other ores containing compounds of zinc, tin, and copper; in rocks such as limestone, sandstone, calcareous shales, and basalt; and in fossil fuels such as coal. Mercury is present in trace amounts in all environmental-media. The bulk of global atmospheric mercury is elemental mercury in vapor form. From the atmosphere, mercury elements are removed through precipitation, resulting in deposition to water bodies, the soil, and vegetation. The ultimate depository of mercury is the sediment of oceans, seas, and lakes, where inorganic mercury is readily transformed into highly toxic organic methylmercury through bacterial synthesis and other enzymic and nonenzymic processes. Organic mercury rapidly accumulates in the aquatic biota and biomagnifies upward through the aquatic food chain, attaining its highest concentrations in fish, especially in large predatory species, where it often exceeds 2.0 $\mu g/g$ and in such species as dolphins, reaching

10 μg/g. Average levels of 0.07 to 0.17 μg/g mercury are found in fish, largely (over 70%) in the form of organic methylmercury. Atmospheric mercury concentrations range from a few ng/m^3 to 0.05 μg/m^3, averaging 0.002 μg/m^3. Near stationary sources such as mines, however, concentrations may reach 0.6-1.5 μg/m^3. Typical concentrations of mercury in water bodies range from below 0.001 to 0.003 μg/m^3. Normal levels in soil range from 0.05 to 0.08 μg/m^3. Mercury tends to bond strongly to particulate matter in fresh water, largely in inorganic mercuric form. Mercury concentrations in soil normally do not exceed 0.1 μg/m^3. Total human daily intake of all forms of mercury from all sources has been estimated at between 5 and 80 μg.

The natural emissions of mercury, mainly a result of the degassing of the Earth's crust and evaporation from water bodies, are two to four times larger than those from anthropogenic sources. About half of the atmospheric mercury generated by anthropogenic sources can be attributed to fossil fuel combustion. Emissions from fossil fuel combustion vary according to the mercury content of the fuel. Mercury levels in coal tend to be one to four orders of magnitude greater than those in fuel oil and natural gas. Waste incineration and the mining and smelting of ores are also important contributors to anthropogenic air pollution. Additional sources include mercury-cell chlor-alkali production and coke ovens. The accumulation, processing, and incineration of mercury-containing waste (for example, batteries and various industrial wastes such as scrubber sludge) contribute to mercury contamination of all environmental media.

Pollution Prevention Practices

The main use of mercury has been as a cathode in the electrolysis of sodium chloride solution to produce caustic soda, which is used by various industries. The mercury-cell chlor-alkali industry has been the largest anthropogenic discharger of mercury to water bodies. The use of liquid metallic mercury in the extraction of gold also contributes to the contamination of rivers, with several major catastrophes occurring within the last few years in British Guiana and Romania. The use of mercury in caustic soda production is being phased out and replaced with membrane technology. The agricultural use of organic mercury in pesticides and fungicides has been banned in many countries to prevent human exposure. Agricultural applications are of particular concern because of the extreme toxicity of the mercury compounds used, the limited control over dispersed use and exposure, and the potential for misuse that could contribute to direct poisoning through the diet. Uses of mercury in electric switches, batteries, thermal sensing instruments, cosmetics, pharmaceuticals and dental preparations have been similarly decreasing.

The main human health hazard of mercury has been associated with exposure to highly toxic organic methylmercury through food, primarily through the ingestion of aquatic organisms, mainly fish. Methylmercury in the human diet is almost completely absorbed into the bloodstream and distributed to all tissues, the main accumulation taking place in the brain, liver, and kidneys. Methylmercury poisoning affects the central nervous system and the areas associated with the sensory, visual, auditory, and coordinating functions. Increasing doses result in paresthesia, ataxia, visual changes, dysarthria, hearing defects, loss of speech, coma, and death. The effects of methylmercury poisoning are, in most cases, irreversible because of the destruction of neuronal cells. Methylmercury shows significant and efficient transplacental transfer and contributes to severe disruptions in the development of the child's brain. Thus, prenatal life is more sensitive to methylmercury exposure than adult life. Not enough evidence exists, however, to establish a no-observed-effect or a dose-response function. According to the WHO, daily intake of 3-7 μg/kg body weight can be connected to an incidence of paresthesia of about 5%. Human intake of mercury through drinking water is generally low, representing only a fraction of the amount of methylmercury intake through diet. The main form of mercury in drinking water is inorganic mercuric mercury with low (7-10%) absorption rates and very low penetration rates to the brain and fetus. The lethal oral dose of metallic and other inorganic mercury compounds for humans is estimated at 1-4 grams. Atmospheric mercury, largely in vapor form, poses a less significant health risk to the general population than exposure to more toxic organic mercury compounds through the diet. About 80% of inhaled vapor is retained and absorbed in the bloodstream. In addition to direct exposure, the indirect impacts of atmospheric mercury on human health through deposition in lakes and rivers are of concern.

NITROGEN OXIDES

Nitrogen oxides (NO_x) in the ambient air consist primarily of nitric oxide (NO) and nitrogen dioxide (NO_2). These two forms of gaseous nitrogen oxides are significant pollutants of the lower atmosphere. Another form, nitrous oxide (N_2O), is a greenhouse gas. At the point of discharge from man-made sources, nitric oxide, a colorless, tasteless gas, is the predominant form of nitrogen oxide. Nitric oxide is readily converted to the much more harmful nitrogen dioxide by chemical reaction with ozone present in the atmosphere. Nitrogen dioxide is a yellowish-orange to reddish-brown gas with a pungent, irritating odor, and it is a strong oxidant. A portion of nitrogen dioxide in the atmosphere is converted to nitric acid (HNO_3) and ammonium salts. Nitrate aerosol (acid aerosol) is removed from the atmosphere through wet or dry deposition processes similar to those that remove sulfate aerosol. Only about 10% of all NO_x emissions come from anthropogenic sources. The rest is produced naturally by anaerobic biological processes in soil and water, by lightning and volcanic activity, and by photochemical destruction of nitrogen compounds in the upper atmosphere. About 50% of emissions from anthropogenic sources comes from fossil-fuel-fired heat and electricity generating plants and slightly less from motor vehicles. Other sources include industrial boilers, incinerators, the manufacture of nitric acid and other nitrogenous chemicals, electric arc welding processes, the use of explosives in mining, and farm silos. Worldwide annual emissions of anthropogenic nitrogen oxides are estimated at approximately 50 million metric tons. The United States generates about 20 million metric tons of nitrogen oxides per year, about 40% of which is emitted from mobile sources. Of the 11 million to 12 million metric tons of nitrogen oxides that originate from stationary sources, about 30% is the result of fuel combustion in large industrial furnaces and 70% is from electric utility furnaces.

Annual mean concentrations of nitrogen dioxide in urban areas throughout the world are in the range of 20-90 mg/m^3. Maximum half-hour values and maximum 24-hour values of nitrogen dioxide can approach 850 and 400 mg/m^3, respectively. Hourly averages near very busy roads often exceed 1,000 mg/m^3. Urban outdoor levels of nitrogen dioxide vary according to time of day, season, and meteorological conditions. Typically, urban concentrations peak during the morning and afternoon rush hours. Levels are also higher in winter in cold regions of the world than in other seasons because of the increased use of heating fuels. Finally, since the conversion of nitrogen dioxide from nitric oxide depends on solar intensity, concentrations are often greater on warm sunny days. Nitrogen oxides decay rapidly as polluted air moves away from the source. Concentrations of nitrogen oxides in rural areas without major sources are typically close to background levels. However, nitrogen oxides can travel long distances in the upper atmosphere, contributing to elevated ozone levels and acidic depositions far from sources of emissions.

Concentrations of nitrogen dioxide in homes may considerably exceed outdoor levels and may therefore be more important for human health. Large sources of indoor nitrogen dioxide include cigarette smoke, gas-fired appliances, and space heaters. Nitrogen dioxide concentrations in kitchens with unvented gas appliances can exceed 200 mg/m^3 over a period of several days. Maximum 1-hour concentrations during cooking may reach 500-1,900 mg/m^3, and 1,000-2,000 mg/m^3 where a gas-fired water heater is also in use. The smoke from one cigarette may contain 150,000-225,000 mg/m^3 of nitric oxide and somewhat less nitrogen dioxide. Epidemiologic studies have rarely detected effects on children or adults from exposure to outdoor nitrogen dioxide. Available data from animal toxicological experiments rarely indicate effects of acute exposure to nitrogen dioxide concentrations of less than 1,880 mg/m^3. Asthmatics are likely to be the group most sensitive to exposure to nitrogen oxides.

Some studies have reported reversible effects on pulmonary function of asthmatics exercising intermittently after 30 minutes of exposure to nitrogen dioxide concentrations as low as 560 mg/m^3. However, the health impact of the change in pulmonary function is unclear; the change of about 10% is within the range of physiological variation and is not necessarily adverse. At levels above 3,760 mg/m^3, normal subjects have demonstrated substantial changes in pulmonary function. Studies with animals have found that several weeks to months of exposure to nitrogen dioxide concentrations less than 1,880 mg/m^3 causes both reversible and irreversible lung effects and biochemical changes. Animals exposed to nitrogen dioxide levels as low as 940 mg/m^3 for six months may experience destruction of cilia, alveolar tissue disruption, obstruction of the respiratory bronchioles, and increased susceptibility to bacterial infection of the lungs. Rats and rabbits exposed to higher levels experience more severe tissue damage, resembling emphysema. The available data suggest that the physiological effects of nitrogen dioxide on humans and animals are due more to peak concentrations than to duration or to total dose.

In addition to the health risks, nitrogen dioxide in reaction to textile dyes can cause fading or yellowing of fabrics. Exposure to nitrogen dioxide can also weaken fabrics or reduce their affinity for certain dyes. Industry has devoted considerable resources to developing textiles and dyes resistant to nitrogen oxide exposure.

Nitrogen oxides are precursors of both acid precipitation and ozone, each of which is blamed for injury to plants. While nitric acid is responsible for only a smaller part of hydrogen ion (H^+) concentration in wet and dry acid depositions, the contribution of nitrogen oxide emissions to acid deposition could be more significant. It is nitrogen oxide that absorbs sunlight, initiating the photochemical processes that produce nitric acid. Approximately 90-95% of the nitrogen oxides emitted from power plants is nitric oxide; this slowly converts to nitrogen dioxide in the presence of ozone.

The extent and severity of the damage attributable to acid depositions is difficult to estimate, since impacts vary according to soil type, plant species, atmospheric conditions, insect populations, and other factors that are not well understood. Nitrates in precipitation may actually increase forest growth in areas with nitrogen-deficient soils.

However, the fertilizing effect of nitrates (and sulfates) may be counterbalanced by the leaching of potassium, magnesium, calcium, and other nutrients from forest soils. There is little evidence that agricultural crops are being injured by exposures to nitrates in precipitation. The amount of nitrates in rainwater is almost always well below the levels applied as fertilizer.

The most evident damage from acid depositions is to freshwater lake and stream ecosystems. Acid depositions can lower the pH of the water, with potentially serious consequences for fish, other animal, and plant life. Lakes in areas with soils containing only small amounts of calcium or magnesium carbonates that could help neutralize acidified rain are especially at risk. Few fish species can survive the sudden shifts in pH (and the effects of soluble substances) resulting from atmospheric depositions and runoff of contaminated waters; affected lakes may become completely devoid of fish life. Figures 1 and 2 illustrate the mechanisms responsible for acid deposition and rain in the environment.

Acidification also decreases the species variety and abundance of other animal and plant life. "Acid pulses" have been associated with the fish kills observed in sensitive watersheds during the spring meltdown of the snowpack. The atmospheric deposition of nitrogen oxides is a substantial source of nutrients that damage estuaries by causing algal blooms and anoxic conditions. Emissions of nitrogen oxides are also a precursor of ground-level ozone (O_3), which is potentially a more serious problem.

Plant scientists blame tropospheric ozone for 90% of the injury to vegetation in North America. Ozone can travel long distances from the source and can contribute to elevated ozone concentrations even in rural areas. Since the meteorological and climatic conditions that favor the production of ozone-abundant sunshine-are also good for agriculture, ozone has the potential to cause large economic losses from reductions in crop yields.

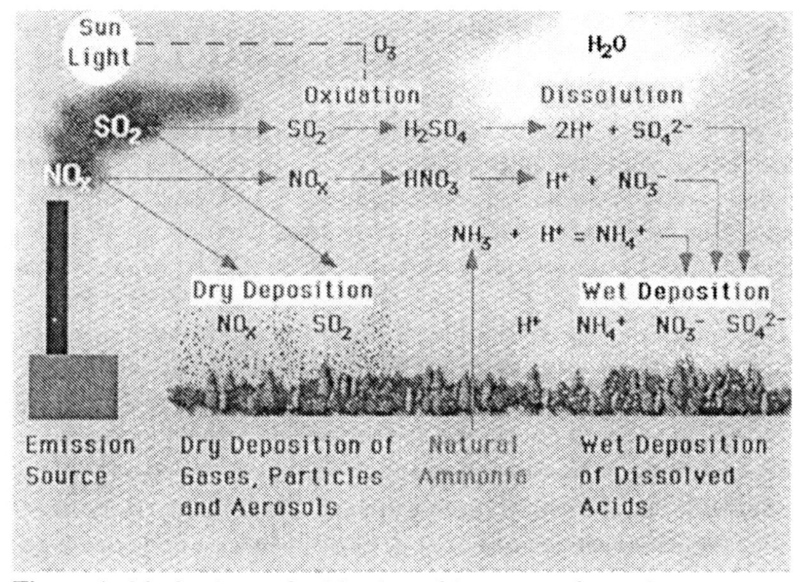

Figure 1. *Mechanisms of acid rain and impact on the environment.*

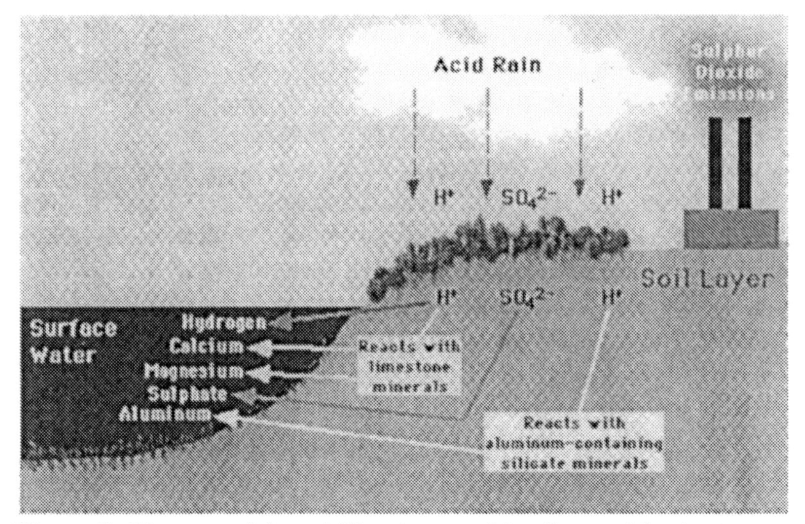

Figure 2. *Illustrates lake acidification resulting from acid rain.*

Nitrogen dioxide affects visibility by absorbing short-wavelength blue light. Since only the longer wavelengths of light are visible to the eye, nitrogen dioxide appears yellowish to reddish brown in color. Nitrogen oxides can also combine with photochemical oxidants to form smog.

Pollution Prevention Practices

The first priority in designing a strategy to control nitrogen oxides is to protect human health. Human health impacts appear to be related to peak exposures to nitrogen oxides (NO_x). In addition to potentially damaging human health, nitrogen oxides are precursors to ozone (O_3) formation, which can harm human health and vegetation. Finally, nitrogen oxides contribute to acid deposition, which damages vegetation and aquatic ecosystems. The extent to which NO_x emissions harm human health depends on ground-level concentrations and the number of people exposed. Source location can affect these parameters. Gases emitted in areas with meteorological, climatological, and topographical features that favor dispersion will be less likely to concentrate near the ground. However, some meteorological conditions, such as inversion, may result in significantly higher ambient levels. Sources away from population centers will expose fewer people to harmful pollution. Plant siting is a critical feature in any air pollution management strategy. However, due to the dispersion of nitrogen oxides that may contribute to ozone formation and acid deposition far from the source, relying on plant siting alone is not a recommended strategy. The long-term objective must be to reduce total emissions. Effective control of NO_x emissions require controls on both stationary sources and mobile transport sources. Each requires different strategies.

Nitrogen oxides are produced in the combustion process by two different mechanisms: (a) the burning the nitrogen in the fuel, primarily coal or heavy oil fuel NO_x and (b) high-temperature oxidation of the molecular nitrogen in the air used for combustion (thermal NO_x). Formation of fuel NO_x depends on combustion conditions, such as oxygen concentration and mixing patterns, and on the nitrogen content of the fuel. Formation of thermal NO_x depends on combustion temperature. Above 1,538 °C, NO_x formation rises exponentially with increasing temperature. The relative contributions of fuel NO_x and thermal NO_x to emissions from a particular plant depend on the combustion conditions, the type of burner, and the type of fuel. Approaches for controlling NO_x from stationary sources can address fuel NO_x, thermal NO_x, or both. One way of controlling NO_x emissions is to use low-nitrogen fuels. Another is to modify combustion conditions to generate less NO_x. Flue gas treatment techniques, such as selective catalytic reduction (SCR) processes, can remove NO_x. Coals and residual fuel oils containing organically bound nitrogen contribute to over 50% of total emissions of NO_x according to some estimates. The nitrogen content of U.S. coal ranges between 0.5% and 2% and that of residual fuel oil between 0.1% and 0.5%. In many circumstances, the most cost-effective means of reducing NO_x emissions will be to use low-nitrogen fuels such as natural gas. Natural gas used as fuel can emit 60% less NO_x than coal and virtually no particulate matter or sulfur oxides.

Combustion control may involve any of three strategies: (a) reducing peak temperatures in the combustion zone; (b) reducing the gas residence time in the high-temperature zone; and (c) reducing oxygen concentrations in the combustion zone. These changes in the combustion process can be achieved either through process modifications or by modifying operating conditions on existing furnaces. Process modifications include using specially designed low-NO_x burners, reburning, combustion staging, gas recirculation, reduced air preheat and firing rates, water or steam injection, and low excess air (LEA) firing. These modifications are capable of reducing NO_x emissions by 50 to 80%. The method of combustion control used depends on the type of boiler and the method of firing fuel.

New low-NO_x burners are effective in reducing emissions from both new power plants and existing plants that are being retrofitted. Low NO_x burners limit the formation of nitrogen oxides by controlling the mixing of fuel and air, in effect automating low-excess-air firing or staged combustion. Compared with older conventional burners, low-NO_x burners reduce emissions of NO_x by 40-60%. Because low-NO_x, burners are relatively inexpensive, power utilities have been quick to accept them; in fact, low-NO_x burners are now a standard part of new designs. Capital costs for low-NO_x burners with overfire air (OFA) range between US$20 and US$25 per kilowatt. Unfortunately, low-NO_x burners are not suitable for reducing NO_x emissions from cyclone fired boilers, which emit large quantities of NO_x due to their high operating temperatures. Because combustion takes place outside the main furnace, the use of low-NO_x burners is not suitable for these applications. However, reburning technology can reduce NO_x emissions.

Reburning is a technology used to reduce NO_x emissions from cyclone furnaces and other selected applications. In reburning, 75-80% of the furnace fuel input is burned in the furnace with minimum excess air. The remaining fuel (gas, oil, or coal) is added to the furnace above the primary combustion zone. This secondary combustion zone is operated substoichiometrically to generate hydrocarbon radicals that reduce to nitrogen the nitrogen oxides that are formed. The combustion process is then completed by adding the balance of the combustion air through overfire air ports in a final burnout zone at the top of the furnace.

Staged combustion (off-stoichiometric combustion) burns the fuel in two or more steps. Staged combustion can be accomplished by firing some of the burners fuel-rich and the rest fuel-lean, by taking some of the burners out of service and allowing them only to admit air to the furnace, or by firing all the burners fuel-rich in the primary combustion zone and admitting the remaining air over the top of the flame zone. Staged combustion techniques can reduce NO_x emissions by 20-50%. Conventional OFA alone can reduce emissions of NO_x by 30%, and advanced OFA has the potential of reducing them still further, although potential for corrosion and slagging exists. Capital costs for conventional and advanced OFA range between US$5 and $10 per kilowatt.

Flue gas recirculation (FGR) is the rerouting of some of the flue gases back to the furnace. By using the flue gas from the economizer outlet, both the furnace air temperature and the furnace oxygen concentration can be reduced. However, in retrofits FGR can be very expensive. Flue gas recirculation is typically applied to oil- and gas-fired boilers and reduces NO_x emissions by 20-50%. Modifications to the boiler in the form of ducting and an energy efficiency loss due to the power requirements of the recirculation fans can make the cost of this option higher.

Reduced air preheat and reduced firing rates lower peak temperatures in the combustion zone, thus reducing thermal NO_x. This strategy, however, carries a substantial energy penalty. Emissions of smoke and carbon monoxide need to be controlled, which reduces operational flexibility.

Water or steam injection reduces flame temperatures and thus thermal NO_x. Water injection is especially effective for gas turbines, reducing NO_x emissions by about 80% at a water injection rate of 2%. For a gas turbine, the energy penalty is about 1%, but for a utility boiler it can be as high as 10%. For diesel-fired units, 25-35% reductions in NO_x emissions can be achieved using water fuel mixtures.

Low-excess-air firing (LEA) is a simple, yet effective technique. Excess air is defined as the amount of air in excess of what is theoretically needed to achieve 100% combustion. Before fuel prices rose, it was not uncommon to see furnaces operating with 50-100% excess air. Currently, it is possible to achieve full combustion for coal-fired units with less than 15-30% excess air. Studies have shown that reducing excess air from an average of 20% to an average of 14% can reduce emissions of NO_x by an average of 19%. Techniques involving low-excess-air firing staged-combustion, and flue gas recirculation are effective in controlling both fuel NO_x and thermal NO_x. The techniques of reduced air preheat and reduced firing rates (from normal operation) and water or steam injection are effective only in controlling thermal NO_x. These will therefore not be as effective for coal-fired units since about 80% of the NO_x emitted from these units is fuel NO_x.

Flue gas treatment (FGT) is more effective in reducing NO_x emissions than are combustion controls, although at higher cost. FGT is also useful where combustion controls are not applicable. Pollution prevention measures, such as using a high-pressure process in nitric acid plants, is more cost-effective in controlling NO_x emissions. FGT technologies have been primarily developed and are most widely used in Japan. The techniques can be classified as selective catalytic reduction, selective noncatalytic reduction, and adsorption.

Selective catalytic reduction (SCR) is currently the most developed and widely applied FGT technology. In the SCR process, ammonia is used as a reducing agent to convert NO_x to nitrogen in the presence of a catalyst in a converter up stream of the air heater. The catalyst is usually a mixture of titanium dioxide, vanadium pentoxide, and tungsten trioxide. SCR can remove 60-90% of NO_x from flue gases. Unfortunately, the process is very expensive (US$40-$80/kilowatt), and the associated ammonia injection results in an ammonia slip stream in the exhaust. In addition, there are safety and environmental concerns associated with anhydrous ammonia storage.

Selective noncatalytic reduction (SNCR) using ammonia- or urea-based compounds is still in the developmental stage. Early pilot studies indicate that SNCR systems can reduce NO_x emissions by 30-70%. Capital costs for SNCR are expected to be much lower than for SCR processes, ranging between US$10 and US$20 per kilowatt. Several dry adsorption techniques are available for simultaneous control of NO_x and sulfur oxides (SO_x). One type of system uses activated carbon with ammonia (NH_3) injection to simultaneously reduce the NO_x to nitrogen (N_2) and oxidize the SO_2 to sulfuric acid (H_2SO_4). If there is no sulfur in the fuel, the carbon acts as a catalyst for NO_x reduction only. Another adsorption system uses a copper oxide catalyst that adsorbs sulfur dioxide to form copper sulfate. Both copper oxide and copper sulfate are reasonably good catalysts for the selective reduction of NO_x with NH_3. This process, which has been installed on a

40-megawatt oil-fired boiler in Japan, can remove about 70% of NO_x and 90% of SO_2 from flue gases.

For *coal-fired boilers* the most widely applied control technologies involve combustion modifications, including low-excess-air firing, staged combustion, and use of low-NO_x burners. For *oil-fired* boilers, the most widely applied techniques include flue gas recirculation, in addition to the techniques used for coal-fired units. For *gas-fired units,* which in any case emit 60% less NO_x than coal-fired units, the primary control technologies include flue gas recirculation and combustion modifications. Finally, for *diesel plants,* the common technologies are water-steam injection, and SCR technology.

The most cost-effective methods of reducing emissions of NO_x are the use of low-NO_x burners and the use of low nitrogen fuels such as natural gas. Natural gas has the added advantage of emitting almost no particulate matter or sulfur dioxide when used as fuel. Other cost-effective approaches to emissions control include combustion modifications. These can reduce NO_x emissions by up to 50% at reasonable cost. Flue gas treatment systems can achieve greater emissions reductions, but at a much higher cost.

GROUND-LEVEL OZONE

Ozone (O_3) is a colorless, reactive oxidant gas that is a major constituent of atmospheric smog. Ground-level ozone is formed in the air by the photochemical reaction of sunlight and nitrogen oxides (NO_x), facilitated by a variety of volatile organic compounds (VOCs), which are photochemically reactive hydrocarbons. The relative importance of the various VOCs in the oxidation process depends on their chemical structure and reactivity. Ozone may be formed by the reaction of NO_x and VOCs under the influence of sunlight hundreds of kilometers from the source of emissions. Ozone concentrations are influenced by the intensity of solar radiation, the absolute concentrations of NO_x and VOCs, and the ratio of NO_x to VOCs. Diurnal and seasonal variations occur in response to changes in sunlight. In addition, ground-level ozone accumulation occurs when sea breezes cause circulation of air over an area or when temperature-induced air inversions trap the compounds that produce smog. Peak ground-level ozone concentrations are measured in the afternoon. Mean concentrations are generally highest during the summer. Peak concentrations of ground-level ozone rarely last longer than two to three hours. Registered average natural background concentrations of ground-level ozone are around 30-100 $\mu g/m^3$. Short-term (one-hour) mean ambient concentrations in urban areas may exceed 300-800 $\mu g/m^3$. Both natural and anthropogenic sources contribute to the emission of ground-level ozone precursors, and the composition of emissions sources may show large variations across locations. VOCs occurring naturally due to emissions; from trees and plants may account for as much as two thirds of ambient VOCs in some locations. Anaerobic biological processes, lightning, and volcanic activity are the main natural contributors to atmospheric NO_x occasionally accounting for as much as 90% of all NO_x emissions.

Motor vehicles are the main anthropogenic sources of ground-level ozone precursors. Other anthropogenic sources of VOCs include emissions from the chemical and petroleum industries and from organic solvents in small stationary sources such as dry cleaners. Significant amounts of NO_x originate from the combustion of fossil fuels in power plants, industrial processes, and home heaters.

The main health concern of exposure to ambient ground-level ozone is its effect on the respiratory system, especially on lung function. Several factors influence these health impacts, including the concentrations of ground-level ozone in the atmosphere, the duration of exposure, average volume of air breathed per minute (ventilation rate), and the length of intervals between short-term exposures. Most of the evidence on the health impacts of ground-level ozone comes from animal studies and controlled clinical studies of humans focusing on short-term acute exposure. Clinical studies have documented an association between short-term exposure to ground-level ozone at concentrations of 200-500 $\mu g/m^3$ and mild temporary eye and respiratory irritation as indicated by symptoms such as coughing, throat dryness, eye and chest discomfort, thoracic pain, and headache. Temporary decrements in pulmonary function have been found in children at hourly average ground-level ozone concentrations of 160-300 $\mu g/m^3$. Similar impacts have been observed after 2.5-hour exposure of heavily exercising adults and children to concentrations of 240 $\mu g/m^3$. Lung function losses, however, have been reversible and relatively mild even at concentrations of 360 $\mu g/m^3$, with a great variety of personal responses. Full recovery of respiratory functions normally occurs within 24 to 48 hours after exposure.

Animal studies have also demonstrated an inflammatory response of the respiratory tract following exposure to ground-level ozone at 1,000 $\mu g/m^3$ for four hours. Although biochemical and morphological alterations in the red blood cells were found in several animal species after exposure to ground-level ozone concentrations of 400 $\mu g/m^3$ for four hours, no consistent changes have been demonstrated in humans, even at concentrations as high as 1,200 $\mu g/m^3$, and extrapolation of such impacts to humans has not been supported. Exposure to elevated concentrations of ground-level ozone has been shown to reduce physical performance, since the increased ventilation rate during physical exercise increases the effects of exposure to ground-level ozone. There is no evidence that smokers, children, older people, asthmatics, or individuals with chronic obstructive lung disease are more responsive to ground-level ozone exposure than others. Ground-level ozone may, however, make the respiratory airways more responsive to other inhaled toxic substances and bacteria. In addition, a synergistic effect of ground-level ozone and sulfur dioxide has been found, indicating that sulfur dioxide potentiates the effects of ground level ozone.

Besides short-term impacts, the potential for irreversible damage to the lungs from repeated exposure over a longer period of time has been a health concern. Some studies have found an association between accelerated loss of lung function over a longer period of time (five years) and high oxidant levels in the atmosphere. The WHO has noted that the length of the recovery period between successive episodes of high ground-level ozone concentrations and the number of episodes in a season may be important factors in the nature and magnitude of health impacts, since prolonged acute exposure to ground level ozone concentrations of 24-360 $\mu g/m^3$ result in progressively larger changes in respiratory function. However, a cross-sectional analysis based on large samples from multiple locations in the United States found no correlation between chronic ground-level ozone pollution and reduced lung function except for the highest 20% of ground-level ozone exposures, suggesting the possibility of a lower threshold for effects of chronic ground-level ozone exposure. No evidence has been found of an association between peak oxidant concentrations and daily mortality rates of the general population.

Elevated ground-level ozone exposures affect agricultural crops and trees, especially slow growing crops and long-lived trees. Ozone damages the leaves and needles of sensitive plants, causing visible alterations such as defoliation and change of leaf color. In North America, tropospheric ozone is believed responsible for about 90% of the damage to plants. Agricultural crops show reduced plant growth and decreased yield. According to the U.S. Office of Technology Assessment (OTA), a 120 $\mu g/m^3$ seasonal average of seven-hour mean ground-level ozone concentrations is likely to lead to reductions in crop yields in the range of 16-35% for cotton, 0.9-51% for wheat, 5.3-24% for soybeans, and 0.3-5.1% for corn. In addition to physiological damage, ground-level ozone may cause reduced resistance to fungi, bacteria, viruses, and insects, reducing growth and inhibiting yield and reproduction. These impacts on sensitive species may result in declines in agricultural crop quality and the reduction of biodiversity in natural ecosystems. The impact of the exposure of plants to ground-level ozone depends not only on the duration and concentration of exposure but also on its frequency, the interval between exposures, the time of day and the season, site-specific conditions, and the developmental stage of plants. Additionally, ground-level ozone is part of a complex relationship among several air pollutants and other factors such as climatic and meteorological conditions and nutrient balances. For example, the presence of sulfur dioxide may increase the sensitivity of certain plants to leaf injury by ground-level ozone. Also the presence of ground-level ozone may increase the growth-suppressing effects of nitrogen dioxide.

Pollution Prevention Practices

Since ground-level ozone is formed by the photochemical reaction of nitrogen oxides and certain hydrocarbons, abatement strategies should focus not only on reduction of emissions of these substances but also on their ratio and balance. In areas where NO_x concentrations are high relative to VOCs, the abatement of VOC emissions can reduce the formation of ground level ozone, while reduction in nitrogen oxides may actually increase it. In areas where the relative concentration of VOCs is high compared with nitrogen oxides, ground level ozone formation is "NO_x-limited," and NO_x reductions generally work better than VOC abatement.

To recap, surrounding the earth at a height of about 25 kilometers is the stratosphere, which is rich in ozone, and its purpose is to prevent the sun's harmful ultraviolet (UV) rays from reaching the earth. UV rays have an adverse effect on all living organisms, including marine life, crops, animals and birds, and humans. In humans, UV is known to affect the immune system; to cause skin cancer, eye damage, and cataracts; and to increase susceptibility to infectious diseases such as malaria. In 1974, it was hypothesized that chlorinated compounds were able to persist in the atmosphere long enough to reach the stratosphere, where solar radiation would break up the molecules and release chlorine atoms that would destroy the ozone layer. Mounting evidence and the discovery of the Antarctic ozone hole in 1985 led to the global program to control chlorofluorocarbons (CFCs) and other ozone-destroying chemicals. In addition to Antarctica, ozone loss is now present over New Zealand, Australia, southern Argentina and Chile, North America, Europe, and Russia. The ozone-depleting substances (ODSs) of concern are CFCs, halons, methyl chloroform (1,1,1,-trichloroethane; MCF), carbon tetrachloride (CTC), hydrochlorofluorocarbons (HCFCs), and methyl bromide. CFC-11 was assigned an ODP of 1.0; all other chemicals have an ODP relative to that of CFC-11. An ODP higher than 1.0 means that the chemical has a greater ability than CFC-11 to destroy the ozone layer; an ODP lower than 1.0 means that the chemical's ability to destroy the ozone layer is less than that of CFC-11. In September 1987, the Montreal Protocol on *Substances that Deplete the Ozone Layer* (the *Protocol*) was signed by 25 nations and the European Community. The Protocol was the first international environmental agreement, and its signing by so many nations represented a major accomplishment, and a major shift in the approach to handling global environmental problems. The Protocol called for a freeze on the production of halons and a requirement to reduce the production of CFCs by 50% by 1999. However, new scientific evidence surfaced after the entry into force of the Protocol, indicating that ozone depletion was more serious than originally thought. Accordingly, in 1990 (London), 1992 (Copenhagen), and 1995 (Vienna), amendments were made to the Protocol to regulate the phaseout of the original chemicals and the control and phase-out of additional chemicals. The principal provisions of the Montreal Protocol as it now stands are as follows:

- Production of CFCs, halons, methyl chloroform, and CTC ceased at the end of 1995 in industrial countries and will cease by 2010 in developing countries. Developing countries are defined in the Protocol as those that use less than 0.3 kilograms (kg) of ODS per capita per year. These are called Article 5 countries in reference to the defining article in the Montreal Protocol.
- HCFCs, originally developed as a less harmful class of CFC alternatives, will be phased out by 2020 in industrial countries, with some provisions for servicing equipment to 2030. Developing countries are to freeze consumption by 2016 (base year 2015) and phase out use by 2040.
- Consumption and production of methyl bromide will end in 2005 in industrial countries (subject to phase-out stages and exemptions) and in 2015 in developing countries.

It was early recognized that undue hardships might be experienced by industry in developing countries as they implemented replacement technologies. Therefore, a fund was established under the Montreal Protocol to pay for incremental costs such as technical expertise and new technologies, processes, and equipment associated with the phase-out. The Multilateral Fund of the Montreal Protocol is managed by an executive committee consisting of delegates from seven developing countries and seven industrial countries.

In general, ODSs are most often used in the following applications: as propellants in aerosols (CFCs and HCFCs); in refrigeration, air conditioning, chillers, and other cooling equipment (CFCs and HCFCs); to extinguish fires (halons);in the manufacture of foams (CFCs and HCFCs); as solvents for cleaning printed circuit boards and precision parts and degreasing metal parts (CFCs, HCFCs, methyl chloroform, and CTC); in a variety of other areas, such as inks and coatings and medical applications (CFCs, HCFCs, methyl chloroform, and CTC); as a fumigant (methyl bromide). The following discussion provides a brief overview of the alternatives to ODSs that have been developed in various industry sectors. It is not intended to be an exhaustive listing of all alternatives, but it does summarize some proven alternatives. The selection of any alternative should be made with due consideration of other issues that could affect the final choice. Identification, development, and commercialization of alternatives to ODSs are going on constantly. For this reason it is important to seek information on the latest alternatives. For any alternative, consideration needs to be given to, for example, its compatibility with existing equipment, its health and safety aspects, its direct global-warming potential, whether it increases or decreases energy consumption, and the costs that may be incurred in eventual conversion to a non-ODS technology if an interim HCFC alternative is chosen. New ways of doing business may also develop in the course of selecting alternatives. For example, many electronics companies have now converted their manufacturing plants to "no-clean" technology. The benefits include elimination of circuit board cleaning after soldering, savings in chemical costs and waste disposal costs, savings in maintenance and energy consumption, improved product quality, and advances toward new technologies such as fluxless soldering. The selection of any alternative should not be made in isolation from the factors listed above.

Flexible and Rigid Foams

Zero-ODP alternatives are the substitutes of choice in many foam-manufacturing applications. However, the use of HCFCs is sometimes necessary in order to meet some product specifications. The viability of liquid hydrofluorocarbon (HFC) isomers in this industry remains to be proved, and hydrocarbon alternatives need to be better qualified, as well. The issues in these evaluations are safety as related to toxicity and flammability, environmental impact as related to the generation of volatile organic compounds and global warming, product performance as related to insulating properties, conformity to fire codes, and the like, cost and availability, and regulatory requirements.

Some of the alternatives for specific products of the foam manufacturing sector are described briefly below. The alternatives are listed as short-term and long-term options, without an elaboration of the merits of each.

Rigid polyurethane foams used in refrigerators and freezers - Alternatives include hydrocarbons (pentane) and HCFC-141b; long-term alternatives include HFCs (-134a, -245, -356, -365). Vacuum panels may be used in the future.

Rigid polyurethane for other appliances. Alternatives include HCFC-141b, HCFC-22, blends of -22 and HCFC-142b, pentane, and carbon dioxide/water blowing. In the long term, the alternatives include HFCs.

Rigid polyurethane used for boardstock and flexible-faced laminations - Alternatives include HCFC-141b and pentane; in the long term, the use of HFCs should be developed.

Sandwich panels of rigid polyurethane - HCFC-141b, HCFC-22, blends of HCFC-22 and -141b, pentane, and HFC-134a are now used as alternatives to CFCs in this application. In the long term, HFCs and carbon dioxide/water will be the replacement technologies.

Spray applications of rigid polyurethane - Alternatives currently in use for spray applications include carbon dioxide/water and HCFC-141b. Long-term alternatives will be HFCs.

Slabstock of rigid polyurethane - Alternatives include HCFC-141b; long-term alternatives include HFCs and carbon dioxide/water. Pentane may also be used.

Rigid polyurethane pipe construction - CFCs in this application are being replaced by carbon dioxide/water, HCFC-22, blends of HCFC-22 and -142b, HCFC-141b, and pentanes. Long-term alternatives will include HFCs and carbon dioxide/water. For district central heating pipes, pentane and carbon dioxide/water are the preferred technologies.

Polyurethane flexible slab - Many alternatives now exist for flexible slab construction, including extended range polyols, carbon dioxide/water, softening agents, methylene chloride, acetone, increased density, HCFC-141b, pentane, and other alternative technologies such as accelerated cooling and variable pressure. The long term will probably see the use of injected carbon dioxide and alternative technologies.

Molded flexible polyurethane - The standard now is carbon dioxide/water blowing.

Integral-skin polyurethane products - The current alternatives for these products include HCFC-22, hydrocarbons, carbon dioxide/water, HFC-134a, pentanes, and HCFC-141b. The long-term alternate is expected to be carbon dioxide/water.

Phenolic foams - Phenolic foams can now be made using HCFC-141b, hydrocarbons, injected carbon dioxide, or HFC-152a instead of CFCs. In the long term, HFCs may be the predominant alternative.

Extruded polystyrene sheet - Alternatives currently include HCFC-22, hydrocarbons, injected carbon dioxide, and HFC-152a. In the long term, these same alternatives (except for HCFC-22) will be used, along with possible use of atmospheric gases.

Extruded polystyrene boardstock - HCFC-22 and -142b and injected carbon dioxide are the current alternatives. Long-term alternatives will be HFCs and injected carbon dioxide.

Polyolefins - Polyolefins are now manufactured using alternatives such as hydrocarbons, HCFC22 and -142b, injected carbon dioxide, and HFC152a. Hydrocarbons and injected carbon dioxide will be long-term alternatives.

Refrigeration, Air Conditioning, and Heat Pumps - Immediate replacements for many applications include hydrocarbons, HFCs, and HCFCs. Some of these will also be candidates for long-term replacement of the currently used CFCs. This following briefly describes the alternatives that are available for specific refrigeration, air conditioning, and heat pump applications.

Refrigeration, Air Conditioning and Heat Pump Application

Domestic refrigeration - Two refrigerant alternatives are predominant for the manufacture of new domestic refrigerators. HFC-134a has no ozone depletion potential and is nonflammable, but it has a high global-warming potential (GWP). HC-600a is flammable, has a zero ODP, and has a GWP approaching zero. Other alternatives for some applications include HFC-152a and binary and ternary blends of HCFCs and HFCs. Retrofitting alternatives may include HCFC/HFC blends, after CFCs are no longer available. However, the results obtained so far are still not satisfactory. Neither HC-600a nor HFC-134a is considered an alternative for retrofitting domestic refrigeration appliances, but preliminary data indicate that a combination of the two may be a retrofit, or "servicing," candidate.

Commercial refrigeration - Alternatives to CFCs for new commercial refrigeration equipment include HCFCs (including HCFC mixtures) and HFCs and HFC mixtures. Retrofit of existing equipment is possible by using both HCFCs and HFCs, in conjunction with reduced charges and more efficient compressors. Hydrocarbons are, to a limited extent, applied in hermetically sealed systems.

Cold storage and food processing - Although there has been a return to the use of ammonia for some cold storage facilities, there are safety issues, and some regulatory jurisdictions restrict its use. Other alternatives to CFCs in cold storage and large commercial food preservation facilities include HCFC-22 and HFC blends. Hydrocarbons and HCFC-22 will continue to be the favored alternatives until equipment using other alternatives is developed; ammonia will be used in selected applications.

Industrial refrigeration - New industrial refrigeration systems that are used by the chemical, petrochemical, pharmaceutical, oil and gas, and metallurgical industries, as well for industrial ice making and for sports and leisure facilities, can use ammonia and hydrocarbons as the refrigerant. Although the product base concerned is small, existing CFC equipment can be retrofitted to use HCFC-22, HFCs and HFC blends, and hydrocarbons.

Air conditioning and heat pumps (air-cooled systems) - Equipment in this category generally uses HCFC-22 as the refrigerant. Alternatives under investigation include HFCs and HCs (in particular, propane). The most promising of these are the nonflammable, nontoxic HFC compounds, although there is more interest in propane in various regions. HCFs have been criticized for their global warming potential, but their total equivalent warming impact (TEWI), a measure that combines GWP and energy efficiency is equal to or lower than that of the other alternatives.

Air conditioning (water chillers) - HCFC-22 has been used in small chillers, and CFC-11 and -12 have been used in large chillers that employ centrifugal compressors. HFC blends are now beginning to be introduced to replace HCFC-22 in small chillers; HCFC-123 and HFC-134a are the preferred replacements for large units. Chillers that have used CFC-114 can be converted to use HCFC-124 or can be replaced by HFC-134a units.

Transport refrigeration - HCFC-22 and CFC-502 have been the refrigerants of choice for transport refrigeration units, although some applications are using ammonia as the refrigerant. The alternatives include various HFC blends.

Automotive air conditioning - The manufacturers of new automobiles have chosen HFC-134a as the fluid for air conditioning units, and retrofit kits are now available to allow older automobiles to convert to this alternative.

Heat pumps (heating-only and heat recovery) - New heating-only heat pumps use HCFC-22. HFC-134a is an alternative for retrofitting existing heat pumps, and investigation into the use of ammonia for large-capacity heat pumps is continuing. Other alternatives being explored include propane, other hydrocarbons, and hydrocarbon blends.

Solvents, Coatings, Inks, and Adhesives Applications

There now exist alternatives or sufficient quantities of controlled substances for almost all applications of ozone-depleting solvents. Exceptions have been noted for certain laboratory and analytical uses and for manufacture of space shuttle rocket motors. HCFCs have not been adopted on a large scale as alternatives to CFC solvents. In the near term, however, they may be needed as the conventional substances in some limited and unique applications. HCFC-141b is not a good replacement for methyl chloroform (1,1,1-trichloroethane) because its ODP is three times higher. Alternatives for specific uses of ozone-depleting solvents are briefly described below.

Electronics cleaning - For most uses in the electronics industry, ozone-depleting solvents can be replaced easily and, often, economically. A wide choice of alternatives exists. If technical specifications do not require postsolder cleaning, no-clean is the preferred technology. If cleaning is required, the use of water-soluble chemistry has generally proved to be reliable. There are however limitations, whereby water-soluble chemistry is not suitable for all applications.

Precision cleaning - Precision cleaning applications are defined as requiring a high level of cleanliness in order to maintain low-clearance or high-reliability components in working order. To meet rigorous specifications, the alternatives that have been developed include solvent and nonsolvent applications. Solvent options include alcohols, aliphatic hydrocarbons, HCFCs and their blends, and aqueous and semi-aqueous cleaners. Nonsolvent options include supercritical cleaning fluid (SCF), ultraviolet (UV)/ozone cleaning, pressurized gases, and plasma cleaning.

Metal cleaning - Oils and greases, particulate matter, and inorganic particles are removed from metal parts prior to subsequent processing steps such as further machining, electroplating, painting. Alternatives to ozone-depleting solvents that have been developed include solvent blends, aqueous cleaners, emulsion cleaners, mechanical cleaning, thermal vacuum deoiling, and no-clean alternatives.

Dry cleaning - Several solvents exist to replace the ozone-depleting solvents that have traditionally been used by the dry cleaning industry. Perchloroethylene has been used for more than three decades. Petroleum solvents, while flammable, can be safely used when appropriate safety precautions are taken. They include white spirit, Stoddard solvent, hydrocarbon solvents, isoparaffins, and n-paraffin. A number of HCFCs can also be used but should be considered only as transitional alternatives.

Adhesives - Methyl chloroform has been used extensively by the adhesives manufacturing industry because of its characteristics - it is nonflammable and quick drying, and it does not contribute to local air pollution. One alternative for some applications is water-based adhesives.

SULFUR OXIDES

Sulfur oxides (SO_x) are compounds of sulfur and oxygen molecules. Sulfur dioxide (SO_2) is the predominant form found in the lower atmosphere. It is a colorless gas that can be detected by taste and smell in the range of 1,000 to 3,000 $\mu g/m^3$. At concentrations of 10,000 $\mu g/m^3$, it has a pungent, unpleasant odor. Sulfur dioxide dissolves readily in water present in the atmosphere to form sulfurous acid (H_2SO_3). About 30% of the sulfur dioxide in the atmosphere is converted to sulfate aerosol (acid aerosol), which is removed through wet or dry deposition processes. Sulfur trioxide (SO_3), another oxide of sulfur, is either emitted directly into the atmosphere or produced from sulfur dioxide and is readily converted to sulfuric acid (H_2SO_4).

Most sulfur dioxide is produced by burning fuels containing sulfur or by roasting metal sulfide ores, although there are natural sources of sulfur dioxide (accounting for 35-65% of total sulfur dioxide emissions) such as volcanoes. Thermal power plants burning high-sulfur coal or heating oil are generally the main sources of anthropogenic sulfur dioxide emissions worldwide, followed by industrial boilers and nonferrous metal smelters. Emissions from domestic coal burning and from vehicles can also contribute to high local ambient concentrations of sulfur dioxide.

Periodic episodes of very high concentrations of sulfur dioxide are believed to cause most of the health and vegetation damage attributable to sulfur emissions. Depending on wind, temperature, humidity, and topography, sulfur dioxide can concentrate close to ground level. During the London fog of 1952, levels reached 3,500 $\mu g/m^3$ (averaged over 48 hours) in the center of the city and remained high for a period of 5 days. High levels have been recorded during temperature inversions in Central and Eastern Europe, in China, and in other parts of the world.

Exposure to sulfur dioxide in the ambient air has been associated with reduced lung function, increased incidence of respiratory symptoms and diseases, irritation of the eyes, nose, and throat, and premature mortality. Children, the elderly, and those already suffering from respiratory ailments, such as asthmatics, are especially at risk. Health impacts appear to be linked especially to brief exposures to ambient concentrations above 1,000 $\mu g/m^3$ (acute exposures measured over 10 minutes). Some epidemiologic studies, however, have shown an association between relatively low annual mean levels and excess mortality. It is not clear whether long-term effects are related simply to annual mean values or to repeated exposures to peak values.

Health effects attributed to sulfur oxides are likely due to exposure to sulfur dioxide, sulfate aerosols, and sulfur dioxide adsorbed onto particulate matter. Alone, sulfur dioxide will dissolve in the watery fluids of the upper respiratory system and be absorbed into the bloodstream. Sulfur dioxide reacts with other substances in the atmosphere to form sulfate aerosols. Since most sulfate aerosols are part of $PM_{2.5}$, they may have an important role in the health impacts associated with fine particulates. However, sulfate aerosols can be transported long distances through the atmosphere before deposition actually occurs. Average sulfate aerosol concentrations are about 40% of average fine particulate levels in regions where fuels with high sulfur content are commonly used. Sulfur dioxide adsorbed on particles can be carried deep into the pulmonary system. Therefore, reducing concentrations of particulate matter may also reduce the health impacts of sulfur dioxide. Acid aerosols affect respiratory and sensory functions.

Sulfur oxide emissions cause adverse impacts to vegetation, including forests and agricultural crops. Studies have shown that plants exposed to high ambient concentrations of sulfur dioxide may lose their foliage, become less productive, or die prematurely. Some plant species are much more sensitive to exposure than others. Plants in the immediate vicinity of emissions sources are more vulnerable. Studies have shown that the most sensitive species of plants begin to demonstrate visible signs of injury at concentrations of about 1,850 $\mu g/m^3$ for 1 hour, 500 $\mu g/m^3$ for 8 hours, and 40 $\mu g/m^3$ for the growing season. Canadian studies showed that chronic effects on pine forest growth were prominent where concentrations of sulfur dioxide in air averaged 44 $\mu g/m^3$, the arithmetic mean for the total 10 year measurement period; the chronic effects were slight where annual concentrations of sulfur dioxide averaged 21 $\mu g/m^3$.

Trees and other plants exposed to wet and dry acid depositions at some distance

Regenerable FGDs generally have higher capital costs than throwaway systems but they tend to have lower waste disposal requirements and costs.

In wet FGD processes, flue gases are scrubbed in a liquid or liquid/solid slurry of lime or limestone. Wet processes are highly efficient and can achieve SO_x removal of 90% or more. With dry scrubbing, solid sorbents capture the sulfur oxides. Dry systems have 70-90% sulfur oxide removal efficiencies and often have lower capital and operating costs, lower energy and water requirements, and lower maintenance requirements, in addition to which there is no need to handle sludge. However, the economics of the wet and dry (including "semi-dry" spray absorber) FGD processes vary considerably from site to site. Wet processes are available for producing gypsum as a by product. Table 1 provides some comparisons between the different pollution control methods that dominate today.

Table 1. *SO_x Pollution Control Process Comparisons.*

Process	% SO_x Reduction	Capital Costs ($/kilowatt)
Sorbent injection	30-70	50-100
Dry flue gas desulfurization	70-90	80-170
Wet flue gas desulfurization	>90	80-160

POLLUTION CONTROL EQUIPMENT

This section provides an overview of conventional pollution control equipment. Historically these equipment were intended as end-of-pipe treatment technologies, and even today, there are many emissions issues that require the use of these conventional devices. Pollution prevention practices do not necessarily displace these equipment, but rather oftentimes incorporate them into cleaner technologies. Therefore, an understanding of the major pollution control devices and hardware is necessary for implementing a pollution prevention program.

Air Pollution Control Equipment

There are several types of conventional equipment used for air pollution control purposes. These include filtration devices, settling chambers, cyclone separators, electrostatic precipitators, venturi scrubbers. Many are briefly described below.

Fabric Filters

Fabric filters, more commonly called baghouses or dust collectors, have been in use since the early 1900s in the mining industry. Dry dust filters are available in sizes ranging from a few square feet up to several hundred thousand square feet of cloth.

Gas flows that can be handled by individual units range from under 100 cfm to over 1,000,000 cfm. The fabric filter's design is similar to that of a large vacuum cleaner. It consists of bags of various shapes constructed from a porous fabric.

Filter bags are available in two major configurations, namely, flat (envelope) bags and round (tubular) bags. Figure 3 illustrates the operation of a baghouse. The dust-laden gas enters the module through an inlet diffuser that breaks up the gas stream and evenly disperses the dust. The heavier dust particles settle into the hopper and the fine particles rise through the tube sheet into the bags.

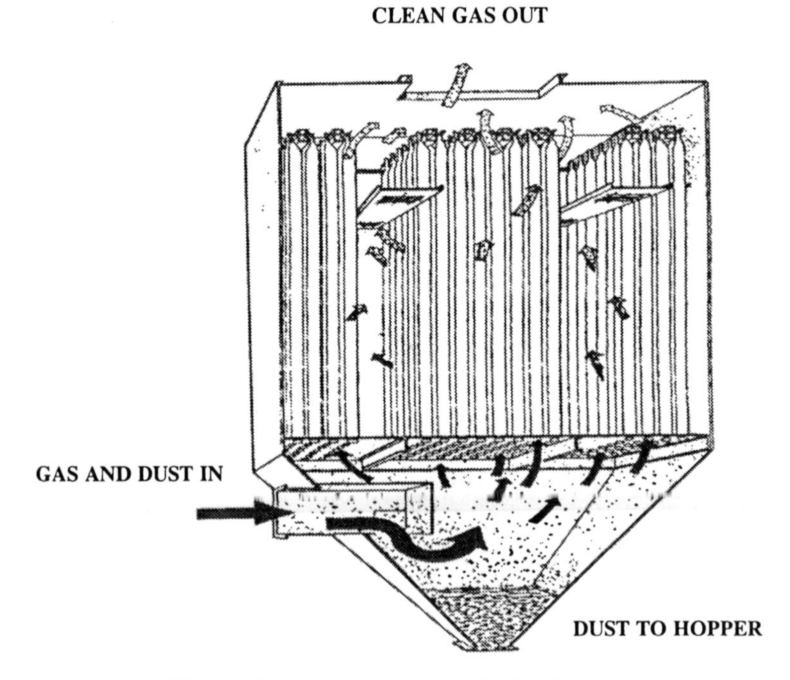

Figure 3. *Basic operation of a baghouse.*

Particles typically as small as 0.5 μ in diameter are collected on the inside of the bags, while the cleaned gas passes through the fabric. Dust is removed from the bags by periodic shaking accomplished by an automatic shaker. The frequency of cleaning depends on the type of dust, the concentration, and the pressure drop which must be overcome. The dust shaken from the bags falls into the hopper below and is removed by a rotary airlock, screw conveyor, or other devices.

Series modules can be joined to provide any desired capacity. When two or more modules are joined together, a single module can be shut down for bag cleaning and then returned to service. The simple closing of an inlet or outlet damper diverts the dirty gas stream to other modules. Thus, the gas is filtered continuously. Multi-module installations typically employ a large single fan or small individual fans mounted on each module. Small fan arrangements are more flexible and eliminate the need for outlet ductwork and the foundation that are required for a large fan. Individual fans often simplify maintenance and permit fan, motor, drive, or other components to be changed readily without interrupting normal service. Any one module can be shut down and isolated from the rest of the system while still maintaining full operation and efficiency levels. The particles to be removed play an important role in the selection of a fabric and filter efficiency. Specifically, particle density, concentration, velocity, and size are important. Each of these properties is interrelated to the pressure drop of the system, which has a direct impact on operating or particle capturing efficiency. Principle variables directly related to pressure drop are gas velocity, the cake resistance coefficient, the weight of cake per unit area, and the air-to-cloth ratio. The cake resistance coefficient is dependent on the particle size and shape, range of the particle sizes, and humidity. Weight of the cake per unit area is related to the concentration of particulate matter.

Large scale air filtration systems for air pollution control applications are generally packaged systems. Commercially available packages are available which include both the reverse air and compressed air types. Reverse air baghouses have typical diameters that range from 8 to 18 ft, from 6,000 cfm to 94,000 cfm. Compressed air baghouses usually include both bottom bag removal (i.e., through bin vents) as well as top bag removal units. Commercial systems can be round or square/rectangular in configuration. Round (plan view) body sections are usually retrofitted with square plenum sections.

Depending on the accessibility afforded by adjacent equipment, staged ladders and access platforms are often quoted as options by vendors. Square body configuration systems (from a plan view) have square body and plenum sections. Safety handrails surrounding the plenum are standard. Structural supports are usually optional, the height of which depends on system location. Figures 4 and 5 provide illustrations of the two basic configurations.

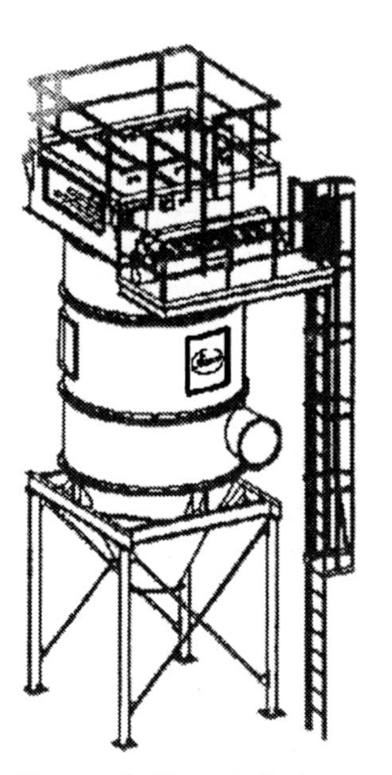

Figure 4. *Round design.*

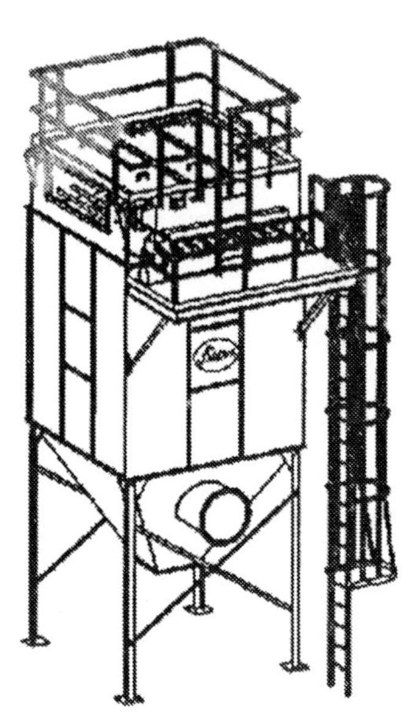

Figure 5. *Square design.*

The design of a fabric filter system must satisfy two criteria, namely, high particle capture efficiency and low pressure drop. Prediction of the pressure drop and knowledge of its dependence on operating conditions of the filter are necessary for proper design. Fabric filters composed of thicker felt materials have a complex orientation of fibers which can achieve a high collection efficiency with less dust buildup. Consequently, felt filters cannot be cleaned by mechanical shaking due to embedment of fine particles. Instead, a high pressure air stream (60 to 120 psi) is used to disengage the dust particles. The specific application dictates the type of fiber. For example, fiberglass filters can withstand higher temperatures than wool; nylon is a poor fiber to use for chemical resistance. The fabric is designed to withstand thermal, chemical, and mechanical action.

Fabric filters can be categorized according to the particular cleaning method, the filter capacity, the type of filter media, the temperature capability, and the type of service (either intermittent or continuous). There are three major cleaning methods employed: shakers, reverse air, and pulse jet. The oldest and most widely used cleaning method is mechanical shaking. The casing is divided into an upper and lower portion by a tube sheet. The woven fabric tubular bags are located in the upper portion with a pyramid-shape hopper in the lower end. Each bag is supported between a flexible cap and a fixed thimble. Gas velocity entering the hopper is reduced, causing the coarse particles to settle out. The gas enters the tube on the inside causing the fine particles to be collected, with the clean gas passing through the fabric into a common outlet manifold. After a certain amount of dust buildup, the flexible support mechanically shakes the particle loose from the fabric into the hopper. The reverse air cleaning baghouse operates in the same manner as the mechanical shaking arrangement, except a reverse air flow replaces the shaking process. An air vent located in the outlet manifold is opened allowing atmospheric air to enter the casing, thereby collapsing the bags and dislodging the dust particles. Baghouses incorporating the pulse jet cleaning method are constructed with an upper and lower compartment separated by a tube sheet. The upper portion serves as the discharge manifold. The felted filter bags are supported by a venturi-shaped thimble attached to the tube sheet. A compressed air jet is located above each filter bag to facilitate cleaning. Internal frames (mesh cages) with a closed bottom prevent the collapse of the bags during the cleaning cycle. Dirty gas enters the hopper and is then directed into the casing, passing through the filter bags. The dust is collected on the outside surface, allowing the clean air to pass through the fabric and out the discharge manifold. The filter bags are cleaned by the force of the pulse jet expanding the bags. Figure 6 illustrates the operation.

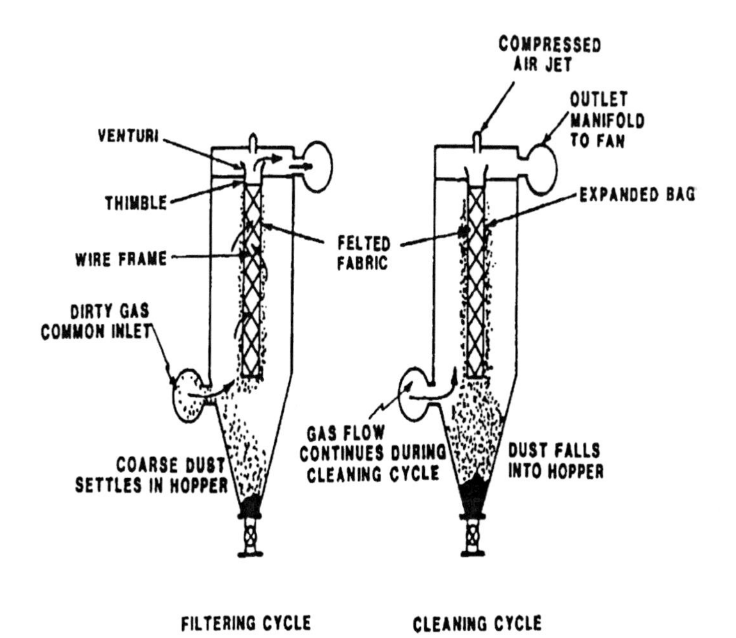

Figure 6. *Automatic baghouse with pulse jet cleaning.*

Filters are grouped according to the capacity by volume as follows: small volumes (i.e., below 10,000 acfm), medium volumes (i.e., 10,000 to 100,000 acfm), and large volumes (i.e., $>$100,000 acfm). The filter-media types include woven and felted media. Temperature capabilities of the media range from higher temperatures ($>$ 400 °F), to medium temperatures (200 to 400 °F) and low temperatures ($<$ 200 °F). The ability of the fabric to collect fine particles and maintain a good cleaning process should serve as the basis in selection of a fabric. As the dust layer or so-called filter cake layer builds up, flow resistance increases. Cleaning reduces the gas flow resistance and maintains the proper pressure drop across the filter.

Cloth filters are designed to remove three types of particles, and each type incorporates the basic principles of air filtration. Particles in the submicron size are collected as a result of the Brownian motion and bridging. As the particles build-up on the surface of the media, the collection surface areas increase causing particles to be captured. Collection efficiencies in excess of 99.95 % are possible. Particles having diameters in the 1 to 10 μ range and coarse particles (above 10 μ) rely on inertial collection. Efficiencies of 90 % to 95 % are achieved with particles under 10 μ in size. Efficiencies of coarse particles fluctuate from 50 % to 99.9 %. The reason for such variations in efficiency is a result of re-entrainment caused by gas flows at high velocities. High velocities can force the particles through the pores of the fabric.

Fabric filters are made of natural fibers, such as cotton and wool, or from synthetic fibers, depending upon their application. Cotton and wool are available in spun form, which limits the individual fibers to a few inches in length. Spun fibers can produce characteristics not found in filament (continuous) fibers. Filters composed of spun fibers are thicker, bulkier, heavier and provide a higher permeability to air flow. Synthetic fibers offer greater versatility such as higher operating ranges and corrosion resistance, but are more costly. Synthetic fibers are available as multi-filament fabrics, which are the most widely used and produced characteristics not common on spun forms. Multi-filament fabrics are light weight, of high tensile strength and high dimensional stability, abrasion resistant, and easy to clean.

Collection efficiency is affected by thread count, yarn size, and twist of the yarn. Permeability is increased by decreasing the thread count (either warp or fill), and subsequently increasing the pore area. Collection efficiency is reduced by increased permeability. A balanced weave is optimum for efficient operation. Permeability is also a function of the yarn size. The larger the yarn size, the lower the permeability. Yarn diameter, which can be altered by the twist of the yarn, also affects permeability; the smaller the diameter, the higher the permeability.

The ability of the gas stream to permeate the filter is also affected by the shrinking and elongation of the fabric. Elongation of the fibers results in increased pore space, and conversely shrinkage decreases the pore volume. Fabrics of good dimensional qualities are essential to fabric life and efficiency.

Fabric finishing steps enhance the collection efficiency of a system. The most commonly applied finishing steps include calendering, napping, singeing, glazing, and coating. Calendering is where surface fibers are pushed down onto the fabric medium by high pressure pressing on the fabric. Napping refers to scraping the filter medium, which raises the surface fibers. Singeing involves separate surface fibers that are removed by passing the filter medium over an open flame. Glazing involves surface fibers fused to the filter medium by high pressure pressing at elevated temperatures. Finally, coating involves a surface preparation over the fibers that reduces self-abrasion.

In order to specify a fabric filter, the properties of the incoming gas and particulate matter must be well defined. The properties of the gas essential to the problem definition are volume, temperature, moisture content, and acid gas concentration. The dust properties of importance are the density, particle size, particle size distribution, and the dust loading. The size of the fabric filter dust collector is directly related to the gas-to-cloth ratio as previously noted, which can be calculated by dividing the total gas volume by the total area of cloth or filter medium. Corrections to the gas-to-cloth ratio are required for the volume of air introduced to the unit and to account for the area of cloth not exposed to the incoming gas. The type of fabric selected is a function of the cleaning process employed. For example, the pulse-jet cleaning method would be used in conjunction with the felt fabric to achieve proper cleaning and efficiency. The filter medium type is the next major consideration following the gas/cloth ratio and cleaning method.

Cotton bags are used in standard installations and are the most economical. A maximum operating temperature of 180 °F is recommended for continuous use with 225 °F allowed for surge conditions. Wool bags are used for applications with dust particles of a combustible nature, or with operating temperatures of 200 °F and an allowable surge temperature of 250 °F. Nylon has a greater tensile strength than cotton or wool and provides excellent abrasion resistance. Fiberglass is most resistant to high temperatures, with a maximum operating temperature of 500 °F. To increase the allowable temperature, fiberglass filters are silicone treated to permit their use in applications such as in carbon black production plants. Replacement of bag filters generates the highest maintenance and cost of the system. Typical causes of bag failure include too high of a gas to cloth ratio, metal-to-cloth abrasion problems, chemical attack by the gas stream or particulates, inlet velocity abrasion, and excessive gas temperatures. The quality of the fabric and method of cleaning are additional factors to consider in evaluating service and maintenance costs. If a filter bag tears, it is important to repair the bag as quickly as possible to prevent abrasion to adjacent bags by jet streams of dust discharging out of the damaged bag. This type of bag failure is limited to inside bag collection types of dust collectors. The speed of repair is determined by the opacity of the outlet bag. In a compartmentalized system, broken bags can be found by monitoring the emissions while isolating one compartment at a time. To prevent a higher filter velocity, damaged filter bags within a compartment should not be replaced with clean bags. The higher velocity could create greater pressure drop or failure due to dust abrasion. An alternative is to plug or tie off the flow.

To recap, the first step in selecting a fabric filter is to define the magnitude of the particulate loading. Knowledge of the particulate matter collected, properties of the gas stream, and the cleaning method are essential to proper design. Improper design leads to low efficiency and unscheduled maintenance. Prior to selection, results from a related application should be investigated. An alternative is to operate a pilot unit to ensure the most optimum gas-to-cloth ratio for a specified pressure drop. With proper design, operation, and maintenance, better than 99.9% efficiency can be achieved, depending on the application. A major advantage of fabric filters, is their ability to operate at a high efficiency at all loads from maximum down to very low gas flow. Some disadvantages of the system are the space requirements and high maintenance costs. Other problems associated with fabric filters are plugging of the fabric due to operation below the dew point or break down of the filter bags, resulting from high temperatures.

Fabric filters can be more costly to operate and maintain than electrostatic precipitators, cyclones, and scrubbers; however, fabric filters are more practicable for filtration of specific dusts. For example: fabric systems are the typical control method for toxic dusts from insecticide manufacturing processes, salt fumes from heat treating, metallic fumes from metallurgical processes, and other applications. Any other control method may not be as efficient, nor economically feasible for such applications.

Up to this point, we have described the operations and features of an industrial baghouse dust collector, which is essentially a system containing filter fabric of some kind which removes dust and particulate from a gas streams. The cleaned gas can then be vented to the atmosphere. If the dust is valuable, it can be recycled back into the process. In contrast, a cartridge dust collector uses gravity together with a downward airflow pattern to provide extremely high filtration efficiency while reducing energy consumption. The cartridge is mounted in a horizontal and slightly downward sloping position. This design increases filter life and improves airflow because both gravity and airflow are pushing the dust downward into the collection hopper. Filters in baghouses typically need to be replaced after approximately one year of operation. Users should consider the cost of maintenance and filter replacement when purchasing or specifying a dust collector. This is an important cost factor to consider when comparing to cartridge filters which generally are less expensive. In addition, cartridge filters tend to be more efficient than a baghouse. A cartridge collector collects 99.9% of submicron particulate as compared to more typically 99.0% for a baghouse. This characteristic can be particularly important in applications involving toxic dust or when air is being recirculated.

On smaller filtration devices such as those used in fume control in indoor air applications, the real objective is to purify air through materials that trap contaminants which may include both particulate matter and gases. In such systems the fabric is impregnated with activated carbon or zeolite. Gases and odors filter through activated carbon and/or zeolite, an organic mineral. HEPA (high efficiency particulate air filter) is an effective medium for screening particles developed by the US Atomic Energy Commission. True medical rated HEPA is 99.97% effective in removing particles above 0.3 μ, or one three-hundredth the size of a human hair.

The following is a partial list of Web sites that can be consulted for specific vendor information on air filtration devices. A brief description of each site is provided. Many of these sites will link the reader to other sites containing additional information on product information. This list is not intened to be a recommendation of any one vendor or product. It is provided only for the purpose of providing the reader with additional sources of information.

- **Astec Microflow**: Manufacturers of fume cupboards, fume hoods and fume cabinets. Leaders in filtration and containment technology. *http://www.astec-microflow.co.uk*
- **TMS Air Filtration Systems Limited**: Specialist in air filtration, extraction and ventilation systems. *http://www.westmids.co.uk/tmsairfiltration*
- **Munktell Filter AB**: Highly qualified producer of filter media in medical, industrial, environmental controlling and analytical, and purifying filtration of air, fumes and fluids. *http://www.munktell.se*
- **Coppus**: Provider of portable ventilators which help fume and vapor removal, compressed air filtration, personnel cooling & confined space ventilation with propane heaters, cooling mist & ventilation blowers. *http://www.coppus.com*
- **Airflow Systems, Inc**: Resource for the collection and filtration of dust, smoke, mist, fumes and other airborne contaminants generated during industrial and commercial manufacturing and processing applications. *http://www.airflowsystems.com*
- **Clean Air Ltd**: Fume cabinets (or cupboards, UK) designed and produced

quickly and efficiently for a wide range of Research, Pharmaceuticals, Chemicals, or educational settings. *http://www.cleanairltd.co.uk*

- **Campbell Environmental Systems:** Sells Austin air HEPA air filters and Miele HEPA vacuum cleaners *http://www.airwaterbestprices-comlindex.html*
- **Air Cleaning Systems, Inc.**: Sells, services and provides equipment and parts for air filtration systems. *http://www.aircleaningsystems.com*
- **Conquest Equipment Corporation**: Industrial filtration products and services to clients in the wood, paper, plastics, metal, aerospace, and mining industries. *http://www.conquestequipment.com*
- **Hoffman & Hoffman, Inc.**: Manufacturers sales representatives in the selection and application of commercial heating, ventilation, air conditioning, filtration and DDC systems. *http://www.hoffman-hoffman.com*
- **Clean Air Machine Corp.**: Air purifiers - commercial site offers customized filters for specialty applications aimed at removal of microscopic particulates and gases. *http://aco.ca/tibbits/index.html*
- **Aircon Corp.**: Site markets air pollution systems including dust control systems and equipment. *http://aircon-corporation.com*
- **CECO Environmental**: Experts in improving air quality through the use of fiber bed filter systems, high temperature baghouse filter fabrics, scrubbing technology, and on-site air quality monitoring. *http://www.cecofilters.com*
- **Fortress Designs**: Supplier of dust control equipment, baghouses, dust collectors, fabric filters, dust filters, bin vents. *http://www.fabricfilters.com*
- **Schrader Environmental Systems, Inc.**: Air pollution control technologies such as catalytic oxidizers, packed tower wet scrubbers, dust collectors. *http://www.angelfire.com*

Despite the versatility of baghouses, there are numerous applications where other gas cleaning equipment are the favored technologies. Selection most often depends on the intended application, the degree of removal efficiency needed, available floor and headspace, operational and maintenance costs, and the trend in future regulations. More stringent air quality regulations are often the basis for technology selection from the standpoints of either pollution control equipment or pollution prevention. Table 2 provides an equipment selection guide that can be used in selecting the most appropriate piece of equipment for a gas cleaning application based upon general vendor literature reported in terms of removal efficiency and applicability to a gas cleaning.

Table 2. *Equipment Selection Guide.*

Particle Size	Small	Tiny	Invisible	Submicron
Particle Type	>1/64", 20 mesh	4/1000", 140 mesh	4/100,000" - 8/10,000"	<4/100,000"
Size	850 μ	51 - 850 μ	2 - 50 μ	<1 μ
Solids Particulates	Gravity settler	Gravity settler, Cylcone	Coarse (Grit) filter, Cyclone	Cartridge filter
Liquid Particulates	Cyclone	Gravity settler	Cyclone, Baghouse	Scrubber
Combination	Fume filters, Scrubber	Fume filters, Scrubber	Mist and fume filters	Scrubber

Cyclone Separators, Hydroclones, and Multiclones

Among industry jargon, the terms cyclone and hydroclone are often used interchangeably. In the strict sense, however, a cyclone is restricted to gas - particulate separations, whereas the hydroclone has been more traditionally applied to solid-liquid or slurry separations. Nonetheless, the design configurations, operating principles, and selection and sizing criteria are nearly identical. Both devices are the simplest and most economical separators (also called solids collectors or, simply- centrifugal collectors). Their operations are identical, in which forces both of inertia and gravitation are capitalized on, and their primary advantages are high collection efficiency in certain applications, adaptability and economy in power. The main disadvantage lies in their limitation to high collection efficiency of large-sized particles only. In general, cyclones are not capable of high efficiencies when handling gas streams containing large concentrations of particulate matter less than 10 μ in size. Cyclones generally are efficient handling devices for a wide range of particulate sizes. They can collect particles ranging in size from 10 to above 2,000 μ with varying degrees of removal efficiency, with inlet loadings ranging from less than 1 gr/scfm to greater than 100 gr/scfm. There are many design variations of the basic cyclone configuration. Because of the cyclone's/hydroclone's simplicity and lack of moving parts, a wide variety of construction materials can be used to cover relatively high operating temperatures of up to 2,000° F. Cyclones are employed in the following general applications:

collecting coarse dust particles; handling high solids concentration gas streams between reactors such as Flexicokers in refinery operations (typically above 3 gr/scf); for classifying particulate sizes; in operations in which extremely high collection efficiency is not critical; and as precleaning devices in line with high-efficiency collectors for fine particles. Because of the similarities between cyclones and hydroclones, the discussions for these equipment are not separated. Figure 7 shows a cutaway view of a typical industrial cyclone. For many years cyclones have been extensively utilized in the classification of particles in comminution circuits. The practical range of classification for cyclones is 40 μ to 400 μ, with some remote applications as fine as 5 μ or as coarse as 1,000 μ.

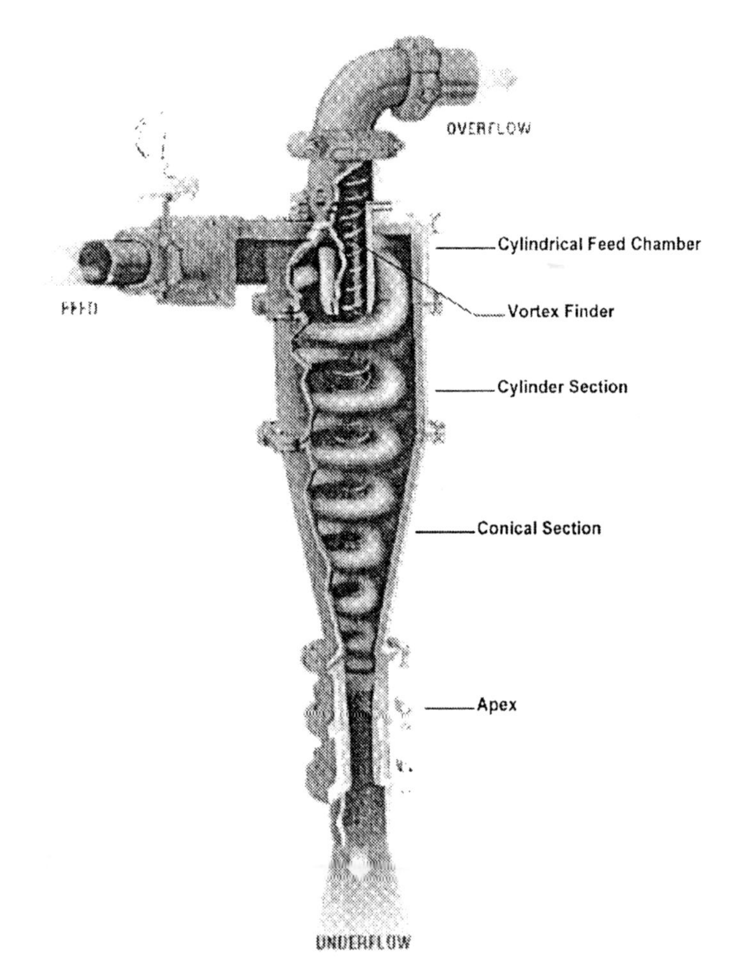

Figure 7. *Cutaway view of a cyclone separator. (Courtesy of Krebs Engineers, Menlo Park, CA).*

Cyclones are used in both primary and secondary grinding circuits as well as regrind circuits. Generally, it is recommended that cyclone suppliers be consulted for sizing confirmation. Some cyclone suppliers employ digital computers to aid in the sizing and selection of cyclones. Following Figure 7, during operation, the polluted feed enters the cyclone or hydroclone under pressure through the feed pipe into the top of the cylindrical feed chamber. This tangential entrance is accomplished by two types of design, as shown in Figure 8. As the feed enters the chamber, a rotation of the two phases inside of the cyclone begins, causing centrifugal forces to accelerate the movement of the particles towards the outer wall. The particles migrate downward in a spiral pattern through the cylindrical section and into the conical section. At this point the smaller mass particles migrate toward the center and spiral upward and out through the vortex finder, discharging through the overflow pipe. This product, which contains the finer particles and the majority of the carrier gas, is termed the overflow and should be discharged at or near atmospheric pressure. The higher mass particles remain in a downward spiral path along the walls of the conical section and gradually exit through the apex orifice. This product is termed the underflow and also should be discharged at or near atmospheric pressure.

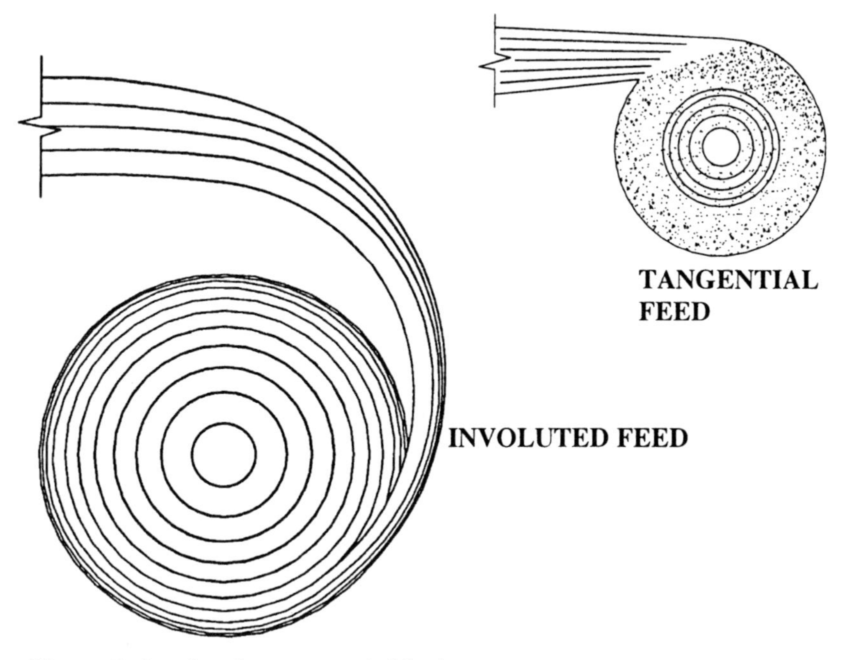

Figure 8. *Involuted vs. tangential feeds.*

The definition of a "standard cyclone" is that cyclone which has the proper geometrical relationship between the cyclone diameter, inlet area, vortex finder, apex orifice, and sufficient length providing retention time to properly classify particles. As with the involuted type design, the graphs and mathematical relationships shown for proper selection and sizing of cyclones apply to the "standard cyclone" geometry. The main parameter is the cyclone diameter. This is the inside diameter of the cylindrical feed chamber. The next parameter is the area of the inlet nozzle at the point of entry into the feed chamber. This is normally a rectangular orifice, with the larger dimension parallel to the cyclone axis. The basic area of the inlet nozzle approximates 0.05 times the cyclone diameter squared. The next important parameter is the vortex finder. The primary function of the vortex finder is to control both the separation and the flow leaving the cyclone. Also, the vortex finder is sufficiently extended below the feed entrance to prevent short circuiting of material directly into the overflow. The size of the vortex finder equals 0.35 times the cyclone diameter. The cylindrical section is the next basic part of the cyclone and is located between the feed chamber and the conical section. It is the same diameter as the feed chamber and its function is to lengthen the cyclone and increase the retention time. For the basic cyclone, its length should be 100% of the cyclone diameter. The next section is the conical section, typically referred to as the cone section. The included angle of the cone section is normally between 10° and 20° and, similar to the cylinder section, provides retention time.

The termination of the cone section is the apex orifice. The critical dimension is the inside diameter at the discharge point. The size of this orifice is determined by the application involved and must be large enough to permit the solids that have been classified to underflow to exit the cyclone without plugging. The normal minimum orifice size would be 10% of the cyclone diameter and can be as large as 35%. Below the apex is normally a splash skirt to help contain the underflow slurry in the case of a hydroclone.

Cyclones and hydroclones can be arranged in bundles, referred to as multiclones. This type of configuration enables considerably higher removal efficiencies, but generally at the cost of increased pressure drop, and hence higher operating costs due to energy. A typical installation for an air pollution application at a coke chemical plant is shown in Figure 9. These designs, though highly efficient, often have significant maintenance problems associated with them. Designs are modular in nature, and hence damaged or worn out units can be taken out of service while the remaining cyclones continue to operate, but at reduced overall collection efficiency. Even so, these systems tend to be more prone to repair and maintenance issues, in part because they are often undersized for intended applications. Vendors should be carefully consulted when considering multiclones.

Figure 9. *Multiclone installation.*

Electrostatic Precipitators

Particles typically less than 10 μ in size generally form a highly stable colloidal suspension in gas (called aerosols). These particles either will not settle out or will require an exceptionally long settling time. Recall that inertia forces are proportional to particle mass and, hence, in the case of aerosols these small forces lead to very low settling velocities.

Particle separation from the gas can be achieved on a practical basis by passing the heterogeneous system between a pair of high-voltage electrodes that ionize the gas. The discharge electrode, responsible for ionizing the gas, consists of a small cross section that generates a high electrical field at its surface.

The collecting electrode serves to precipitate the charged particles. Particles migrate according to an emf (electromotive force) gradient and are attracted to the appropriately charged electrodes, i.e., the negatively charged particles are attracted to the positive electrode and the positively charged particles to the negative electrode. The direction of the velocity vector of the charged particles is determined by their sign and speed and, consequently, their kinetic energy, as determined by the field intensity. A photograph of an installation is shown in Figure 10.

The electrical force field responsible for particle precipitation may exceed inertia forces by orders of magnitude. There are several types or configurations of electrostatic precipitators. Normally, the flow section of a precipitation chamber (simply called a precipitator) consists of a bundle of vertical tubes arranged in parallel fashion (usually of round or hexagonal cross section), or as a packet of vertical parallel plates. Thin wires, approximately 2 mm in diameter, are suspended and stretched along the axes of the tubes, which are 150-300 mm in diameter. The same wires are stretched between the plate configurations. The wires and tubes are connected to a source of direct electrical current of high voltage (up to 90,000 V). Cylindrical wires are connected to the negative terminal where a zone of ionized gas is generated. Gas molecules in this vicinity comprise both positively and negatively charged ions.

Figure 10. *Photograph of an electrostatic precipitator.*

The negative ions are repelled to the walls of the tubes and plates and fill the total volume of the precipitator. The dispersed particulate matter are contacted by the ions and entrained to the surface of the tubes and plates. Here, the particles lose their charge and descend downward along the electrode surfaces. The ionized gas layer in the vicinity of the wires dissipates an energy field in the form of a thin layer called the corona, hence the name "wire corona-forming electrode." The plates and tubes in which the particles are precipitated are called collecting electrodes. Depending on the shape of the electrodes, the electrical field may be either homogeneous or heterogeneous. A homogeneous electrical field is observed in the gaseous space between flat parallel electrodes. The lines of force uniformly fill the volume and are located in a parallel arrangement. A heterogeneous field is generated when the surface of one of the electrodes is reduced greatly by geometry changes. A thickening of the lines of force occurs around these electrodes, which corresponds to an increase in the field intensity.

Water Pollution Equipment

Sedimentation Equipment

There are a large number of operations particularly in the chemical processing and allied industries that handle a variety of suspensions of solid particles in liquids. The application of filtration techniques for the separation of these heterogeneous systems is sometimes very costly. If, however, the discrete phase of the suspension largely contains settleable particles, the separation can be effected by the operation of sedimentation. The process of sedimentation involves the removal of suspended solid particles from a liquid stream by gravitational settling. This unit operation is divided into *thickening,* i.e., increasing the concentration of the feed stream, and *clarification,* removal of solids from a relatively dilute stream.

A thickener is a sedimentation machine that operates according to the principle of gravity settling. Compared to other types of liquid/solid separation devices, a thickener's principal advantages include simplicity of design and economy of operation; its capacity to handle extremely large flow volumes; and versatility, as it can operate equally well as a concentrator or as a clarifier. In a batch-operating mode, a thickener normally consists of a standard vessel filled with a suspension. After settling, the clear liquid is decanted and the sediment removed periodically.

The operation of a continuous thickener is also relatively simple. A drive mechanism powers a rotating rake mechanism. Feed enters the apparatus through a feed well designed to dissipate the velocity and stabilize the density currents of the incoming stream. Separation occurs when the heavy particles settle to the bottom of the tank. Some processes add flocculants to the feed stream to enhance particle agglomeration to promote faster or more effective settling. The clarified liquid overflows the tank and is sent to the next stage of a process. The underflow solids are withdrawn from an underflow cone by gravity discharge or pumping.

Thickeners can be operated in a countercurrent fashion. Applications are aimed at the recovery of soluble material from settleable solids by means of continuous countercurrent decantation (CCD). The basic scheme involves streams of liquid and thickened sludge moving countercurrently through a series of thickeners. The thickened stream of solids is depleted of soluble constituents as the solution becomes enriched. In each successive stage, a concentrated slurry is mixed with a solution containing fewer solubles than the liquor in the slurry and then is fed to the thickener. As the solids settle, they are removed and sent to the next stage. The overflow solution, which is richer in the soluble constituent, is sent to the preceding unit. Solids are charged to the system in the first-stage thickener, from which the final concentrated solution is withdrawn. Wash water or virgin solution is added to the last stage, and washed solids are removed in the underflow of this thickener.

Continuous clarifiers handle a variety of process wastes, domestic sewage and other dilute suspensions. They resemble thickeners in that they are sedimentation tanks or basins whose sludge removal is controlled by a mechanical sludge-raking mechanism. They differ from thickeners in that the amount of solids and weight of thickened sludge are considerably lower.

Figure 11 shows one of the various types of cylindrical clarifiers. In this type of sedimentation machine, the feed enters up through the hollow central column or shaft, referred to as a siphon feed system. The feed enters the central feed well through slots or ports located near the top of the hollow shaft. Siphon feed arrangements greatly reduces the feed stream velocity as it enters the basin proper. This tends to minimize undesirable cross currents in the settling region of the vessel. Most cylindrical units are equipped with peripheral weirs; however, some designs include radial weirs to reduce the exit velocity and minimize weir loadings. The unit shown also is equipped with adjustable rotating overflow pipes.

Pre-settling operations are often performed in wastewater treatment applications. These can be as simple as rectangular settling basins, like the ones shown in the photograph of Figure 12. These basins serve to remove grit and other heavy particle prior to clarification. Chemical treatment may or may not be used at this stage of the water treatment operation. In general, this can be thought of as a pretreatment stage, prior to clarification and thickening operations.

Figure 11. *Clarifier used in a wastewater treatment plant.*

Figure 12. *Rectangular settling basins in operation.*

The total solids in municipal wastewaters exist in a distribution of sizes from individual ions up to visible particles. Specific analytical procedures have been established to distinguish the suspended fraction of the total solids and to further distinguish the settleable fraction within the suspended solids. A typical concentration of SS (suspended solids) for raw domestic wastewaters is 200 mg/ℓ, but this can vary substantially from system to system. The lower limiting size for the SS fraction (about 1.5 μ) is arbitrarily defined by the test procedures and it should be noted that variations in test procedures themselves can also lead to widely varying results, especially at the low solids levels characteristic of treated effluents.

The settleable and supracolloidal fractions together are essentially equivalent to the suspended fraction referred to above. Dividing lines between fractions again are somewhat arbitrary depending on tests applied, and overall concentrations in different fractions can vary substantially between systems depending on factors such as water use, travel time in sewers, ground-water infiltration, and prevalence of home garbage grinding. Contributions of dissolved, colloidal and suspended solids from individual homes, multi-family dwellings or other point sources often have concentrations two or more times the average for a whole system. In addition to particle size, specific gravity and strength or shear resistance of wastewater solids may affect solids separation performance.

The three basic types of solids separation processes are gravity separation, physical straining, and granular media filtration. Wastewater solids characteristics can be altered to enhance performance of the separation processes. The processes of chemical treatment (precipitation and/or coagulation) and physical treatment (flocculation) are aimed at alteration of solids characteristics. In addition, during the separation processes themselves, agglomeration and compaction of solids generally continues, increasing separation efficiency and reducing the volume of separated solids.

Biological wastewater treatment processes also affect solids characteristics and hence solids separation. Activated sludge solids have been found to have a distinct bimodal distribution with one mode in the supracolloidal to settleable range and another near the border between the colloidal and supracolloidal fractions. The concentrations and size limits in each range are affected by conditions in the biological reactor. Bacteria cellular debris, etc. fall into the finer (colloidal-supracolloidal) range. Agglomeration of these finer solids generally increases the efficiency of subsequent separation processes.

Processes for SS separation may fill three distinct functions in wastewater treatment, namely, pretreatment to protect subsequent processes and reduce their loadings to required levels, treatment to reduce effluent concentrations to required standards, and separation of solids to produce concentrated recycle streams required to maintain other processes. In the first two functions effluent quality is the prime consideration, but where the third function must be fulfilled along with one of the others, design attention must be given to conditions for both the separated solids (sludge) and the process effluent.

Wedge-wire screens can operate at very high hydraulic and solids loadings, but do not greatly reduce SS (suspended solids). Hence, wedge wire screens are limited to pretreatment applications where subsequent processes will assure production of a satisfactory final effluent. They can be considered as an adjunct to primary sedimentation or, where conditions prescribe, as an alternative. Sedimentation units must operate at relatively low hydraulic loadings (large space requirements), but can accept high solids loadings. With proper chemical or biological pretreatment and design, they can produce good quality effluents.

Microscreens and granular-media filters, operating at significantly higher hydraulic loads than sedimentation units, can produce an effluent with lower SS than is possible with sedimentation alone. In general they are not designed to accept high solids loadings, and are normally used following other processes which put out relatively low effluent SS concentrations. Selection of one of the alternative processes can be based on cost only where all factors not reflected in cost are equivalent. Direct cost comparison of individual solids removal processes usually proves impossible because of differences in factors such as: effluent quality, pretreatment requirements, effects on sludge processing, housing, space and head requirements. Meaningful cost comparisons usually involve practically the entire process configuration of the treatment facility, including processes for disposal of solid residues, and reflect how the individual unit processes affect one another. Some important cost considerations include:

- Where chemical treatment is used to remove BOD or phosphates or improve SS removals, significant quantities of chemical sludge are produced. The cost of disposal of this sludge must be considered in process selection unless configurations being compared involve similar chemical treatment. The actual cost involved will depend greatly on the particular method of sludge disposal to be used;
- Head Requirements - Some of the processes employed for SS separation (sedimentation, microscreens, etc.) require relatively small head (only 2 to 3 ft. to overcome losses at inlet and effluent controls and in connecting piping). Others, such as granular-media filters, and wedge-wire screens, require greater differential head (10 ft or more). Differences in head requirements are most significant where they necessitate capital outlay for an extra pumping step. The costs for pumping, however, even with lifts above 10 ft are usually not large in relation to the overall costs for treatment facilities;
- Flow Variation - Both the rate and characteristics of the inflow to most treatment plants vary significantly with time. Diurnal cycles are found in all domestic discharges. Weekly and seasonal cycles are common in municipal systems as are variations between wet and dry weather. Even where only domestic flows are involved, the magnitude of variations can differ widely between different systems depending on system configuration, water use habits of the population and opportunities for groundwater infiltration or direct inflow of surface or subsurface drainage.

Industrial and institutional flows where significant, can further alter domestic patterns. Because of these wide differences, design of treatment facilities should be based, whenever possible, on measurements of actual flow variations in existing systems. Projected flow variations from existing systems should reflect elimination of excessive flows. Flows tend to be less variable in larger systems, due chiefly to differing times of travel from different sections and to damping effects of flow storage in large sewers. Equalization storage can be used to reduce diurnal variations in flow and in concentration of SS or other wastewater characteristics. Storage may also be used to handle peaks caused by direct inflow to the sewers during wet weather. Assuming that equivalent performance can be obtained either by increasing the size of treatment facilities or by providing equalizing basins, selection between these approaches can be based on their relative costs and environmental impacts. In plants using processes involving large, short-term recycle flows-such as for backwashing granular media filters-equalization is almost always justified.

Chemical Treatment and Settling Characteristics

As noted earlier, chemical coagulation and flocculation are common operations in wastewater treatment applications that are accomplished by a combination of physical and chemical processes which thoroughly mix the chemicals with the wastewater and promote the aggregation of wastewater solids into particles large enough to be separated by sedimentation, flotation, media filtration or straining. The strength of the aggregated particles determines their limiting size and their resistance to shear in subsequent processes. For particles in the colloidal and fine supra colloidal size ranges (< 1 to $2\ \mu$) natural stabilizing forces (electrostatic repulsion, physical separation by absorbed water layers) predominate over the natural aggregating forces (van der Waals) and the natural mechanism (Brownian movement) which tends to cause particle contact. Coagulation of these fine particles involves both destabilization and physical processes which disperse coagulants and increase the opportunities for particle contact. Chemical coagulants used in wastewater treatment are generally the same as those used in potable water treatment and include: alum, ferric chloride, ferric sulfate, ferrous chloride, ferrous sulfate and lime. The effectiveness of a particular coagulant varies in different applications, and in a given application each coagulant has both an optimum concentration and an optimum pH range. In addition to coagulants themselves, certain chemicals may be applied for pH or alkalinity adjustment (lime, soda ash) or as flocculating agents (organic polymers). For full effectiveness chemical coagulation requires initial rapid mixing to thoroughly disperse the applied chemicals so that they can react with suspended and colloidal solids uniformly.

Destabilization Mechanisms: The destabilizing action of chemical coagulants in wastewater may involve any of the following mechanisms: electrostatic charge reduction by adsorption of counter ions, inter-particle bridging by adsorption of specific chemical groups in polymer chains, and physical enmeshment of fine solids in gelatinous hydrolysis products of the coagulants.

Electrostatic Charge Reduction: Finely dispersed wastewater solids generally have a negative charge. Adsorption of cations from metal salt coagulants (in the case of iron and aluminum from their hydrolysis products), or from cationic polymers can reduce or reverse this charge. Where electrostatic charge reduction is a significant destabilization mechanism, care must be taken not to overdose with coagulant. This can cause complete charge reversal with restabilization of the oppositely charged coagulant-colloid complex.

Interparticle Bridging: When polymeric coagulants contain specific chemical groups which can interact with sites on the surfaces of colloid particles, the polymer may adsorb to and serve as a bridge between the particles. Coagulation using polyelectrolyte of the same charge as the colloids or nonionic polymers depends on this mechanism. Restabilization may occur if excessive dosages of polymer are used. In this case all sites on the colloids may adsorb polymer molecules without any bridging. Excessive mixing ran also cause restabilization by fracture or displacement of polymer chains.

Enmeshment in Precipitated Hydrolysis Products: Hydroxides of iron, aluminum or, at high pH, magnesium form gelatinous hydrolysis products which are extremely effective in enmeshing fine particles of other material are formed by reaction of metal salt coagulants with hydroxyl ions from the natural alkalinity in the water or from added alkaline chemicals such as lime or soda ash. Sufficient natural magnesium is frequently present in wastewater so that effective coagulation is obtained merely by raising the pH with lime. Organic polymers do not form hydrolysis products of significance in this mechanism. At a pH value lower than that required to precipitate magnesium, the precipitates produced by lime treatment are frequently ineffective in enmeshing the colloidal matter in wastewater. The remedy for this generally involves the addition of low dosage of iron salts or polymers as coagulant aids.

Use of Coagulants: Coagulants may also react with other constituents of the wastewater, particularly anions such as phosphate and sulfate, forming hydrolysis products containing various mixtures of ions. The chemistry of the reactions is extremely complex and highly dependent on pH and alkalinity. The presence of high concentrations of these anions may require increased doses of coagulants or pH adjustment to achieve effective removal of suspended solids (SS). The design of chemical treatment facilities for SS removal must take into account:

- the types and quantities of chemicals to be applied as coagulants, coagulant aids and for pH control, and
- the associated requirements for chemical handling and feeding, and for mixing and flocculation after chemical addition. The selection of coagulants should be based on jar testing of the actual wastewater to determine dosages and effectiveness, and on consideration of the cost and availability of different coagulants. Where expected changes in waste characteristics or market conditions may favor different coagulants at different times, chemical feed and handling should be set up to permit a switchover. In developing a testing program, general information on experience at other locations and on costs should be considered to aid in selection of processes and coagulants to be tested. Aluminum or iron salts tend to react with soluble phosphate preferentially so that substantial phosphorus removal must be involved before organic colloids can be destabilized. Required dosages will be affected by phosphorus content. Similarly, lime treatment to a pH at which coagulation is effective precipitates substantial phosphorus. Because chemical dosage and pH range for optimum SS removal may differ somewhat from those for optimum phosphorus removal, coagulant requirements may be determined by the effluent criteria for either pollutant, depending on wastewater characteristics and the choice of chemical.

Sludge Production: Chemical coagulation increases sludge production in sedimentation units due both to greater removal of influent suspended solids and to insoluble reaction products of the coagulation itself.

The weight of sludge solids can be estimated by calculation of the sum of the expected SS removal and of the precipitation products expected from the coagulant dosages applied. Usually jar tests can be employed to obtain the necessary information for this calculation.

pH Control and Alkalinity: The critical factor in the control of lime reactions is pH. The pH for optimum effectiveness of lime coagulation, determined from jar testing and process operating experience can be used as a set point for a pH control of lime dosing. Alum and iron salt coagulation are much less sensitive to pH. Testing can determine optimum dosages for coagulation and whether natural alkalinity is adequate for the reactions. If supplemental alkalinity is needed either regularly or on an intermittent basis (e.g. during high wet weather flows) provisions should be included for feeding necessary amounts of lime or soda ash.

Points of Chemical Addition: In independent physical-chemical treatment or in phosphate removal in the primary clarifier ahead of biological treatment, chemicals are added to raw sewage. In tertiary treatment for phosphate removal and suspended solids (SS) reduction, they are added to secondary effluent. In both cases, proper mixing and flocculation units are needed. For phosphate removal or improvement of SS capture in biological secondary treatment, chemicals are often added directly to aeration units or prior to secondary settling units, without separate mixing and flocculation. In some phosphate removal applications coagulants are added at multiple points, e.g. prior to primary settling and as part of a secondary or tertiary treatment step.

Supplementary Coagulants: Addition of the hydrolyzing metal coagulants to wastewater often results in a small slow-settling floc or precipitate of phosphorus. Additional treatment is required to produce a water with low residual suspended solids. Polymeric coagulants are beneficial in aggregating the precipitation products to a settleable size and increasing the shear strength of the floc against hydraulic breakup.

Coagulation Control: Because coagulation represents a group of complex reactions, laboratory experimentation is essential to establish and maintain the optimum coagulant dosage and to determine the effects of important variables on the quality of coagulation of the wastewater under investigation. With alum and iron coagulants two procedures are generally followed for this purpose: the jar test and measurement of zeta potential. Proper control of lime coagulation may be maintained by measuring the pH or automatically titrating alkalinity after lime addition.

The single, most widely used test to determine coagulant dosage and other parameters is the jar test, which attempts to simulate the full scale coagulation-flocculation process and has remained the most common control test in the laboratory since its introduction in 1918. Since the intent is to simulate an individual plant's conditions, it is not surprising that procedures may vary but generally have certain common elements. The jar test apparatus consists of a series of sample containers, usually six, the contents of which can be stirred by individual mechanically operated stirrers. Wastewater to be treated is placed in the containers and treatment chemicals are added while the contents are being stirred. The range of conditions, for example, coagulant dosages and pH, are selected to bracket the anticipated optima. After a 1 to 5 minute period of rapid stirring to ensure complete dispersion of coagulant, the stirring rate is decreased and flocculation is allowed to continue for a variable period, 10 to 20 minutes or more, depending on the simulation. The stirring is then stopped and the flocs are allowed to settle for a selected time. The supernatant is then analyzed for the desired parameters. With wastewater, the usual analyses are for turbidity or suspended solids, pH, residual phosphorus and residual coagulant. If desired, a number of supernatant samples may be taken at intervals during the settling period to permit construction of a set of settling curves which provide more information on the settling characteristics of the floc than a single sample taken after a fixed settling period. A dynamic settling test may also be used in which the paddles are operated at 2 to 5 rpm during the settling period. This type of operation more closely represents settling conditions in a large horizontal basin with continuous flow. A simple apparatus can be constructed from tubing, rubber stoppers and small aquarium valves to permit rapid sampling of the supernatant. The unit is placed next to the sample jars at the beginning of the settling period with the curved stainless steel tubes dipping into the jars. At desired intervals the vent valve is covered with a finger, permitting vacuum to draw samples into the small sample bottles. The needle valves are adjusted so that supernatant is drawn into all the bottles at the same rate. When sufficient sample is obtained, the vent is uncovered and the bottles are replaced with empties. The maximum sampling rate is about once per minute.

Filtration Equipment

Liquid filtration equipment is commercially available in a wide range. Proper selection must be based on detailed information of the wastewater to be handled, cake properties, anticipated capacities and process operating conditions. Operational modes include batch, sem-batch and continuous). Continuous filters are comprised of essentially a large number of elemental surfaces, on which different operations are performed. These operations performed in series are solids separation and cake formation, cake washing, cake dewatering and drying, cake removal, and filter media washing. The specific equipment used can be classified into two groups: stationary components (which are the supporting devices such as the suspension vessel), and scraping mechanisms and movable devices (which can be the filter medium, depending on the design). Either continuous or batch filters can be employed in cake filtration. In filter-medium filtration, however, where particulate matter are retained within the framework of the filter medium, batch systems are the most common. Batch filters may be operated in any filtration regime, whereas continuous filters are most often operated under constant pressure. It is important to note that equipment used in liquid filtration applications fall into three general categories based on the difference in orientation between gravity force and filtrate motion. These orientations are; forces acting in the opposite direction (i.e, countercurrent systems), forces acting in the same direction (cocurrent), and forces acting normal to each other (cross-flow).

Solid-liquid separation is a fundamental unit operation that exists in almost every flowscheme related to the chemical process industries, ore beneficiation, pharmaceutics, food or drinking water, as well as wastewater treatment. The separation techniques are very diverse. As a unit operation, there are many types of equipment that are used, not just filtration systems.

In addition to filtration by such equipment as vacuum and pressure filters, there are centrifugation by filtering and sedimenting centrifuges, sedimentation by conventional, storage and high-rate thickeners, clarification by conventional, solids-contact and sludge-blanket clarifiers, polishing by precoat filters, pressure and deep-bed filters, and upward separation by dissolved-air flotation.

Among the most common filter device is the vacuum filter. Vacuum filters are relatively simple and reliable machines. The various types of vacuum filters are summarized in the chart provided in Figure 13. For a sense of physical dimensions, the following are some typical sizes: drum and disc filters are up to 100 m^2 in filtration area; horizontal belt filters up to 120 m^2; tilting pan and table filters up to 200 m^2. There are several advantages that vacuum filtration has over other solid-liquid separation methods. Some of these advantages include:

- Continuous operation (with the exception of the Nutsche type filter),
- Intensive soluble recovery or removal of contaminants from the cake as accomplished by countercurrent washing operations. This is especially the case with horizontal belt, tilting pan and table filters, which are described later in this subsection,
- Polishing of solutions (as performed on a precoat filter operation).

Additional advantages that vacuum filters have over other separation methods is the ability to produce relatively clean filtrates by using a sedimentation basin, as in the cases with the horizontal belt, tilting pan and table filters; they provide easy access to cake for sampling purposes; they allow easy control of operating variables such as cake thickness and wash rations; commercial systems are available in a wide range of materials of construction.

The principle disadvantages of vacuum filters include higher residual moisture in the cake, untight construction so it is difficult to contain process vapors, and high power requirements as demanded by the vacuum pump. One of the oldest filter designs is the rotary vacuum drum filter. This machine belongs to the group of bottom feed configurations. Rotary drum filters are typically operated in the countercurrent mode of operation. The principle advantage of these machines is the continuity of their operation.

The key features of the a rotary drum filter are as follows. First, the machine has a drum that is supported by a large diameter trunnion on the valve end and a bearing on the drive end, and its face is divided into circumferential sectors each forming a separate vacuum cell. The internal piping that is connected to each sector passes through the trunnion and ends up with a wear plate having openings which correspond to the number of sectors.

The drum deck piping is arranged so that each sector has a leading pipe to collect the filtrate on the rising side of the drum and a trailing pipe to collect the remaining filtrate from the descending side to ensure complete evacuation prior to cake discharge. The drum is driven with a variable speed drive at speeds that normally range from 1 to 10 rpm.

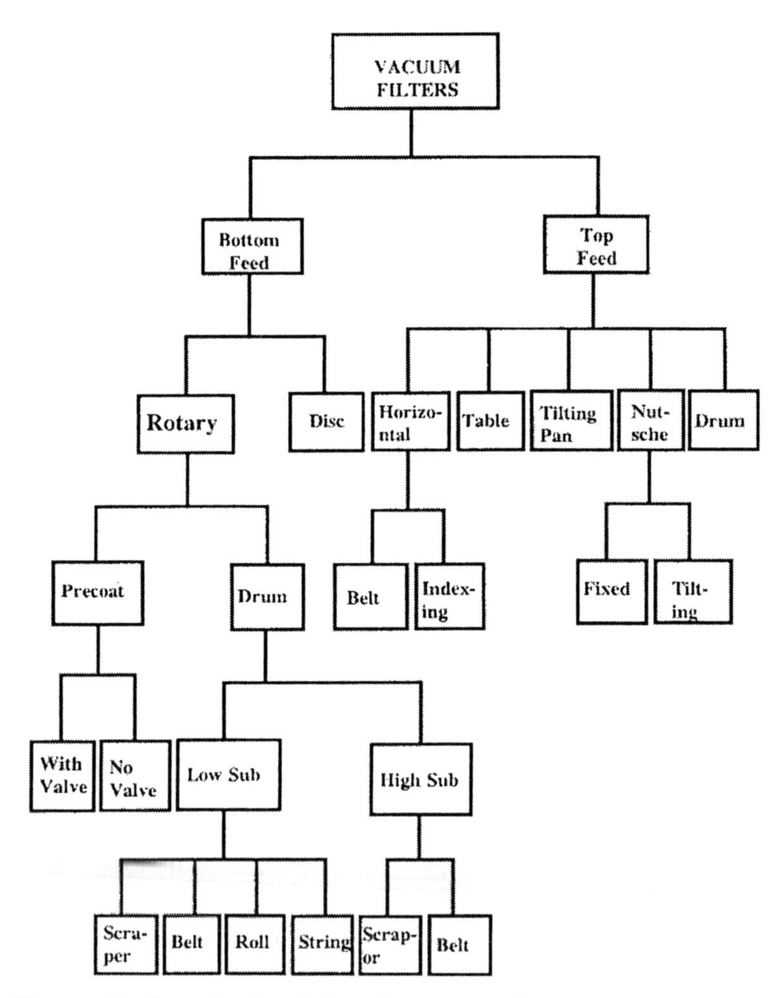

Figure 13. *Organizational chart for vacuum filters.*

The machine is equipped with a valve with a bridge setting which controls the sequence of the cycle so that each sector is subjected to vacuum, blow and a dead zone. When a sector enters submergence, vacuum commences and continues to a point that it is cut-off and blow takes place to assist in discharging the cake.

A cake discharge mechanism is available in several forms, such as a scraper, belt, roll and in very rare cases a string discharge. Blow is applied only to filters with scraper and roll discharge mechanisms but not to filters with a belt or string discharge. The various types of cake discharge mechanisms are illustrated in Figures 14 through 17.

A mild agitator is used to keep the slurry in suspension and reciprocates between the drum face and tank bottom at 16 or so cpm. The system is equipped with a tank with baffled slurry feed connections, an adjustable overflow box to set a desired drum submergence and a drain connection.

The tanks are normally designed for an apparent submergence of 33 to 35%, however on certain applications 50% and more is possible. With these special designs the tank ends are higher in order to accommodate stuffing boxes on both the drive shaft and valve end trunnion. Refer to references 4 through 6 for further descriptions of these and other processing equipment.

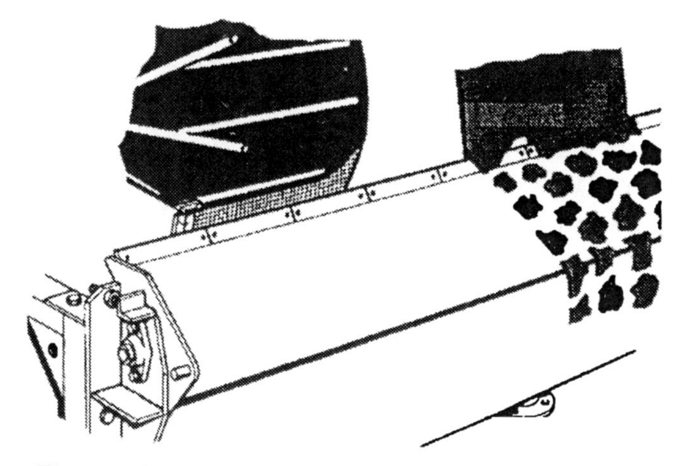

Figure 14. *Shows scraper discharge.*

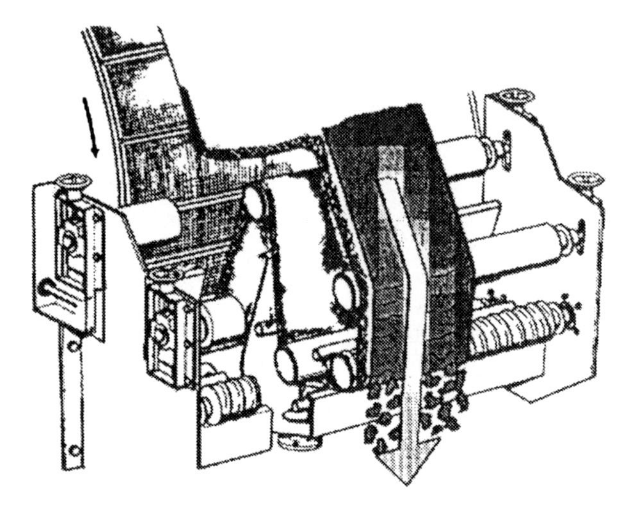

Figure 15. *Shows a belt discharge.*

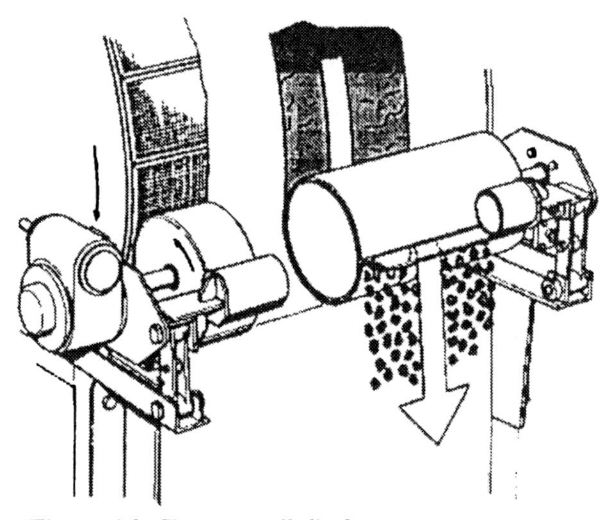

Figure 16. *Shows a roll discharge.*

On applications where cake washing is required, 2 or 3 manifolds with overlapping nozzles are mounted to a pair of splash guards that are bolted to the tank ends.

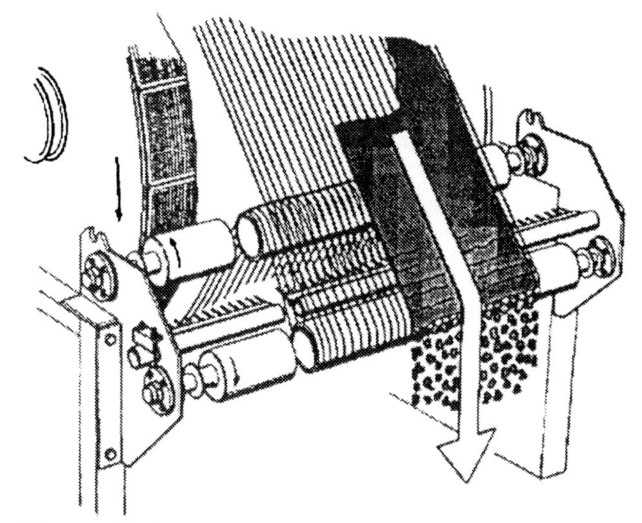

Figure 17. *Shows a string discharge.*

The position of the manifolds and the quantity of wash liquid are adjustable depending on the wash characteristics of the cake. Optional controls may be used to automate settings such as drum speed, applied wash liquid and drum submergence for a desired cake thickness or throughput. The monitoring of drum submergence controls the slurry feed valves so an adjustable overflow weir is not necessary except for a fixed connection in case of an emergency. In broad terms, drum filters are suitable to the following process requirements:

- Slurries with solids that do not tend to settle rapidly and will remain in a uniform suspension under gentle agitation,
- Cakes which do not require long drying times to reach asymptotic moisture values,
- Cakes when a single washing stage is sufficient to remove residual contaminants from the cake or yield maximum recovery of filtrate,
- Filtrates that generally do not require a sharp separation between the mother and wash filtrates. Some complex valves, however, enable atmospheric purging of the sectors and internal piping to facilitate a sharp separation of filtrates,
- Filtrates that are acceptable with a low quantity of fines that pass trough the filter cloth in the first few seconds of cake formation. Broadly, and depending on particle size and cloth permeability, the filtrate may contain 1,000 to 5,000 ppm insolubles.

The entire filtration cycle on a rotary drum filter must be completed within a geometry of 360 degrees. The stages of the cycles are cake formation, cake washing and drying, cake discharge, and the dead zone. Each is briefly described below.

Cake Formation: With the overflow weir set to a maximum the "apparent submergence" is normally 33 to 35% so the slurry levels are between 0400 and 0800 hours. Once a sector is completely submerged vacuum is applied and a cake starts to form up to a point where the sector emerges from the slurry. The portion of the cycle available for formation is the "effective submergence" and its duration depends on the number of sectors, the slurry level in the tank and the bridge setting which controls the form to dry ratio.

Cake Washing and Drying: After emerging from submergence the drying portion of the cycle commences and for nonwash applications continuous to about 0130 hours where the vacuum is cut-off. If cake washing is required the wash manifolds will be located from about 1030 to 1130 hours and the remaining time to vacuum cut- off at 0130 is the portion allocated to final cake drying.

Cake Discharge: After vacuum for the entire sector is cut-off air blow commences at about 0200 hours in order to facilitate the discharge of the cake. The blow, depending on the position of the tip of the scraper blade, will cut-off at approximately 0300 hours. Drum filters are normally operated with a low pressure blow but on certain applications a snap blow is applied and to avoid the snapping out of the caulking bars or ropes wire winding of the cloth is recommended. Blow is used on scraper and roll discharge mechanisms but on belt discharge filters vacuum cuts-off when the filter media leaves the drum.

Dead Zone: Once the blow is cut-off the sector passes through a zone blocked with bridges so that no air is drawn through the exposed filter media which might cause the loss of vacuum on the entire drum surface.

The slow rotation of the drum and reciprocation of the agitator reduce maintenance requirements to a minimum but the following should be inspected periodically. The strip liner of the trunnion bearing at the valve end will normally wear at the lower half. However, in situations when the slurry has a high specific gravity, the drum has a tendency to become buoyant causing wear to the upper half.

One way to remove the lower half of the liner, when hoisting facilities are not available or operational, is to float the drum by filling the tank with a sufficiently concentrated solution. Other components that should be periodically checked are:

- The stuffing boxes on high submergence filters should be inspected for leakage and, if necessary, the stud nuts should be tightened. It should be noted that excess tightening can increase substantially the load on the drum drive so the use of a torque wrench is recommended.

- The face of the wear plate should be checked periodically and remachined if necessary. A whistling noise during operation is an indication the wear plate is worn out or the valve spring requires tensioning.
- The drum has a bailer tube that protrudes from the drive end shaft and must be kept open at all times since its blockage may cause the collapse of the drum. The bailer tube is a tell-tale indication to the following: (a) if a lighter flame is drawn through the bailer tube to the inside of the drum it indicates that a vacuum leak exists in the drum shell or the internal piping; (b) if leakage is observed from the bailer tube it indicates that a hole exists in the drum head causing penetration of slurry from the tank into the drum.
- The on-line filter on the wash headers manifold should be checked periodically for pressure buildup due to progressive blockage. Likewise, the nozzles on the wash headers should be kept clean in order to ensure overlapping for full coverage of the washed cake.

Internal Rotary-Drum Filters: An example of an internal rotary-drum filter is illustrated in Figure 18. The filter medium is contained on the inner periphery. This design is ideal for rapidly settling slurries that do not require a high degree of washing. Tankless filters of this design consist of multiple-compartment drum vacuum filters. One end is closed and contains an automatic valve with pipe connections to individual compartments. The other end is open for the feed entrance.

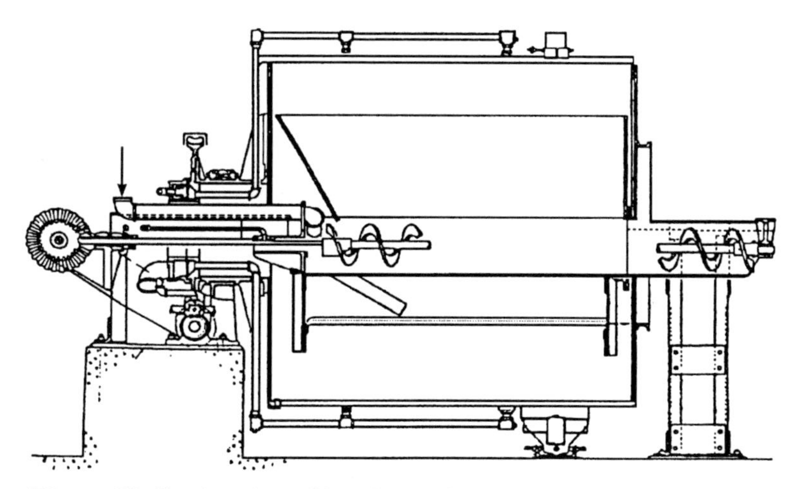

Figure 18. *Section view of interior medium rotary-drum vacuum filter.*

The drum is supported on a tire with rigid rollers to effect cake removal. The drum is driven by a motor and speed-reducer connected to a riding roll shaft. The feed slurry is discharged to the bottom of the inside of the drum from the distributor and is maintained as a pool by a baffle ring located around the open end and the closed portion of the outer end. As the drum revolves, the compartments successively pass through the slurry pool, where a vacuum is applied as each compartment becomes submerged. Slurry discharge is accomplished at the top center where the vacuum is cut off and gravity (usually assisted by blowback) allows the solids to drop off onto a trough. From there, a screw or belt conveyor removes the solids from the drum. This filter is capable of handling heavy, quick-settling materials.

Nutsche Filters: Nutsche filters are one design type with a flat filtering plate. This configuration basically consists of a large false-bottomed tank with a loose filter medium. Older designs employ sand or other loose, inert materials as the filtering medium, and are widely employed in water clarification operations. In vacuum filtration, these false-bottom tanks are of the same general design as the vessels employed for gravity filtration. They are, however, less widely used, being confined for the most part to rather small units, particularly for acid work. Greater strength and more careful construction are necessary to withstand the higher pressure differentials of vacuum over gravity. This naturally increases construction costs. However, when high filtering capacity or rapid handling is required with the use of vacuum, the advantages may more than offset higher costs. Construction of the vacuum false-bottom tank is relatively simple; a single vessel is divided into two chambers by a perforated section. The upper chamber operates under atmospheric pressure and retains the unfiltered slurry. The perforated false bottom supports the filter medium. The lower chamber is designed for negative pressure, and to hold the filtrate. Nutsche filters are capable of providing frequent and uniform washings. A type of continuous filter that essentially consists of a series of Nutsche filters is the rotating-tray horizontal filter.

Horizontal Rotary Filters: An example of a horizontal rotary filter is illustrated in Figure 19. These machines are well suited to filtering quick-draining crystalline-like solids. Due to its horizontal surface, solids are prevented from falling off or from being washed off by the wash water. As such, an unusually heavy layer of solids can be tolerated. The basic design consists of a circular horizontal table that rotates about a center axis

The table is comprised of a number of hollow pie-shaped segments with perforated or woven metal tops. Each of the sections is covered with a suitable filter medium and is connected to a central valve mechanism that appropriately times the removal of filtrate and wash liquids and the dewatering of the cake during each revolution. Each segment receives the slurry in succession. The wash liquor is sprayed onto each section in two applications. Then the cake is dewatered by passing dry air over it. The cake is finally removed by a scraper.

Belt Filters: Belt filters consist of a series of Nutsche filters moving along a closed path. Nutsche filters are connected as a long chain so that the longitudinal edge of each unit has the shape of a baffle plate overlapping the edge of the neighboring unit. Each unit is displaced by driving and tensioning drums. Nutsche filters are equipped with supporting perforated partitions covered with the filtering cloth. The washed cake is removed by turning each unit over. Sometimes a shaker mechanism is included to ensure more complete cake removal. A belt filter can be thought of as consisting of an endless supporting perforated rubber belt covered with the filtering cloth. The basic design is illustrated in Figure 20.

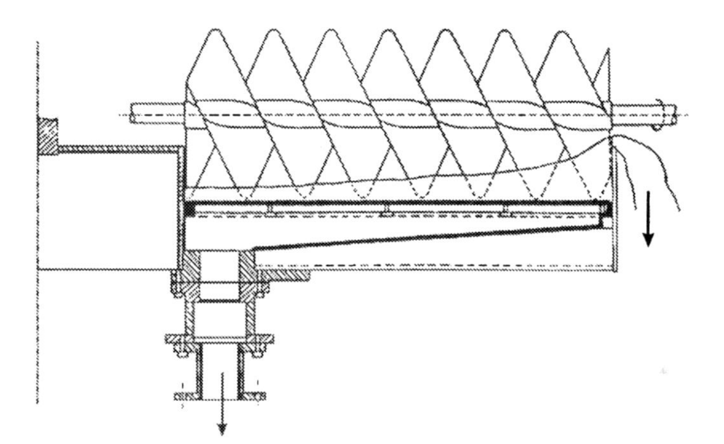

Figure 19. *Cross section of a rotary horizontal vacuum filter.*

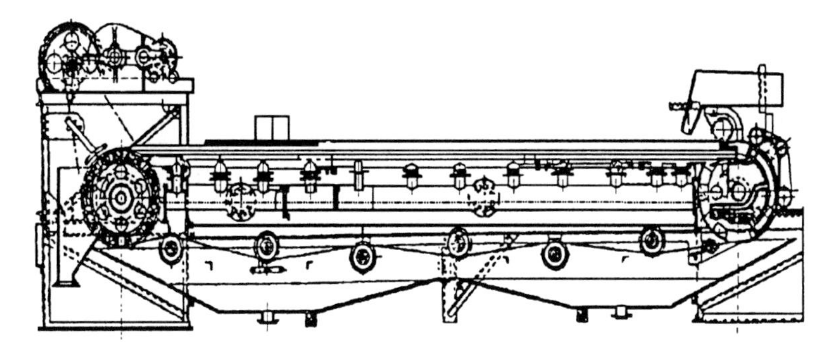

Figure 20. *Basic design features of a belt filter.*

Supporting and filtering partitions are displaced by a driving drum and maintained in a stretched condition by tensioning the drum, which rotates due to friction against the rubber belt. Belt edges (at the upper part of their path) slide over two parallel horizontal guide planks. The elongated chamber is located between the guide planks. The chamber in the upper part has grids with flanges adjoining the lower surface of the rubber belt. The region under the belt is connected by nozzles to the filtrate collector, which is attached to a vacuum source. The chamber and collector are divided into sections from which filtrate and washing liquid may be discharged. The sludge is fed by the trough. The cake is removed from the drum by gravity or blowing air, or sometimes it is washed off by liquid from the distributor nozzle. The washing liquid is supplied from a dike tank, which can move along the filtering partition. It can be washed during the belt's motion along the lower path. The filtering partition consists of a riffled rubber belt with slots, grooves and the filter cloth, which is fixed in a set of grooves by cords. Slots through which the filtrate passes are located over the grids of the elongated chamber. The edges of the rubber belt are bent upward by guides forming a gutter on the upper path of the belt. The velocity of the filtering partition depends on the physical properties of the sludge and the filter length. The cake thickness may range from 1 to 25 turns. The advantages of belt filters are their simplicity in design compared to filters with automatic valves, and the abilities to provide countercurrent cake washing and removal of thin layers of cake. Their disadvantages include large area requirements, inefficient use of the total available filter area, and poor washing at the belt edges.

Cross Mode Filters: Filters of this group have a vertical flat or cylindrical filtering partition. In this case, filtrate may move inside the channels of the filtering elements along the surface of the filtering partition downward under gravity force action, or rise along this partition upward under the action of a pressure differential. In the separation of heterogeneous suspensions, nonuniform cake formation along the height can occur because larger particles tend to settle out first. This often results in poor cake washing due to different specific resistance over the partition height. The cake may creep down along the partition due to gravity; this is almost inevitable in the absence of a pressure gradient across the filtering partition. The vertical filtering partition makes these filters especially useful as thickeners, since it is convenient to remove cake by reverse filtrate flow.

Filter Presses: The most common filter press is the plate-and-frame design, consisting of a metal frame made up of two end supports rigidly held together by two horizontal steel bars. Varying numbers of flat plates containing cloth filter media are positioned on these bars. The number of plates depends on the desired capacity and cake thickness. The plates are clamped together so that their frames are flush against each other, forming a series of hollow chambers. The faces of the plates are grooved, either pyramided or ribbed. The entire plate is covered with cloth, which forms the filtering surface. The filter cloth has holes that register with the connections on the plates and frames, so that when the press is assembled these openings form a continuous channel over the entire length of the press and register with the corresponding connections on the fixed head. The channel opens only into the interior of the frames and has no openings on the plates. At the bottom of the plates, holes are cored so that they connect the faces of the plates to the outlet cocks. As the filterable slurry is pumped through the feed channel, it first fills all of the frames. As the feed pump continues to supply fluid and build up pressure, the filtrate passes through the cloth, runs down the face of the plate and passes out through the discharge cock. When the press is full, it is opened and dumped. The cake cannot be washed in these units and is therefore discharged containing a certain amount of filtrate with whatever valuable or undesirable material it may contain. Each plate discharges a visible stream of filtrate into the collecting launder. Hence, if any cloth breaks or runs cloudy, that plate can be shut off without spoiling the entire batch. If the solids are to be recovered, the cake is usually washed. In this case, the filter has a separate wash feed line. The plates consist of washing and nonwashing types arranged alternately, starting with the head plate as the first nonwashing plate. The wash liquor moves down the channels along the side of each washing plate, and moves across the filter cake to the opposite plate and drains toward the outlet. To simplify assembly, the nonwashing plates are marked with one button and the washing plates with three buttons. The frames carry two buttons. In open-delivery filters the cocks on the one-button plates remain open and those on the three-button plates are closed. In closed-delivery filters a separate wash outlet conduit is provided. Initial capital investment and floor space requirements are low in comparison to other types of filters. They can be operated at full capacity (i.e., all frames in use) or at reduced capacity by blanking off some of the frames by dummy or blank plates. They can deliver reasonably well washed and relatively dry cakes. However, the combination of labor charges for removing the cakes and fixed charges for downtime may constitute a high percentage of the total cost per operating cycle.

Leaf Filters: Leaf filters are similar to plate-and-frame filters in that a cake is deposited on each side of a leaf and the filtrate flows to the outlet in the channels provided by a coarse drainage screen in the leaf between the cakes. The leaves are immersed in the sludge when filtering, and in the wash liquid when washing. Therefore, the leaf assembly may be enclosed in a shell, as in pressure filtration, or simply immersed in sludge contained in an open tank, as in vacuum, filtration. When operating a pressure leaf filter, the sludge is fed under pressure from the bottom and is equally distributed. The clear filtrate from each leaf is collected in a common manifold and carried away. In filters with an external filtrate manifold the filtrate from each leaf is visible through a respective sight glass. This is not possible when the leaves are mounted on a hollow shaft that serves as an internal filtrate collecting manifold. The filter cakes build on each side of the leaves as filtration continues, until a specified cake thickness is achieved. During the cake washing phase, excess sludge is first drained while compressed air is introduced at about 3 to 5 psi. The role of the compressed air is to prevent the cake from peeling off of the leaves during the washing stage.

Disk Filters: Disk filters consist of a number of concentric disks mounted on a horizontal rotary shaft. The operating principle is the same as that of rotary-drum vacuum filters. The basic design is illustrated in Figure 21. The disks are formed by using V-shaped hollow sectors assembled radially about a central shaft. Each sector is covered with filter cloth and has an outlet nipple connected to a manifold extending along the length of the shaft and leading to a port on the filter valve. Each row of sectors is connected to a separate manifold. The sludge level in the tank should provide complete submergence to the lowest sector of the disks. Compared to drum vacuum filters, disk filters require considerably less floor space for the same filter area. However, because these machines have vertical filtering surfaces, cake washing tends not to be as efficient as in the case of a drum filter. The disk filter is best suited for cases when the cake does not require washing and floor space is critical.

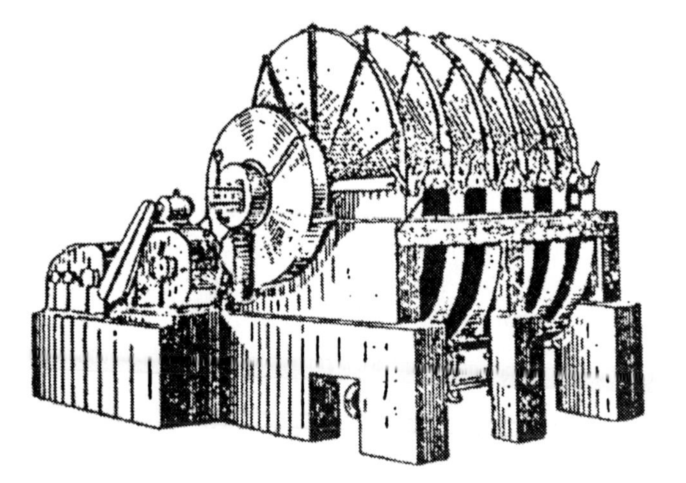

Figure 21. *Rotary-disc vacuum filter.*

Cartridge Filters: Cartridge filters are used in a multitude of solid-liquid filtration applications ranging from laboratory scale operations to industrial flows in excess of 5,000 gpm. These units are typically operated in the countercurrent mode. Common configurations consist of a series of thin metal disks that are 3 to 10 inches in diameter, set in a vertical stack with very narrow uniform spaces between them. The disks are supported on a vertical hollow shaft, and fit into a closed cylindrical casing. Liquid is fed to the casing under pressure, whence it flows inward between the disks to openings in the central shaft and out through the top of the casing. Solid particles are captured between the disks and remain on the filter media. Since most of the solids are removed at the periphery of the disks, the unit is referred to as an edge filter. The accumulated solids are periodically removed from the cartridge. As with any filter, careful media selection is critical. Media that are too coarse, for example, will not provide the needed protection. However, specifying finer media than necessary can add substantially to both equipment and operating costs. Factors to be considered in media selection include the solids loading, the nature and properties of the particles, particle size, shape and size distribution, the amount of solids to be filtered, fluid viscosity, slurry corrosiveness, abrasiveness, adhesive qualities, liquid temperature, and flowrate. Typical filter media are wire mesh (typically 10 to 700 mesh), fabric (30 mesh to 1 μ), slotted screens (10 mesh to 25 μ) and perforated stainless steel screens (10 to 30 mesh). Multiple filters are also common, consisting of two or more single filter units valved in parallel to common headers. The distinguishing feature of these filters is the ability to sequentially backwash each unit in place while the others remain on stream. Hence, these systems are continuous filters. These units can be fully automated to eliminate manual backwashing. Backwashing can be controlled by changes in differential pressure between the inlet and outlet headers. One possible arrangement consists of a controller and solenoid valves that supply air signals to pneumatic valve actuators on each individual filter unit. As solids collect on the filter elements, flow resistance increases. This increases the pressure differential across the elements and, thus, between the inlet and outlet headers on the system. When the pressure drop reaches a preset level, an adjustable differential pressure switch relays information through a programmer to a set of solenoid valves, which in turn sends a signal to the valve actuator. This rotates the necessary valve(s) to backwash the first filter element. When the first element is cleaned and back on stream, each successive filter element is backwashed in sequence until they are all cleaned. The programmer is then automatically reset until the rising differential pressure again initiates the backwashing cycle. Filter cartridges or tubes are made from a variety of materials. Common designs are natural or synthetic fiber wound over a perforated plastic or metal core. A precision winding pattern covers the entire depth of the filter tube with hundreds of funnel-shaped tunnels, which become gradually finer from the outer surface to the center of the tube and trap progressively finer particles as the fluid travels to the center. This provides greater solids retention capacity than is associated with surface filter media of the same dimensions. Typical cartridge materials are cotton, Dynel, polypropylene, acetate, porous stone and porous carbon filter lubes. Supporting perforated cores for cotton, Dynel or polypropylene are stainless steel, polypropylene or steel. Supporting cores for acetate tubes are tin-plated copper with voile liner. Porous stone and porous carbon filter tubes do not require supporting cores. Stainless steel cores are recommended for mildly acid and all alkaline solutions, i.e., pH 4-14. Polypropylene cores are used where all metal contact must be eliminated or where stainless steel is attacked, such as high chloride and sulfuric acid solutions.

Two types of polypropylene cores are commonly used: mesh polypropylene and rigid perforated polypropylene. Mesh polypropylene is satisfactory for temperatures below 140° F. The more expensive rigid polypropylene cores are used for temperature applications over 140° F, and for double- and triple-tiered filter chambers because greater strength is needed here. Perforated steel cores are used for dilute alkaline solutions, solvents, lacquers, oils, emulsions, etc.

Strainers and Filter Baskets: Strainer filter baskets and filter bag baskets are used as prefiltering devices. This prestraining or prefiltering prevents larger contaminated particles from continuing through the filtration process and thus extends the life of the entire system. Single-stage strainers and bag filters differ only in the basket design. Strainer baskets have solid flat bottoms, and baskets for filter bags have perforated bottoms to accept standard size filter bags. Dual-stage straining/ filtering action is achieved by insertion of a second, inner basket. It is supported on the top flange of the outer basket. Both baskets can be strainers (with or without wire mesh linings) or both can be baskets for filter bags. They may also be a combination: one a strainer basket, the other a filter bag basket. Dual-stage action increases strainer or filter life and reduces servicing needs. All baskets are equipped with a seal.

The seal is maintained during operation by a hinged basket bail handle being held down under the closed cover, which holds the basket down against a positive stop in the housing. Fabric bag filter baskets are capable of providing removal ratings from 20 mesh to nominal 1 micron (μ) for both Newtonian and viscous liquids. Wire mesh or fabric baskets can be cleaned and reused in many applications, or are disposable when cleaning is not feasible. Side-entry models feature permanent flanged connections, for line pressures to 150 psi. These filters are fabricated to American Society of Mechanical Engineers (ASME) codes for applications that must comply with piping standards established in many processing plants. Top-entry models feature the inlet connection as an integral part of the lid.

The inlet can be equipped with different types of quick disconnects for fast basket removal. Strainers should be selected so that the pressure drop incurred does not exceed a specified limit with a clean strainer basket (typically 2 psi).

Rapid Filtration: Rapid filtration is a standard unit operation in wastewater treatment. The operation is performed either in open gravitational flow filters or in closed pressure filters. Rapid pressure filters have the advantage of being able to be inserted in the pumping system, thus allowing use of a higher effective loadings. Note that pressure filters are not subject to development of negative pressure in a lower layer of the filter. These filters generally support higher speeds, as the available pressure allows a more rapid flow through the porous medium made up by the filter sand. Pressure filtration is generally less efficient than the rapid open type with free-flow filtration. Pressure filters have the following disadvantages. The injection of reagents is complicated, and it is more complicated to check the efficiency of backwashing. Work on the filter mass is difficult considering the assembly and disassembly required. Also, the risk of breakthrough by suction increases. Another disadvantage is that pressure filters need a longer filtration cycle, due to a high loss of head available to overcome clogging of the filter bed. Another option is to use open filters, which are generally constructed in concrete. They are normally rectangular in configuration. The filter mass is posed on a filter bottom, provided with its own drainage system, including bores that are needed for the flow of filtered water as well as for countercurrent washing with water or air. There are several types of washing bottoms. One type consists of porous plates which directly support the filter sand, generally without a layer of support gravel. Even if the system has the advantage of being of simple construction, it nevertheless suffers from incrustation. This is the case for softened water or water containing manganese. Porous filters bottoms are also subject to erosion or disintegration upon the filtration of aggressive water. The filter bottom is often comprised of pipes provided with perforations that are turned toward the underpart of the filter bottom and embedded in gravel. The lower layers are made up of gravel of approximate diameter 35-40 mm, decreasing up to 3 mm. The filter sand layer, located above this gravel layer, serves as a support and equalization zone. Several systems of filter bottoms comprise perforated self-supporting bottoms or false bottoms laid on a supporting basement layer. The former constitutes a series of glazed tiles, which includes bores above which are a series of gravels in successive layers. All these systems are surpassed to some extent by filter bottoms in concrete provided with strainers. The choice of strainers should in part be based on the dimensions of the slits that make it possible to stop the filter sand, which is selected as a function of the filtration goal. Obstruction or clogging occurs only rarely and strainers are sometimes used. Strainers may be of the type with an end that continues under the filter bottom. These do promote the formation of an air space for backwashing with air. If this air space is not formed, it can be replaced by a system of pipes that provide for an equal distribution of the washing fluids. Pressure filters are worth noting. These are usually set up in the form of steel cylinders positioned vertically. Another variation consists of using horizontal filtration groups. This has the drawback that the surface loading is variable in the different layers of the filter bed; moreover, it increases with greater penetration in the filter bed (the infiltration velocity is lowest at the level of the horizontal diameter of the cylinder). The filter bottom usually consists of a number of screens or mesh sieves that decrease in size from top to bottom or, as an alternative, perforated plates supporting gravel similar to that used in the filter bottoms of an open filter system. Filter mass washing can influence the quality of water being filtered. Changes may be consequent to fermentation, agglomeration, or formation of preferential channels liable to occur if backwashing is inadequate. Backwashing requires locating a source that will supply the necessary flow and pressure of wash water. This water can be provided either by a reservoir at a higher location or by a pumping station that pumps treated water. Sometimes an automated system is employed with washing by priming of a partial siphon pumping out the treated water stored in the filter itself. The wash water must have sufficient pressure to assure the necessary flow.

Washing of the filter sands is accomplished followed by washing with water and in most cases including a short intermediate phase of simultaneous washing with air and water. The formation of a superficial crust on the filter sand is avoided by washing with air. After washing with air, water flow is gradually superimposed on the air flow. This operational phase ends at the same time that the wash air is terminated, to avoid the filter mass being blown away. The wash water contains materials that eventually require treatment in a sludge treatment plant. Their concentration varies as a function of the washing cycle. Accounting for the superficial load in filtration, velocity of the wash water, and length of the filtration cycle, it may be assumed that the water used for washing will not attain 5% of the total production.

For new installations the first washing cycles result in the removal of fine sand as well as all the other materials usually undesirable in the filter mass, such as particles of bitumen on the inner surface of the water inlet or other residuals from the crushing or straining devices of the filter media. Consequently, it is normal that at the beginning of operation of a filter sand installation, dark colored deposits appear at the surface of the filter mass. In the long term they have no consequence and disappear after a few filtration and wash cycles. If, after several weeks of filtration, these phenomena have not disappeared, it will be necessary to examine the filter sand. The elimination of fine sand must stop after 1 or 2 months of activity. If this sand continues to be carried away after the first several dozen washes, it is necessary to reexamine the hydraulic criteria of the washing conditions, the granulometry of the filter mass, and the filter's resistance to shear and abrasion.

REFERENCES

1. Cheremisinoff, N. P., *Handbook of Industrial Toxicology and Hazardous Materials*, Marcel Dekker, Inc, New York, 1999.
2. Cheremisinoff, N. P., *Safety Management Practices for Hazardous Materials*, Marcel Dekker, Inc., New York, 1996.
3. Cheremisinoff, N. P., *Practical Guide to Industrial Safety: Methods for Process Safety Professionals*, Marcel Dekker, Inc. New York, 2001.
4. Cheremisinoff, N. P., *Chemical Engineer's Condensed Encyclopedia of Process Equipment*, Gulf Publishing Co., Houston, TX, 2000.
5. Cheremisinoff, N. P. *Liquid Filtration*, 2nd Edition, Butterworth-Heinemann, MA, 1998.
6. Cheremisinoff, N. P., *Handbook of Chemical Processing Equipment*, Butterworth-Heinemann , MA, 2000.

Chapter 3
The Economics of Pollution Prevention

INTRODUCTION

In the United States, the Pollution Prevention Act of 1990 shifted emphasis away from traditional treatment options toward waste avoidance. The USEPA defines pollution prevention as any effort to reduce the quantity of industrial, hazardous, or toxic waste through changes in the waste generating or production process at the source. This means of course, that pollution prevention can encompass all actions, taken prior to the waste being generated, which provide for net reductions in either waste volume or hazard/toxicity. This is not to imply that end-of-pipe techniques such as recycling and volume reduction are not desirable. It does, however, indicate that while these methods can help, there are better, or perhaps - more cost effective approaches. In this volume the definition of pollution prevention is even broader than EPA's, because we use the terms *waste* and *pollution* essentially interchangeably, and "waste" implies any inefficiency in manufacturing, including but not limited to energy losses, low production yields, and even variable product quality.

If we consider pollution in the narrow sense, preparing a financial justification for a pollution prevention project is often limited to declaring that if funding isn't made available, there could be an environmental incident resulting in fines or penalties, and perhaps litigation. Unfortunately, this approach in the past has led to many poor decisions. Projects with limited benefit were often supported by management, and some projects that could have had large impacts on profit and cash flow were not. A pollution prevention investment must be able to stand up to every other funding request and effectively compete for monies on the projects' own merits.

Unfortunately, investment projects in pollution prevention have often been among the first to be postponed in times of budget shortfalls. This has been due in a large part to the inadequate support and defensive posture taken by corporations towards environmental projects on an economic basis. Typically, when a production division requests money, all the necessary documentation, facts, and figures are ready for presentation. The production project is justified by showing how the project will increase revenues and how the added revenue will not only recover costs, but substantially increase the earnings for the company or company's operating division as well. But in fact there is no difference since pollution prevention project justification requires this same emphasis. To be competitive and to get "management buy-in," an understanding of the financial system is essential. Financial tools demonstrate the importance of the pollution prevention investment on a life cycle or total cost basis; in terms of **REVENUES**, **EXPENSES**, and **PROFITS**. In this chapter, we provide principles and practices for cost accounting, and apply these to some industry examples.

TOTAL COST AND COST ACCOUNTING

The term *Total Cost Accounting* (TCA) has come to be known more commonly as Life Cycle Costing (LCC). LCC is a method aimed at analyzing the costs and benefits associated with a piece of equipment or a practice over the entire time the equipment or practice is to be used. The idea actually originated in the federal government and was first applied in procuring weapons systems. Experience showed that the up-front purchase price was a poor measure of the total cost. Instead, costs such as those associated with maintainability, reliability, disposal/ salvage value, as well as employee training and education had to be given equal weight in making financial decisions. By the same token, in justifying pollution prevention, all benefits and costs must be clearly defined in the most concrete terms possible over the life of each option. Well, how is this done? By applying the basic financial terms described below.

Present Worth or Present Value

The importance of present worth, also known as present value, lies in the fact that time is money. The preference between a dollar now versus a dollar a year from now is driven by the fact that a dollar in-hand can earn interest. Present value can be expressed by a simple formula:

$$P = F/(1 + i)^n$$

where P is present worth or present value, F is future value, i is the interest or discount rate, and n is the number of periods. As a simple example, if we have or hold \$1,000 in one year at 6% interest compounded annually would have a computed present value of:

$$P = \$1,000/(1 + 0.06)^1 = \$943.40$$

Because our money can "work," at 6% interest, there is no difference between \$943.40 now and \$1,000 in one year because they both have the same value at the current time. (Economically, there is an additional factor at work in present value, and that factor is *Pure time preference* (or *impatience*) - see Pearce and Turner (1).

However, this issue is generally ignored in business accounting in that the firm has no such emotions and opportunities can be measured in terms of per financial return.) But going back to our $1,000, if the money was received in 3 years, the present value would be:

$$P= \$1{,}000/(1 + 0.06)^3 = \$839.62$$

In considering either multiple payments or cash into and out of a company, the present values are additive. For example, at 6% interest, the present value of receiving both $1,000 in one year and $1,000 in 3 years would be $943.40 + $839.62 = $1,783.06. Similarly, if one was to receive $1,000 in one year, and pay $1,000 in 3 years the present value would be $943.40 - $839.62 = $103.78. This example illustrates Rule #1 to remember:

Rule #1 - ***Present worth calculations allow both costs and benefits which are expended or earned in the future to be expressed as a single lump sum at their current or present value.***

Financial Analysis Factors

It is common practice to compare investment options based on the present value equation shown above. In addition, we may also apply one or all of the following four factors when comparing investment options:

☞ **PAYBACK PERIOD**
☞ **INTERNAL RATE OF RETURN**
☞ **BENEFIT TO COST RATIO**
☞ **PRESENT VALUE OF NET BENEFIT**

What is Payback Period?

The payback period of an investment is essentially a measure of how long it takes to break-even for an investment. In other words - how many weeks, months or years does it take to get back the investment capital that has been laid out for a project or a piece of equipment. Obviously, those projects with the fastest returns are highly attractive. The technique for determining payback period again lies within present value; however, instead of solving the present value equation for the present value (P), the cost and benefit cash flows are kept separate over time. First, the project's anticipated benefit and cost are tabulated for each year of the project lifetime. Then these values are converted to present values by using the present value equation with the firm's discount rate plugged in as the discount factor. Finally the cumulative total of the benefits (at present value) and the cumulative total of the costs (at present value) are compared on a year by year basis. At the point in time when the cumulative present value of the benefits start to exceed the cumulative present value of the costs, the project has reached the payback period. Ranking projects then becomes a matter of selecting those projects with the shortest payback period.

Although straightforward, there are dangers in selecting pollution prevention projects based upon a minimum payback time standard. One danger is that because the pollution prevention benefit stream generally extends far into the future, discounting makes its payoff period very long. Another danger is that the highest costs and benefits associated with most environmental projects are generally due to catastrophic failure, also a far future event. Since the payback period analysis stops when the benefits and costs are equal, the projects with the quickest positive cash flow will dominate. Hence, for a pollution prevention project, with a high discount rate, the long term costs/benefits may be so far into the future that they do not even enter into the analysis. In essence, the importance of life-cycle costing is lost in using the minimum payback time standard because it only considers costs and benefits to the point where they balance instead of considering them over the entire life of the project. Later on we will look at an example, but for now, we state **Rule # 2** - ***The minimum payback time standard is a good financial comparison, but it should not be the deciding basis for project selection.***

What do we mean by the Internal Rate of Return?

Many readers are likely more familiar with the term *return on investment* or ROI. The ROI is defined as the interest rate that would result in a return on the invested capital equivalent to the project's return. For illustration, if we had a pollution prevention project with a ROI of 30%, that's financially equivalent to investing resources in the right Internet stock and having its price go up 30%. As before, this method is based in the net present value of benefits and costs; however, it does not use a predetermined discount rate. Instead, the present value equation is solved for the discount rate i. The discount rate that satisfies the zero benefit is the rate of return on the investment and project selection is based on the highest rate.

From a simple calculation standpoint, the present value equation is solved for i after setting the net present value equal to zero and plugging in the future value obtained by subtracting the future costs from the future benefits over the lifetime of the project. This approach is frequently used in business, however, the net benefits and costs must be determined for each time period and brought back to present value separately. Computationally, this could mean dealing with a large number of simultaneous equations which can complicate the analysis.

What is the Benefit to Cost Ratio?

The benefit to cost (B/C) ratio is a benchmark that is determined by taking the total present value of all of the financial benefits of a pollution prevention project and dividing by the total present value of all the costs of the project. If the ratio is greater than unity, then the benefits outweigh the costs, and we may conclude that the project is economically worthwhile to invest in.

The present values of the benefits and costs are kept separate and expressed in one of two ways. First, as already explained, there is the pure benefit/cost ratio which implies that if the ratio is greater than unity, the benefits outweigh the costs - and the project is acceptable. Second, there is the net ratio which is the net benefit (i.e., benefits less costs) divided by the costs. In this latter case, the decision criteria is that the benefits must outweigh the costs which means the net ratio must be greater than zero (e.g., if the benefits exactly equaled the costs, the net B/C ratio would be zero). In both cases, the highest B/C ratios are considered as the best projects.

The B/C ratio can be misleading. Take for example, if the present value of a project's benefits were $10,000 and costs were $6,000, the B/C ratio would be $10,000/$6,000 or 1.67. Upon further reassessment of the project we find that some of the costs are not "true" costs, but instead simply offsets to benefits. In this case the ratio could be changed considerably. For argument sake, let's say that $4,500 of the $6,000 total cost is for waste disposal, and that $7,000 of the $10,000 in benefits is due to waste minimization, then one coùld use them to offset each other. Mathematically then, both the numerator and denominator of the ratio could be reduced by $4,500 with the following effect: ($10,000 - $4,500) / ($6,000 - $4,500) = 3.7. Without changing the project, the recalculated B/C ratio would make the project seem to be considerably more attractive. This leads us to **Rule # 3 - *Identify and account for true costs and offsets to benefits to better assess the financial attractiveness of a pollution prevention project.***

What is Present Value of Net Benefits?

The present value of net benefits (PVNB) shows the worth of a pollution prevention project in terms of a present value sum. The PVNB is determined by calculating the present value of all benefits, doing the same for all costs, and then subtracting the two totals. The result is an amount of money that would represent the tangible value of undertaking the project. This comparison evaluates all benefits and costs at their current or present values. If the net benefit (i.e., the benefits less costs) is greater than zero, the project is worth undertaking; if the net is less than zero, the project should be abandoned on a financial basis. This technique is firmly grounded in microeconomic theory and is ideal for total cost analysis (TCA) and pollution prevention financial analysis.

Even though it requires a preselected discount rate which can greatly discount long term benefits, it assures all costs/benefits over the entire life of the project are included in the analysis. Once the present value of all options with positive net values are known, the actual ranking of projects using this method is straight forward; those with the highest PVNB's are funded first.

There are no hard and fast rules as to which factors one may apply in performing life cycle costing or total cost analysis, however conceptually the PVNB method is preferred. There are, however, many small-scale pollution prevention projects where the benefits are so well defined and obvious that a comparative financial factor as simple as a ROI or the payback period will suffice.

ESTABLISHING BASELINE COSTS

To properly determine the cost of a project, we first need to establish a baseline for comparative purposes. If nothing else, a baseline defines for management the option of maintaining "status quo." Changes in material consumption, utility demands, manpower, etc., for other options being considered can be measured as either more or less expensive than the baseline. In this book, we follow the methodology of McHugh (2). McHugh defines four tiers of potential costs that are integral to pollution prevention:

- **Tier 0**: Usual or normal costs such as direct labor, raw materials, energy, equipment, etc.;
- **Tier 1**: Hidden costs - examples are monitoring expenses, reporting and recordkeeping, permitting requirements, Environmental Impact Statements, legal;
- **Tier 2**: Future liability costs - examples are remedial actions, personal injury under the OSHA regulations, property damage;
- **Tier 3**: Less tangible costs - examples are consumer response/confidence, employee relations, corporate image.

Tier 0 and Tier 1 costs are direct and indirect costs. They would include the engineering, materials, labor, construction, contingency, etc., as well as waste collection and transportation services, raw material consumption (increase or decrease) and production costs. Tier 2 and Tier 3 represent intangible costs. They are much more difficult to define and include potential corrective actions under the Resource Conservation and Recovery Act (RCRA), possible site remediation at third-party sites under Superfund, liabilities that could arise from third party lawsuits for personal injury or property damages, and benefits of improved safety and work environments. Although these intangible costs often cannot be accurately predicted, they can be very important and should not be ignored when assessing a pollution prevention project.

A Present Value analysis under uncertainty generally requires a little ingenuity in assessing the full merits of a pollution prevention project. This leads us to **Rule # 4 - When it is not possible to analyze the intangible costs and benefits financially, they should be listed as additional factors to consider when making the pollution prevention investment decision.**

Procurement versus Operating Costs

In analyzing the financial impact of projects, it is often useful to further categorize costs as either *procurement costs* or *operations Costs*. This distinction better enables projecting costs over time, since procurement costs are of shorter duration and refer to all costs required to bring a new piece of equipment or a new procedure on line. Conversely, operations costs are long term and represent all costs of operating the equipment or performing the procedure in the post procurement phase.

Establishing the Baseline

The baseline defines the current cost of doing business, and it gives management the option of doing nothing or simply maintaining status-quo. To illustrate this, let's consider a facility that the author worked with in Samara, Russia - a beverage washing and bottling operation. To gain a physical understanding of the operation, photographs of the main production lines are shown in Figures 1 and 2. The plant cleans bottles in a series of automatic washing machines, dries the bottles, inspects them for defects, then passes them on to bottle filling, labeling and corking operations. Washing is done with drinking quality, fresh hot water and a detergent solution. The washing stage is performed as once-through (i.e., there is no recycling of wash water). Because the spent wash water contains a contaminant (organic matter and the detergent), the effluent is subject to a pollution fee, and it must be treated by a local municipal water treatment plant.

To establish the baseline, the current cost of doing business must first be determined. Once the present costs are known, all potential alternatives such as substituting a more biodegradable cleaning fluid or recycling a portion of the spent rinse waters would then be related to this baseline cost. There are three basic steps to establishing a baseline cost:

Figure 1. *Automatic bottling line for vodka.*

Figure 2. *Shows an automatic bottle washing machine.*

Step 1. Add up all the relevant input and output materials for the process and compute their appropriate dollar value. (Note, we are only focusing on mass - not energy or other issues). This is done by balancing the material entering and leaving the operation which contributes to the waste. This is known as a material balance (see Chapter 4 for a discussion on the material balance).

Step 2. Check to make sure that the material balance makes sense. Specifically, is the volume of cleaning solvent purchased accounted for in the losses, product, inventory, and/or the waste? Account for such losses as evaporative. In this example, the losses due to evaporation are on the order of 5%. Once accomplished, determining the baseline costs becomes a matter of pricing each input and output and then multiplying their volumes by the appropriate unit. The baseline costs for this example are tabulated in Table 1.

Step 3. Examine the expected business outlook and most likely changes such as business expansions, new accounts, rising prices, cutbacks, etc. For simplicity, Table 1 costs and volumes will be assumed constant. In other words, the current annual costs will be the same in the out-years with the exception of one very important aspect - the time value of money.

Table 1. *Current Yearly Costs for Bottle Cleaning Operation.*

Item	Cost/Unit	# Units	Cost/year
Cleaning solvent	$1.75/gal.	3,500 gal	$6,125
Water	$1.60/1,000 gal.	20,000 gal	$32
Water disposal	$0.20/gal.	20,000 gal	$7,000
		Total Annual Cost	**$13,157**

Since we are assuming in this example that the bottling plant's costs are constant, the $13,157 annual cost shown in Table 1 will be repeated for each year. The present value calculations shown earlier enable the annual expenditure to be expressed as a single sum which includes the affect of interest. Assuming that the bills are paid at the end of the year, the first year's cost would be the amount of money that would have to be banked starting today, to pay a $13,157 bill in one year. Using a 10% interest rate, the computation is as follows:

$$P = \$13,157/(1 + 0.10)^1 = \$11,960.91$$

This means that if $11,960.91 was banked at 10% interest, it would provide enough monies to pay the $13,157 bill at the end of the year. Similarly, the second, third, fourth, etc., years expenditures can also be expressed in present value. This is shown in Table 2.

Table 2. *Present Value Calculations for the Vodka Bottling Plant.*

Year	Expenditure, $	Present Value, $
1	13,157	11,960.91
2	13,157	10,873.55
3	13,157	9,885.05
4	13,157	8,986.41
5	13,157	8,169.46
6	13,157	7,426.78
7	13,157	6,751.62
8	13,157	6,137.84
9	13,157	5,579.85
10	13,157	5,072.59
	TOTAL	**$80,844.07**

The simple analysis shows that the total cost of the bottle washing system over the next 10 years, given a 10% interest rate, is $80,844.07 in present value terms. In other words, $81,000 (rounding up) invested today at 10% interest would be sufficient to pay the entire material and wastewater discharge costs for the next 10 years.
Hence, any changes to the operation of the firm can now be compared to this $81,000 baseline. Any change which would result in a lower 10 year cost would be a benefit in that it would save money; any option with a higher cost will be more expensive and should not be adopted from a financial or economic standpoint. We have, of course, not taken into account inflationary issues, which for Russian companies could be a serious unknown. In this example, the cost for raw water and for pollution fees are low by Western standards; however the costs of these services could indeed rise significantly over a decade, and hence the decision of status quo ("do-nothing") could be a grave mistake.

Analyzing Present Value Under Uncertainty

Tier 2 and 3 costs are difficult to quantify or predict. A typical tier 3 cost would be cost of lost sales due to adverse public reaction to a pollution incident such as a leaking underground storage tank, a PCB spill, or a fire and explosion incident. The variables would include the types of incidents that could occur, the severity of each incident, the ability of the firm to control or respond to the emergency, the public's reaction to the incident, the company's ability to address the public's concerns, etc. At the very least, a complex situation. In many cases, there is a probability that can be connected with a particular event. This enters into the calculation of *expected value*. The expected value of an event is the probability of an event occurring times the cost or benefit of the event. Once all expected values are determined, they are totaled and brought back to present value as done with any other benefit or expense. Hence, the expected value measures the central tendency or the value that an outcome would have on the average.

For example, there are a number of games at county fairs that involve betting on numbers or colors much like roulette. If the required bet is $1, and the prize is worth $5, and there are 10 selections (e.g., the numbers 0-9) the expected value of the game can be computed as: (Benefit of Success) × (Probability of Success) - (Cost of Failure) × (Probability of Failure).

And so, ($5) × (.1) - ($l) × (.9) = -$.40. Hence, on the average, the player will lose (i.e., the game operator will win) $.40 on every $1 wagered. For tier 2 and 3 expenses, the analysis is the same. For example, there is a great deal of data available from Occupational Safety and Health Administration (OSHA) studies regarding employee injury in the workplace. In justifying a material substitution pollution prevention project, if the probability of injury and a cost could be found, the benefit of the project could be computed.

The concept of expected value is not complicated although the calculations can be cumbersome. For example, even though each individual's chance of injury may be small, given the number of employees, their individual opportunity costs, the various probabilities for each task, etc., could mean a large number of calculations. However, if one considers the effect of the sum of these small costs, or the large potential costs of environmental lawsuits or site remediation under either the Resource Conservation and Recovery Act (RCRA) or the Comprehensive Environmental Response, Compensation, and Liability Act (CERCLA), the expected value computations can be quite important in the financial analysis.

REVENUES, EXPENSES AND CASH FLOW

At first, the relatively simple pollution prevention projects tend to require more financial justification than the savings related to tier 0 or possibly even tier 1 costs. However, as an organization gets more and more sophisticated in its subsequent efforts, the less tangible tier 2 and 3 costs become more important and more easily incorporated into the analysis. Even if these costs cannot be accurately predicted, in cases where two investment options appear to be financially equivalent, if one is a pollution prevention project, the tier 2 and 3 considerations can favor that option. Generally, all that is needed is one or two success stories and companies can become quickly hooked on the P3 concept (*Pollution Prevention Practices*).

Since it is the goal of any business to make a profit, the costs and benefits cash flows for each option can be related to the basic profit equation:

REVENUES - EXPENSES = PROFITS

The most important aspect is that profits can be increased by either an increase in revenues or a decrease in expenses. A benefit of pollution prevention is often lowered expenditures and increased profit. There are different categories of pollution prevention revenues and expenses, and it is important to distinguish between them.

What Are Revenues?

Obviously, revenue is money coming into the firm; from sale of goods or services, rental fees, interest income, etc. From the profit equation, it can be seen that a revenue increase leads to a direct increase in profit and vice versa if all other revenues and expenses are held constant. Note that we are going to assume that the condition of other expenses/revenues are held constant in the discussions below.

An important rule to remember is **Rule # 5** - ***A pollution prevention project has the potential to either increase or decrease production rates.*** As such, revenue impacts must be closely examined. For example, oftentime firms can cut wastewater treatment costs if water utlilization (and in turn the resulting wastewater flows) is regulated to nonpeak times at the wastewater treatment facility. However, this limitation on water use could hamper production. Consequently, even though the company's actions to regulate water use could reduce wastewater charges, unless alternative methods could be found to maintain total production, revenue could also be decreased. Conversely, a change in production procedure as a result of a pollution prevention project could increase revenue. For example, a process change such as moving from liquid to dry paint stripping can not only reduce water consumption, but also affect production output. Since clean up time from dry paint stripping operations (such as bead blasting) is generally much shorter than from using a hazardous, liquid based stripper, it could mean not only the elimination of the liquid waste stream (the direct objective of the pollution prevention project), but less employee time spent in the cleanup operation. In this case, production is enhanced and revenues are also increased by the pollution prevention practice. Another potential revenue effect is the generation of marketable by-products as a result of the pollution prevention practice. In later chapters several case studies are described where certain waste streams were identified as having value in secondary markets. Such opportunities enable incrementally new revenues to the overall operation of the plant. The overall point to remember is that pollution prevention has the potential to either increase or decrease revenues and profits - and that's the reason for doing a financial analysis.

What Are Expenses?

Expenses are monies leaving the firm to cover the costs of operations, maintenance, insurance, etc. There are several major cost categories and effects:

- Insurance Expense;
- Depreciation Expense;
- Interest Expense;
- Labor Expense;
- Training Expense;
- Auditing and Demo Expense;
- Floor Space Expense

Insurance Expenses - Depending upon the pollution prevention project, insurance expense could either increase or decrease. For example, OSHA has set limits on worker's exposure to a number of chlorinated solvents. If one pollution prevention option was to eliminate a hazardous, chlorinated solvent from production operations, there could be savings in employee health coverage, liability insurance, etc. Likewise, using a nonflammable solvent in place of a flammable one could lead to a decrease in the fire insurance premium. Conversely, insurance expense could be increased. For example, if a heat recovery still was added to a process operation, fire insurance premiums could increase. Depending upon the premium change (if any), expenses, and in turn profits, could be increased or decreased by a pollution prevention practice.

Depreciation Expenses - If the pollution prevention project involves the purchase of capital equipment with a limited life (such as storage tanks, recycle or recovery equipment, new solvent bath systems, new fabric dyeing baths, etc.), the entire cost is not charged against the current year. Instead, a system of depreciation spreads that expense over time. Depreciation expense calculations allocate the equipment's procurement costs (including delivery charges, installation, start up expenses, etc.) by taking a percentage of the cost each year over the life of the equipment. For example, if a piece of equipment was to last 10 years, an accounting expense of 10% of the procurement cost for the equipment would be charged each year. This is known as the *method of straight-line depreciation.*

Although there are other methods available, all investment projects under consideration at any given time should use a single depreciation method to allow for accurate comparisons of expense and revenue impacts between the alternatives. Since straight-line depreciation is easy to compute, it is the preferred method. Even though a firm must use a different depreciation system for tax purposes, e.g., the Accelerated Cost Recovery System (ACRS), it is acceptable to use other methods for bookkeeping and analysis. In any event, any pollution prevention capital equipment must be expensed through depreciation.

Interest Expense - Pollution prevention investment implies one of two things must occur; either a company must pay for the project out of its own cash, or it must finance the cost by borrowing money from a bank, issuing bonds, etc. In the case where a firm pays for a pollution prevention project out of its own cash reserves, the action is sometimes called an *opportunity cost*. If cash for the project must be borrowed, there is an interest charge associated with using someone else's money. It is important to recognize that interest is a true expense and must be treated like insurance expense as an offset to the project's benefits. The magnitude of the expense will vary with bank lending rates, returns required on corporate notes issued, etc., however, there will be an expense.

The basis of arguing for accounting for pollution prevention equipment purchases as a cost is that if cash is used for the purpose of pollution prevention, it is unavailable to use for other opportunities or investments. As a result, revenues which could have been generated by the cash (e.g., interest from a Certificate of Deposit at a bank) should be treated as an expense and reduce the value of the pollution prevention project.

Although the reasoning seems sound, opportunity costs are not really expenses. It is true that the cash will be unavailable for other investments; however, opportunity cost should be thought of as a comparison criteria and not an expense. The opportunity forgone by using the cash is considered when the pollution prevention project competes for funds and is expressed by one of the financial analysis factors discussed earlier (e.g., net value of present worth, pay back period, etc.). It is this competition for company funds that encompasses opportunity cost and opportunity cost should not be accounted directly against the project's benefits.

Many companies apply a minimum rate of return or hurdle rate to express the opportunity cost competition between investments. For example, if a firm can draw 10% interest on cash in the bank, then 10% would be a valid choice for the hurdle rate as it represents the company's cash opportunity cost. Then in analyzing investment options under a return on investment criteria, not only would the highest returns be selected, but any project which pays the firm a return less than the 10% hurdle rate would not be considered. It's important to remember that pollution prevention has good investment potential. In reducing or eliminating waste generation and the related disposal/treatment expenses, pollution prevention can have a significant positive impact on the company's bottom line. Even in cases where revenues are not generated, reducing the expenses and liabilities that are associated with generating hazardous waste or an occupational health risk represent a substantial reduction in overall expenses and an increase in profit.

Labor Expenses - In the majority of situations, a company's labor requirements will change due to the pollution prevention project. This change could be a positive effect which increases available productive time, or, if extra man hours were required to run new equipment, perform preventive maintenance, etc., but there could also be a decrease in employee's production time depending upon the pollution prevention practice.

When computing labor expenses, the tier 1 costs could be significant. For example, if a material substitution project eliminated a hazardous input material which eliminated a hazardous waste, there could be a significant decrease in labor required to complete and track manifests, costs of labeling, handling and storing hazardous waste drums would be eliminated, transportation costs reduced, etc. Hence, both direct, tier 0, expenses (e.g., 5 hours per week preventive maintenance on the pollution prevention equipment) and secondary, tier 1, expenses can have an effect on manpower costs. Labor expense calculations can be simplistic or comprehensive. The most direct and basic approach is to multiply the wage rate times the hours of labor. More comprehensive calculations include the associated costs of payroll taxes, administration, and benefits. Many companies routinely track these costs and establish an internal "burdened" labor rate to be used in financial analysis.

Training Expenses - Pollution prevention practices may also involve the purchase of equipment or new, nonhazardous input materials which require additional operator training. In computing the total training costs, both the direct costs and the man hours spent in training must be considered as an expense. In addition, any other costs for refresher training or training for new employees, which is above the level currently needed, must be included in the analysis. Computing direct costs is simply a matter of adding the costs of tuition, travel, per diem, etc. for the employees. Similarly, to compute the labor costs, simply multiply the employee's wage rate by the number of hours spent away from the job in training.

Auditing and Demo Expenses - Labor and other expenses associated with defining a pollution prevention project are often overlooked. Although these tend to be small for low investment projects, some pollution prevention practices may require extensive auditing, pilot or plant trials which can be significant up front costs from production down times, personnel, monitoring equipment, laboratory measurements, plus engineering design time and consultant time charges. Some companies may prefer to absorb these costs as part of R&D budgets. For those organizations which tend to practice pollution prevention on a regular basis, many of these expenses simply are a part of the baseline cost of operations, and are more than paid for by a few successful projects.

Floor Space Expense - As with any opportunity costs, the floor space cost must be based on the value of alternative uses. For example, multiple rinse tanks have long been used to reduce water use in electroplating operations. If a single dip rinse tank of 50 square feet were replaced with a cascade rinse system of 65 square feet, then the floor space expense would be the financial worth of the extra 15 square feet and must be included as an expense in the financial analysis for the pollution prevention project.

Unfortunately, computing this floor space opportunity cost is not always as straightforward as it was with the case of training costs. In instances where little square footage is required, there maybe no other use for the floor space which implies a zero cost. In other cases, the area is currently only being used for storage of extra parts, bench stock, feed materials, etc., the costs may involve determining the worth of having a drum of chemical or an extra part closer to the operator. Alternatively, as square footage increases, calculating floor space costs becomes more straightforward.

For example, if a new building was needed to house the pollution prevention equipment, it would be easy to compute a cost. Similarly, if installing the equipment at the production site displaces enough storage room to require additional sheds be built, the cost would again be easy to compute. As a default, the cost of floor space can be estimated from information available from realtors. The average square foot cost for a new or used warehouse, or administrative, or production space, that would be charged to procure the space on the local market, is the average market worth of a square foot of floor space. Unless there is a specific alternative proposal for the floor space, this market analysis should work as a proxy.

Cash Flow Considerations

Although cash flow does not have a direct effect on a company's revenues or expenses, the concept must be considered with any pollution prevention project. If the pollution prevention project involves procurement costs, they often must be paid upon delivery of the equipment. Conversely, cash recovery could take many months or even years.

Hence, three things can effect a firm's available cash. First, cash is used at the time of purchase. Second, it takes time to realize financial returns from the project through enhanced revenues or decreased expenses. Finally, depreciation expense is calculated at a much slower rate than the cash was spent. As a result of the investment, a company could find itself cash poor. Conversely, pollution prevention efforts can have a very positive effect on cash flow. For example, eliminating a hazardous waste via an input material substitution could result in a large amount of cash available from not having to pay for hazardous waste disposal every three months or so. Hence, even though cash flow does not have a direct impact on revenues and expenses, it may be necessary to consider in analyzing pollution prevention projects.

INTEREST AND DISCOUNT RATES

In determining the value of a pollution prevention project, the discount rate used becomes very important. If pollution prevention project benefits are accrued far into the future, or if a larger discount rate is used, the effect on the present value (and hence the apparent value of the pollution prevention project) could be dramatic.

Figure 3 illustrates the relationship between percent of future worth regained over time at varying interest rates. On the average, companies prefer a return on investment (ROI) or *hurdle rate* in the range of 10-15%. At 10% over half of a future benefit stream can be lost due to the time value of money within the first 10 years. This factor works against the acceptability of projects which provide benefits far in the future. Hence, to justify pollution prevention projects with long term benefit cash flows it is often necessary to move to tier 2 or 3 criteria described earlier.

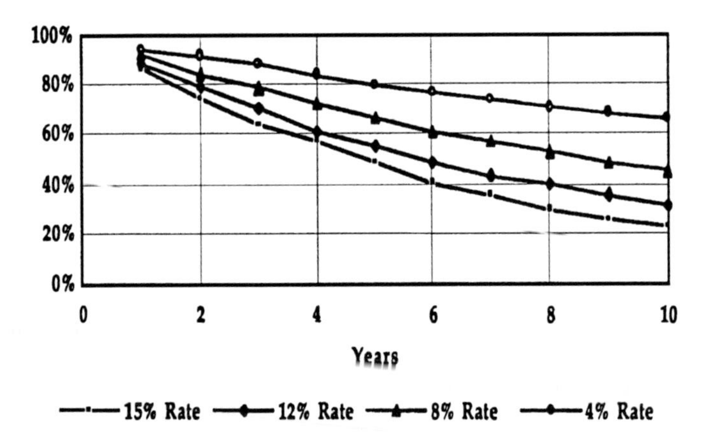

Figure 3. *Comparison of present value to future value.*

INCOME TAXES

Although most companies use only revenue and expense figures in comparing investment projects, income tax effects can enter into each calculation if either revenues or expenses are changed from the baseline values; more expenses mean lower profits and less taxes, and vice versa.

If the effect of income taxes on profit is needed, the computations are simple and can be done during or after the analysis. As with expenses and revenues, the total tax liability for each option does not need to be computed. Instead, only the difference in tax liability resulting the changes in revenues and/or expenses from the baseline due to the options being considered is required. The profit equation reflects gross or pretax profits. Income tax is based on the gross profit figure from this equation and cannot be computed until the changes in revenues/expenses are known.

Refer to Table 3 for the following discussion. For the purposes of illustration, the income tax rate shall be taken as constant at 40% of gross profit. Taxes act to soften the impact on net profit due to changes in revenue/expenses as follows. If revenues increase by $100 with no other changes, pre-tax profits would also increase $100. Since income taxes take $40 of this increase, the effect on net profit would be to soften the $100 revenue increase to a $60 net profit increase. Similarly, if expenses increase $100, pretax income would decrease $100. The tax liability would be $40 less, so in this latter case, the $100 pretax impact would be softened to a $60 net-profit decrease. Table 3 illustrates this and was taken from Stephen (3).

Table 3 shows the profit impact of an increase or decrease in revenues or expenses is limited by 1 minus the tax rate (1 - t). If the tax rate is different from 40%, it can be inserted into (1 - t) and used in calculating the impact. For example, for a 33% tax rate, a $100 increase in revenue would increase profit by (1 - 0.33) or $67.

Tax credits are a special case allowed by the IRS at various times. For example, during the energy crisis of the 1970s, certain capital expenses which reduced energy consumption (such as solar energy projects) were given special treatment as tax credits. Unlike personal tax deductions, tax credits could be deducted directly from the tax obligation of a firm. As a result, in this special tax credit case, capital expenses which would otherwise lower pretax income can be subtracted directly from the tax liability and increase profit.

Unfortunately, there are currently no tax credit pollution prevention projects available in the U.S., which is a sad situation for some heavy industry sectors that are struggling to cope with environmental laws. It is also an area that should be given consideration in those countries which have strong environmental legislation but very weak enforcement. An example of such a country is Russia.

Table 3. *Effect of Changes in Revenues and Expenses on Pre-tax and Net Profits.*

Revenue Increase	
Initial Condition:	
Beginning pre-tax profit:	$100
Tax liability:	$40
Net Profit without pollution prevention project:	**$60**
Post Pollution Prevention:	
Revenue increase subsequent to project:	$100
New pre-tax profit:	$200
New tax liability:	$80
New net profit:	**$120**
Increase in net profit due to +100 in revenues	**+$60**
Expense Increase	
Initial Condition:	
Beginning pre-tax profit:	$100
Tax liability:	$40
Net Profit without pollution prevention project:	**$60**
Post Pollution Prevention:	
Expense increase subsequent to project:	$100
New pre-tax profit:	$00
New tax liability:	$00
New net profit:	$00
Decrease in net profit due to +100 in expenses:	-$60

The reader should refer to the published works of Shim and Siegel (4,5), Purcell (6) and to reference (7) for more in-depth reading materials on the subject areas covered thus far in the chapter.

APPLICATION OF TOTAL COST ASSESSMENT

As explained earlier, the economic analysis for a pollution prevention project involves tabulating the financial costs, revenues and the savings that a project is expected to generate. Such estimates provide necessary information and data needed to evaluate the economic advantages among projects that are competing for fixed funds.

For a variety of reasons, many facilities often fail to apply the same financial tests to a pollution prevention project that they would to other capital intensive projects associated with operations, product quality, debottlenecking, and other more conventional project categories that have direct relationships to revenues and profits. One reason for this is that there are still many companies that simply don't view or understand that pollution prevention practices can indeed have the same positive and direct benefits. Another reason simply boils down to incomplete analyses of factors that fall into Tier 2 and Tier 3 cost parameters. This is particularly true when attempting to assess the economic impacts from reduced risks from occupational safety issues or future and sometimes present environmental liabilities.

And then there are companies that are simply willing to accept risks, and believe that their operations are based on sound environmental management practices and hence, future or off-site liabilities are acceptable risks. For those companies, a history lesson in environmental management is well worthwhile. Companies that operated in the 1960s and 1970s, before the advent and full implementation of many of today's environmental laws were operating well within legals standards and in many cases acceptable industry norms. However, what was legally acceptable in the 1960s and 1970s from and environmental standpoint, later became recognized as illegal under environmental legislation. The environmental laws in the United States became *retroactive* - meaning, that offsite damages, both environmental and those that created dangerous health risks for the public, were now the financial responsibilities of waste generators. Even those companies which innocently disposed of their wastes through third parties, in say - off-site landfill operations, became legally responsible for the costs of cleanup for properties they never owned or managed many years after the fact. The generator of a regulated waste cannot escape the future liability issue. At some point in time, the impacts will be felt in a company, and hence these so-called "hard to identify future costs and liabilities" are best addressed by taking a look at some of the environmental litigations and torts over the past two decades. One can most likely find some analogies for possible costs and situations that fit your company.

To recap - we have introduced some of the concepts as they relate to Total Cost Accounting and the Total Cost Assessment. Total cost accounting is also referred to as Full Cost Environmental Accounting, and is applied in management accounting to represent the allocation of all direct and indirect costs to specific products, the lives of products, or to operations. Total cost assessment has come to represent the process of integrating environmental costs into the capital budgeting analysis. It is generally defined as the long-term, comprehensive analysis of the entire range of costs and savings associated with the investment experienced by the entity making the investment

An additional term that has only been mentioned in passing to this point is the *Life Cycle Cost Assessment*. Life cycle cost assessment represents a methodical process for evaluating the life cycle costs of a product, product line, process, system, or facility starting from raw material acquisition to disposal by identifying the environmental consequences and assigning monetary value. We shall expand on this important subject shortly.For pollution prevention projects to compete fairly with pollution control and other project alternatives, more potential costs and savings must be considered. Additionally, the cost inventory should also include indirect costs, liability costs, and less tangible benefits. The way to accomplish this is by exapnding cost and savings inventories in the analysis. Tables 4 and 5 provide a list of capital and operating costs that environmental managers can use to determine the financial costs and savings associated with a particular project opportunity. The challenge in applying an expanded cost/savings inventory for investment analysis is that some of the cost data associated with a particular piece of equipment or process may be difficult to obtain. Oftentimes many costs may be a challenge because they may be grouped with other cost items in existing overhead accounts. For example, waste disposal costs for existing processes are often placed into a facility overhead account, whereas an expanded cost inventory would call for these costs to be directly allocated to the process that produces them. As such, it is not expected that information for all the cost categories will be identified during analyses. Analysts can use the list of categories provided in Tables 4 and 5 to incrementally expand their existing financial analyses whenever possible.

One approach to uncovering more of the true economic benefits of pollution prevention projects is to expand the evaluation of costs and savings over a longer time horizon, usually five or more years. This is because many of the costs and savings can take years to materialize, and because the savings from pollution prevention projects often occur every year for an extended period of time. For example, some pollution prevention projects may result in recurrent savings as a result of less waste requiring management and disposal every year. Conventional project analysis, however, often confines costs and savings to a three to five year time frame. This time horizon can be shorter than the useful life of the equipment being evaluated. Using this traditional time frame in project evaluation will exclude some of the areas of savings generated by pollution prevention projects.

Table 4. *Partial Inventory of Potential Capital Cost Items.*

PURCHASED EQUIPMENT	**SITE PREPARATION (Labor, Supervision, Materials)**	**SHAKE-DOWN AND START-UP (Labor, Supervision, Materials)**
Equipment Delivery Sales and VAT Tax Insurance Price for initial spare parts	Site studies (EIS, other) Demolition and cleaning Old equipment/garbage disposal Grading/landscaping Equipment rental Ties-ins to existing utilities and infrastructure	In-house Contractor/vendor/consultant fees Trials/manufacturing variance Training
MATERIALS	**CONSTRUCTION/INSTALLATION (Labor, Supervision, Materials)**	**REGULATORY AND PERMITTING (Labor, Supervision, Materials)**
Piping Electrical Instrumentation Structural Insulation Other (e.g., painting, ductwork)	In-house Contractor/vendor/consultant fees Equipment rental	In-house Contractor/vendor/consultant fees Permit fees
UTILITY SYSTEMS AND CONNECTIONS	**PLANNING AND ENGINEERING (Labor, Supervision, Materials)**	**WORKING CAPITAL**
General plumbing Electricity Steam Water (e.g., cooling, process) Fuel (e.g., gas, oil) Plant air Inert gas supplies Refrigeration Sewerage	In-house planning and engineering (e.g, design, shop drawings, cost estimating, etc.) Contractor/vendor/consultant fees Procurement	Raw materials Other materials and supplies Product inventory Protective equipment

CONTINGENCY	**BACK-END**
Future compliance costs Remediation	Closure and decommissioning Inventory disposal Site survey

Table 5. *Partial Inventory of Potential Operating Costs.*

DIRECT MATERIALS	WASTE MANAGEMENT (Labor, Supervision, Materials)	INSURANCE, FUTURE LIABILITY, FINES AND PENALTIES, COST OF LEGAL PROCEEDINGS (e.g., transaction costs), PERSONAL INJURY
Raw materials (e.g., wasted raw materials costs/savings) Solvents Catalysts Transport Storage	Pre-treatment On-site handling Storage Treatment Hauling Insurance Disposal	Property damage Natural resource damage Superfund
DIRECT LABOR	**UTILITIES**	**REVENUES**
Operating (e.g., worker productivity changes) Supervision Manufacturing clerical Inspection/QA/QC	Electricity Steam Water (e.g., cooling, process) Fuel Plant air Inert gas Refrigeration Sewerage	Sale of product (e.g., from changes in manufacturing throughput, market share, corporate image) Marketable by-products Sale of recyclables
REGULATORY COMPLIANCE (labor, Supervision, Materials)	Permitting Training (e.g., Right-to-Know training, Hazmat, etc.) Monitoring and inspections Notifications Testing Labeling and packaging Manifesting Record keeping	Reporting General fees and taxes Closure and postclosure care Financial assurance Value of marketable pollution permits (e.g., SOx) Avoided future regulation (e.g., CAAA)

While expanding cost inventories and time horizons can greatly enhance the ability to accurately portray the economic consequences of a single pollution prevention project, the financial performance indicators - Payback Period, Net Present Value, and Internal Rate of Return, are needed to allow comparisons to be made between competing project alternatives. By way of a short review - the payback period analysis focuses on determining the length of time it will take before the costs of a new project is recouped. A useful formula used to calculate Payback Period is:

Payback Period (in years) = start up costs / (annual benefits - annual costs)

So for example, Payback Period = $800 / ($600/yr - $400/yr) = 4 years. Those investments that recoup their costs before a pre-defined "threshold" period of time are determined to be projects worthy of funding. Remember from earlier discussions that the payback period analysis does not discount costs and savings over future years. Furthermore, costs and savings are not considered if they occur in years later than the threshold time in which a project must pay back in order to be justified. The more thorough analysis is based upon Net Present Value. The NPV method is particularly useful when comparing P2 projects against alternatives that result in higher annual waste management and disposal costs. The increased costs of current operations (or of the investment options that do not reduce wastes) will tend to lower their NPV. The method easily accommodates the expanded cost inventory when analyzing all costs and benefits.

An additional financial analysis term to consider is the *Internal Rate of Return*, or IRR. The purpose of IRR calculations is to determine the interest rate at which NPV is equal to zero. If the rate exceeds the hurdle rate (defined as the minimum acceptable rate of return on a project), the investment is deemed worthwhile. The following formula may be used:

$$\text{Initial Cost} + \text{Cash Flow in Yr.1}/(1 + r)^1 + \text{Cash Flow Yr.2}/(1 + r)^2 + \text{Cash Flow Yr.3}/(1 + r)^3 + \cdots + \text{Cash Flow Yr.n}/(1 + r)^n = 0$$

where "r" is the discount rate (assume IRR), and "n" is the number of years of the investment. A trial and error procedure can be applied to solve the above equation for r.

To help put the theory presented thus far into practice, Table 6 has been devised. Table 6, originally devised by the USEPA (*Federal Facility Pollution Prevention Project Analysis: A Primer for Life Cycle and Total Cost Assessment Concepts*, EPA 300-B-95-008, July 1995) has been modified for the discussions, and constitutes a *pollution prevention project analysis worksheet* (P3AW). Organizing such a worksheet on an Excel, Quatro Pro or other spreadsheet can assist the reader in analyzing the costs and benefits associated with current operations, potential pollution prevention projects, and alternative project opportunities.

Table 6. *Pollution Prevention Project (P3) Analysis Worksheet.*

	Section		ESTIMATED CASH FLOW IN EACH YEAR							
			Start-Up	1	2	3	4	5	6	7
CASH OUTFLOW	1	**CAPITAL COSTS**								
		Equipment								
		Utility Connections								
		Construction								
		Engineering								
		Training								
		Other								
		Subtotal								
	2	**OPERATING COSTS**								
		Materials								
		Labor								
		Utilities								
		Waste Mgmt.								
		Compliance								
		Liability								
		Other								
		Subtotal								
CASH INFLOWS	3	**REVENUES**								
		Sale of Products								
		Sale of By-products								
		Sale of Recyclables								
		Other								
		Subtotal								
	4	**PAYBACK**	**years**	*Equals Section 1 divided by (Sec. 2 - Sec. 3)*						
	5	**CASH FLOW**								
			Cash flow is calculated by subtracting Cash Outflows from Cash Inflows during each year of the investment (i.e., Sec. 3 minus Sec. 2 minus Sec. 1 subtotals)							
	6	**PV FACTORS**								
			For PVs for different investment durations, refer to the text for discussions.							
	7	**CF x PV**								
	8	**NPV**	*Equals the sum of all values from Sec. 7.*							

The use of worksheets such as Table 6 helps to demonstrate ways of capturing more cost categories by better allocating costs to specific activities, expanding the cost areas included in the analysis, and expanding the time horizon over which the project is analyzed. Potential costs and revenues in Table 6 have been abbreviated for ease of use. The reader can readily modify the worksheet to adapt it to his or her needs.

The P3AW enables one to calculate two measures of financial performance: a simple payback analysis and a net present value calculation (which incorporates the time value of money). Both of these calculations can assist in making comparisons between competing project options or in comparing a proposed project against the status quo. IRR calculations are not included on the worksheet, but they may be readily added by the reader.

It is important to note that when completing the worksheet, some data might not be available to complete all the sections. However, by completing even only a few of the sections of the worksheet with data that would not have normally been collected, the accuracy in evaluating project opportunities will be significantly enhanced.

Before attempting to organize information and data needed for the P3AW, it is advisable to define the objective of the analysis. In addition, the analysis ultimately will be used by decision makers - i.e., senior management that decides on whether or not a proposed P2 project will be funded. Like any other sales pitch, presentation means a lot. Therefore, knowing the audience to whom the analysis is to be directed, company decision making criteria for projects, and the format in which the analysis is best presented are important areas to consider. By applying some forward planning, we can assure that the scope of the analysis is appropriate, and that the completed analysis will be presented in a readily understood and accepted manner. When using P3AWs for the purpose of comparing project alternatives, or to compare a potential project to current operations, it is advisable to employ a separate worksheet for each option under consideration. The following guidelines will assist the reader in using the P3AW.

For Sections 1-3: First identify the economic consequences associated with the activity under review. Specific items (such as categories of cash outflows) noted in the P3AW may not necessarily represent a complete list of costs incurred for the facility under review. As such, Tables 4 and 5 should prove helpful in identifying additional capital and operating cost categories. If the focus is on a payback analysis, completing information for only the initial year is acceptable provided that the data are available to describe annual costs and annual savings. If the focus is on analyzing the financial performance using a NPV calculation, then you will need to obtain estimates of future costs and benefits. It is not necessary to make adjustments for inflation if the calculations are addressed through a nominal discount factor. This option is described further on and some nominal discount factors are given in Table 7 for different periods of investments. Note also that to allow comparisons with other project options, there are two measures of economic performance included on the P3AW (*Payback Analysis* and *Net Present Value Analysis*). To conduct a payback analysis, refer to Section 4 discussions below. To conduct a NPV analysis, refer to Sections 5 through 8 in the discussions below.

Table 7. *Present Value Factors for Nominal Discount Rates.*

Year	7.3%	7.6%	7.7%	7.9%	8.1%
1	0.93197	0.92937	0.92851	0.92678	0.92507
2	0.86856	0.86372	0.86212	0.85893	0.85575
3	0.80947	0.80272	0.80048	0.79604	0.79163
4		0.74602	0.74325	0.73776	0.73231
5		0.69333	0.69012	0.68374	0.67744
6			0.64078	0.63368	0.62668
7			0.59496	0.58729	0.57972
8				0.54429	0.53628
9				0.50444	0.49610
10				0.46750	0.45893
11					0.42454
12					0.39273
13					0.36330
14					0.33608
15					0.31090
16					0.28760
17					0.26605
18					0.24611
19					0.22767
20					0.21061
21					0.19483
22					0.18023
23					0.16673
24					0.15424
25					0.14268
26					0.13199
27					0.12210
28					0.11295
29					0.10449
30					0.09666

For Section 4: The section focuses on calculating the number of years that it will take to recoup the initial capital expenditure. This value is obtained by dividing the initial investment needed to establish the project by the net annual benefits (which are obtained by subtracting the annual cash outflows from the expected annual cash inflows). If only the payback analysis is important, then the reader may skip the following discussions.

For Section 5: For each year included in the evaluation, calculate the annual net cash flow by subtracting the capital expenditures subtotal (Section 1) and the annual cash outflows (subtotals from Sections 3, 4, 5) from the annual cash inflows (Section 2).

For Section 6: In order to calculate the NPV, we need to determine the value of future cash flows starting from today. In order to accomplish this, present value factors can be used to discount future cash flows. The discount will be specific to the local region, but for a comparative basis we can use the discount rates used by the U.S. government. Table 7 provides present value factors for nominal discount rates (based on 1995 figures).

For Section 7: Multiply the net cash flows (Section 7) by the PV factors (Section 8) in order to determine the present value today of the cash flow in each year.

Section 8: Sum up all the annual discounted cash to determine the NPV of the project. If the value is positive, the project is cost-beneficial. If more than one investment is being studied, then the project with the greatest NPV is the most attractive.

Once the analysis has been completed, a report may be prepared explaining the results. The report should highlight the economic benefits of the proposed P2 projects. Also discuss the noneconomic benefits as these may tip the scales in favor of the P2 opportunity if the financial analysis is marginal or too close to call.

THE LIFE CYCLE ANALYSIS

Pollution control based on end-of-pipe treatment technologies, as well as early introduction of pollution prevention programs at facilities approached the project review process by considering only those environmental impacts that could be easily translated into financial terms (e.g., permitting costs and pollution control equipment costs). Consequently, these financially-based budgeting tools often did not fully capture the benefits of pollution prevention opportunities, particularly those that reduce environmental concerns for the present and future. Without the tools to completely document environmental benefits, pollution prevention opportunities have often been difficult to support when competing against more easily-quantified environmental projects such as end-of-pipe controls, and nonenvironmental investments such as retooling or plant expansion. Decision makers require analytical tools that accurately and comprehensively account for the environmental consequences and benefits of competing projects. These environmentally-based project review tools must be flexible, easy-to-use, and require limited staff and funding so that they can be easily incorporated into the review process.

This is where we can begin to apply the principles of Life Cycle Assessment (LCA). The LCA provides a means to evaluate environmental consequences and impacts. LCA is a procedure to identify and evaluate "cradle-to-grave" natural resource requirements and environmental releases associated with processes, products, packaging, and services. LCA concepts can be particularly useful in ensuring that identified pollution prevention opportunities are not causing unwanted secondary impacts by shifting burdens to other places within the life-cycle of a product or process. LCA is an evolving tool undergoing continued development. Nevertheless, LCA concepts can be useful in acquiring a broader appreciation of the true environmental impacts of current manufacturing practices and of the proposed pollution prevention opportunities. It has taken a good two decades for many environmental professionals to become more aware that the consumption of manufactured goods and services can have adverse impacts on the supplies of natural resources as well as the quality of the environment. These effects occur at virtually all stages of the life cycle of a product, starting with raw materials harvesting, continuing through materials manufacturing and product fabrication, and concluding with the product consumption and disposal. LCA is essentially a tool that enables us to evaluate the environmental consequences of a product or activity across its entire life. The LCA is comprised of the components listed in Figure 4. These components or stages are defined as follows:

☞ **Goal Definition and Scoping** - This is a screening process which involves defining and describing the product, process or activity; establishing the context in which the assessment is to be made; and identifying the life cycle stages to be reviewed for the assessment.

☞ **Inventory Analysis** - This process involves identifying and quantifying energy, water and materials usage, and the environmental releases (e.g., air emissions, solid waste, wastewater discharges) during each life cycle stage.

☞ **Impact Assessment** - This process is used to assess the human and ecological effects of material consumption and environmental releases identified during the inventory analysis.

☞ **Improvement Assessment** - This process involves evaluating and implementing opportunities to reduce environmental burdens as well as energy and material consumption associated with a product or process.

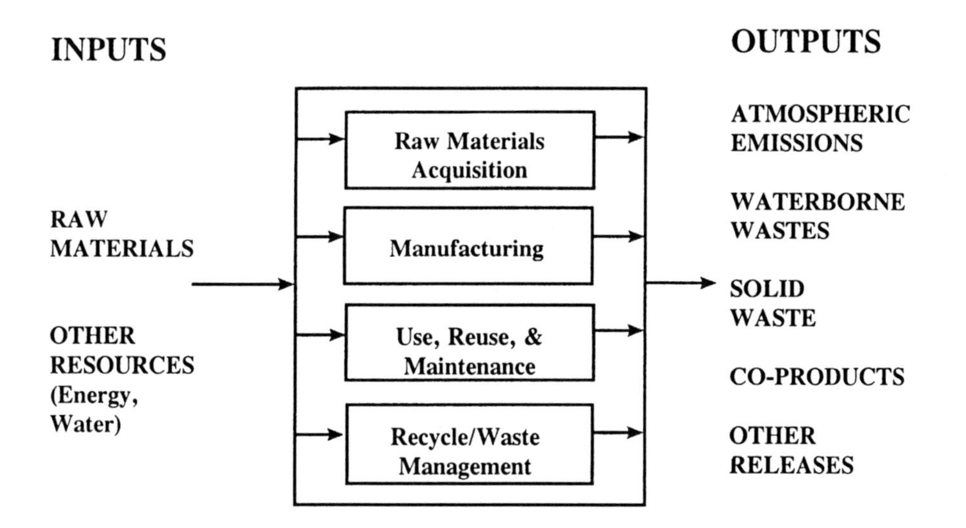

Figure 4. *Life cycle stages.*

The LCA can be used in process analysis, materials selection, product evaluation, product comparison, and even in policy-making. LCA can be used by acquisitions staff, new product design staff, and staff involved in investment evaluations.

What makes this type of assessment unique is its focus on the entire "life cycle," rather than a single manufacturing step or environmental emission. The theory behind this approach is that operations occurring within a facility can also cause impacts outside the facility's gates that need to be considered when evaluating project alternatives. Examining these "upstream and downstream" impacts can identify benefits or drawbacks to a particular opportunity that otherwise may have been overlooked. For example, examining whether to invest in plastic bottle cartons for the beverage bottling facility described earlier, or to use wooden crates for staging and storing incoming bottles should include a comparison of all major impacts, both inside the facility (e.g., disposing of the wooden crates) and "outside the gate" (e.g., additional wastewater discharges from the off-site washing of the reusable plastic cartons).

To gain complete understanding of a proposed project's environmental effects requires identifying and analyzing inputs and releases from every life cycle stage. However, securing and analyzing this data can be frustrating, and perhaps an endless task. Oftentimes process engineers in plants are faced with immediate priorities and may not have the time or resources to examine each life cycle stage or to collect all pertinent data. Despite this shortcoming, it is worthwhile to discuss the steps required to begin applying LCA concepts and principles to project analysis. Examples will demonstrate steps within selected life cycle stages. Before beginning to apply LCA concepts to projects under review, it is important to first determine the purpose and the scope of the study. In determining the purpose, facility managers should consider the type of information needed from the environmental review (e.g., Does the study require quantitative data or will qualitative information satisfy the requirements?). Once the purpose has been defined, the boundaries or the scope of the study should then be determined. What stages of the life cycle are to be examined? Are data available to study the inputs and outputs for each stage of the life cycle to be reviewed? Are the available data of an acceptable type and quality to meet the objectives of the study? Are adequate staff and resources available to conduct a detailed study?

The definition and scoping activity links the purpose and scope of the assessment with available resources and time and allows reviewers to outline what will and will not be included in the study. In some cases, the assessment may be conducted for all stages of the life cycle (i.e., raw materials acquisition, manufacturing, use/reuse/maintenance, and recycling/waste management). In many cases, the analysis may begin at the point where equipment and/or materials enter the facility. In other cases, primary emphasis may be placed on a single life cycle stage, such as identifying and quantifying waste and emissions data. In all cases, managers should ensure that the boundaries of the LCA address the purpose for which the assessment is conducted and the realities of resource constraints. Whenever possible, include in the analysis all life-cycle stages in which significant environmental impacts are likely to occur.

Conducting a LCA that includes all life-cycle stages will provide decision makers with the most complete understanding of environmental consequences. However, if resources are limited and an in-depth, quantitative analysis is not practical, a simplified approach may be taken. This alternative approach makes use of a simple checklist to identify and highlight certain environmental implication's associated with competing projects. A checklist using qualitative data instead of quantitative inputs can be very useful when available information is limited or as a first step in conducting a more thorough LCA. In addition, a *Life Cycle Checklist* (LCC) should include questions regarding the environmental effects of current operations and/or potential projects in terms of materials and resources consumed and wastes/emissions generated.

Table 8 provides a sample checklist. Of course this example is very general, but the reader can expand upon it, making it specific to the facility and nature of the operations under evaluation.

Table 8. *Example of a Life Cycle Checklist.*

Issue	Question	Yes	No
Material Usage	*Does the project minimize the use of raw materials?*		
	Is there a potential to reduce the number of suppliers for certain raw materials?		
	Are there any significant waste/discard/off-spec of raw materials due to inventory and shelf-life issues?		
Resource Conservation	*Does the project minimize energy use?*		
	Does the project minimize water use?		
	Does the project involve fuel switches or alternative energy technologies?		
Local Environmental Impacts	*Does the project eliminate or minimize impacts to the local environment (i.e., air, water, land)?*		
	Does the project have possible impacts on future or trending environmental legislation?		
	Does the project reduce or eliminate dependency on external resources that are polluting (e.g., reduce dependence on a coal fired electric utility plant where electricity is purchased from?		
Global Environmental Impacts	*Does the project eliminate or minimize impacts known to cause global environmental concerns (e.g., global warming, ozone depletion, acid rain)?*		
Toxicity Reduction	*Does the project improve the management of toxic and regulated hazardous materials and/or processes/operations which could result in human/ecological exposure?*		
Off-site Environmental Liabilities	*Does the project eliminate potential third party liabilities (e.g., eliminating the need for regulated wastes to be disposed of off-site by contractors)?*		
OSHA and Health Risk Liabilities	*Does the project eliminate the need for personal protective equipment and confined space operations?*		
	Does the project have the potential to lower insurance premiums by reduced risks of fire, explosion, occupational exposure?		

Application of a LCC has specific advantages and disadvantages when compared to other forms of life cycle assessment. The primary advantage is that completion of a checklist is relatively straightforward and requires only limited resources. However, a LCC does not provide a detailed or complete assessment of the environmental consequences associated with the activity under review. Regardless of how specific or detailed the checklist is, the method only provides general qualitative data.

Detailed Project Reviews

Conducting a more thorough review of the environmental consequences of pollution prevention projects will require more and likely dedicated resources. A more in-depth analysis would be aimed at identifying and evaluating the resource and material inputs and the environmental releases associated with each life cycle stage. This is a resource intensive operation and much more comprehensive than applying a LCC. As first steps, definition of and scoping out the analysis to fit available resources while including all significant areas of environmental impact are recommended.

The first step in identifying and evaluating the inputs and outputs associated with life cycle stages under review is to describe and understand each step in the process. One common method to do this is to construct a system flow diagram for the product, process, or activity being studied (further discussions may be found in Chapter 4). Each step within the relevant life cycle stages is represented by a box. Each box is connected to other boxes that represent the preceding and succeeding step. A simple example of a process flow diagram is as follows. In this example, the life cycle stages covered within the diagram begin at the point a solvent is purchased for use and enters a facility property. Each of these boxes can be further divided into detailed process flow steps.

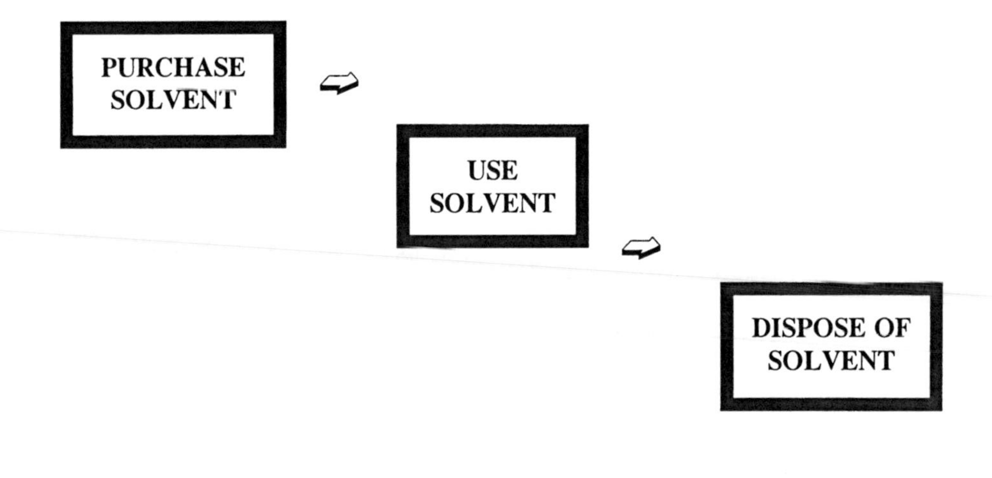

Once all of the relevant steps for each stage of the product, process, or activity under review have been identified, the flow diagram can be expanded to identify specific material and energy inputs, and the specific environmental releases and wastes associated with each box on the diagram. The process flow diagram serves as a roadmap for us to follow the flows of all materials and energy into and out of each stage in the operation. Data on each of these identified inputs and releases will be collected later on in the assessment, and will form the basis for our findings, conclusions and recommendations. In the simple example then, we can note the following:

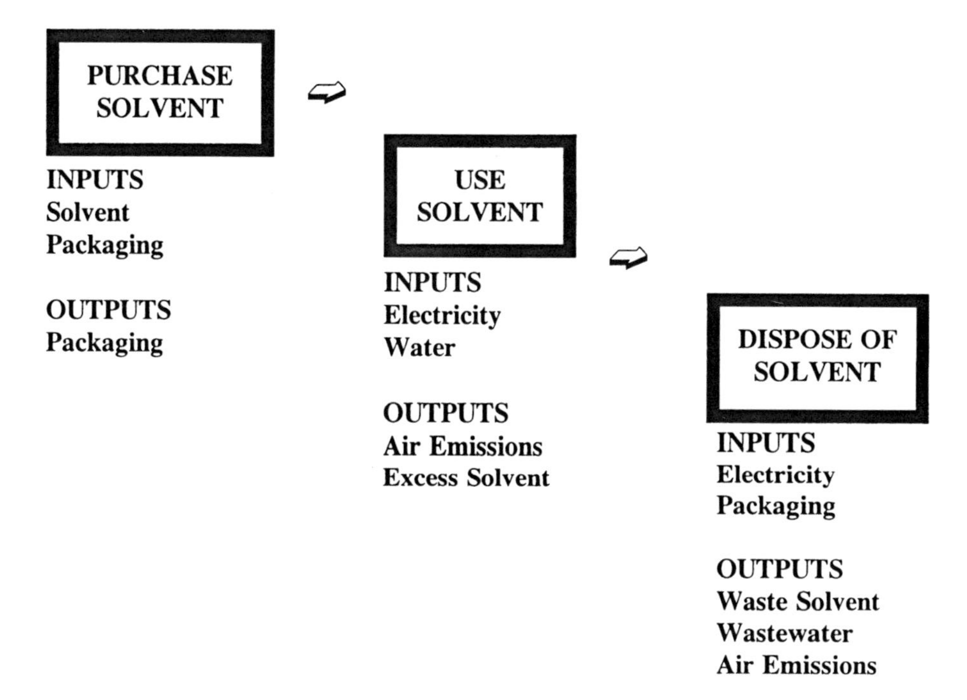

Although quantitative data are preferable (and are necessary to accurately and completely conduct an impact assessment), qualitative data may be acceptable in cases where quantitative data are lacking. A sample LCA worksheet and instructions are provided in Table 9 to help readers complete a sample process flow diagram. The intent of this worksheet is to acquaint the reader with both a suggested form and systematic approach to assessing project options or process changes under consideration. This life cycle-based worksheet is organized into three sections. The first section asks for a flowchart of the process steps/activities to be included in the analysis. The second section asks for inputs (i.e., raw materials, energy, and water), and the third section asks for outputs (i.e., products, air, water, and land releases).

Table 9. *Example of a Life Cycle Worksheet.*

I. PROCESS STEPS *Provide process step name or unit operation in each box (Step 1... Step 4)*	*1*		*2*		*3*		*4*		
II. INPUTS									
2a Raw Materials(units)									
______	___		___		___		___		___
______	___		___		___		___		___
______	___		___		___		___		___
2b Energy Usage									
Electricity (kW-hr)	___	+	___	+	___	+	___	=	___
Natural Gas (cu. ft)	___	+	___	+	___	+	___	=	___
Fuel Oil (gal.)	___	+	___	+	___	+	___	=	___
Other	___	+	___	+	___	+	___	=	___
2c Water Use (gal.)	___	+	___	+	___	+	___	=	___
2d Other Inputs (units)									
______	___	+	___	+	___	+	___	=	___
______	___	+	___	+	___	+	___	=	___
III. OUTPUTS									
3a Products, Useful By-Products (Item and Amount)									
	___		___		___		___		
	___		___		___		___		
	___		___		___		___		
3b Releases to the air (Including gaseous wastes - nonhazardous)									
______	___	+	___	+	___	+	___	=	___
______	___	+	___	+	___	+	___	=	___
______	___	+	___	+	___	+	___	=	___
3c Releases to water (Including liquid wastes - non-hazardous)									
______	___	+	___	+	___	+	___	=	___
______	___	+	___	+	___	+	___	=	___
3d Solid Wastes									
______	___	+	___	+	___	+	___	=	___
	___	+	___	+	___	+	___	=	___

The worksheet provides space for four process steps, however the reader can readily expand this to as many steps as necessary by incorporating the worksheet onto a computerized spreadsheet. Using this or any other life cycle worksheet has specific advantages and disadvantages when compared to conducting a complete LCA. The principal advantage is that it provides a more detailed analysis of the process than the checklist, and it is easier to conduct than a complete LCA. On the other hand, it does not encompass the full environmental impacts of a process or activity life cycle stage. The following are some guidelines for applying Table 9. First, the reader should view this worksheet as an aid or general outline. The worksheet is intended to help managers gain a more complete understanding of the life cycle environmental consequences associated with existing processes, potential pollution prevention projects, and competing project alternatives. When completing the worksheet, do not worry if data are not available to complete all requested information. Even by just completing a few sections of the worksheet, the information on each individual line can still be useful in evaluating and comparing the environmental performance of existing processes and potential projects. However, be aware that completing only certain sections of the worksheet may provide misleading results. For example, completing sections on solid wastes and releases to air without entering data on releases to water may bias the analysis toward projects whose primary environmental consequences result from water pollution. Similarly, collecting and analyzing data on a limited number of life cycle stages may bias the analysis toward projects whose primary environmental effects occur upstream or downstream from stages under analysis.

The information requested on the worksheet can be indicated either numerically or by text description. Descriptive information is sometimes the only information available. Specific instructions follow:

For Line 1: Indicate the process steps that are to be reviewed. For example, a life cycle analysis of a solvent degreaser tank system might examine the following three activities: acquisition of solvent, use of tank, and disposal/recycling of waste materials.

For Line 2a: For each of the process steps indicated in Line 1, identify the raw materials used. Examples of typical materials include chemicals, parts, and minerals. Do not forget to include associated packaging materials such as cans, cardboard, plastic wrap, etc.

For Line 2b: Indicate the energy involved with operating the process activity. Three common energy source categories have been included (i.e., electricity, natural gas, and fuel). Include other categories if needed. If numerical data are available, it is possible to sum together all entries from the same energy source (i.e., electricity usage from each of the process steps examined).

For Line 2c: Indicate the quantity of water consumed in each of the process steps being evaluated. Note that water could be coming from surface sources (e.g., pumped in from a nearby river), from a well, or from purchased city water.

For Line 2d: Indicate other inputs, as needed. Some process steps that can generate additional inputs include pre-process cleaning, process cleaning and maintenance supplies required in the upkeep of the process.

For Line 3a: For each process step, indicate the products that result. Be aware that the products often become the inputs to the next step in the sequence.

For Line 3b: Indicate numerically or by description the air releases associated with the process step. Examples of typical releases from an industrial process include particulates/dust and solvent vapors. Numerical records of air emissions can often be found on permitting applications or in engineering records. If numerical data is not available, provide a narrative list of emissions.

For Line 3c: Indicate the wastewater discharges and liquid hazardous wastes associated with each process step.

For Line 3d: Identify the solid waste generated from each process step. If possible list the type/quantity of solid waste and how it is managed (e.g., 10 pounds of paper products that are recycled or 5 cubic yards of sludge that is landfilled).

FINAL COMMENTS

This chapter has provided an overview of the financial tools, including life cycle assessment techniques needed for evaluating pollution prevention project opportunities. Only very fundamental principles have been presented, but hopefully, when some of the theory presented in this chapter is combined with the information in Chapter 4, along with examples of case studies described in later chapters, the reader will begin putting the tools provided in the handbook to practical use. The reader should review some of the rules to financial analyses highlighted in the chapter before going on to Chapter 4, and also quickly scan the short glossary of terms provided in this section to make sure that the key terms are understood. In addition, there is a brief list of recommended reading materials at the end of this chapter which mostly provide additional discussions on the financial calculations already discussed.

Before moving on, we should make some final comments regarding cost items and on placing value on future costs and benefits. Both of these can pose some difficulties in a project financial analysis since quite often true costs can be buried in overhead items, or masked by other operating costs and categories within an organization. It is important to try and better define these in order to properly assess the financial benefits of the project.

Allocation of Cost Categories

Compared with the traditional project analysis processes, expanding the analysis to include broader cost inventories requires a more detailed data tracking system. Many companies utilize tracking systems that group together inventory categories into facility-wide overhead accounts. These types of tracking methods make it very difficult to identify all of the discreet costs that will be impacted by proposed project alternatives. Pollution prevention activities, in particular, are at a disadvantage because many of the savings that result from these projects (such as energy, sewage, water, permitting, and waste disposal) often occur in areas lumped into overhead accounts. To overcome this, staff performing project analyses must first identify the exact data needs for the project under review. Then, a comparison can be made to information available from traditional record keeping systems in order to identify information gaps resulting from items being lumped together or reported on a facility-wide basis. To eliminate the data gaps, one of several approaches can be employed:

For the simplest of cases where several inventory categories have been combined, a review of the input data developed by each department in a facility may reveal the data for the particular project in question. For example, while the accounting department indicates on its books only the total quantity of valves used at the entire facility, a review of department specific expenses would likely reveal a more detailed account of valve use by location. For categories that are aggregated for the whole facility and not by specific project (e.g., water usage), engineering estimates or a facility walk through may be used to generate an estimate allocation to specific projects. For aggregated categories that cannot be easily allocated on a project specific basis by either of the above two methods, it may be useful to discuss the data needs both with the vendors that supplied the original equipment to see if any baseline consumption data exist and/or auditing professionals to identify what types of measurement devices or meters could be located at the specific project to meet the data needs.

Placing Value on Future Costs and Benefits

Developing reliable estimates of future costs and benefits can be a difficult task. Quantitatively estimating future costs for items such as property clean-up and environmental compliance for a facility decommissioning and for post closure can be extremely difficult. A generalized approach to this problem is to group future costs into one of two categories: recurring costs and contingent costs.

Recurring costs include items that are currently incurring costs and are anticipated to continue incurring costs into the foreseeable future based upon regulatory requirements. These include permits, monitoring costs, and compliance with regulatory requirements. The first step in estimating the future costs of these items is to determine how much the facility is currently paying. Then estimate how much the cost can reasonably be expected to escalate in the future. For example, if monitoring costs are currently $10,000 and are expected to rise with inflation, a conservative estimate would be a 4% annual increase. Consequently, the monitoring costs for the following year would be estimated at $10,400, assuming that monitoring requirements do not become more stringent. Note that when using the P3AW (Table 6), it is not necessary to escalate these values because nominal discount rates from Table 7 can be incorporated into the worksheet. By doing so the worksheet already takes inflation into account when calculating present values.

Contingent costs include catastrophic future liabilities such as remediation and clean-up costs. While current activities can lead to these future costs, quantitative estimates of these liabilities are difficult to obtain. Often the only way to include these future liabilities in the budgeting process is to qualitatively describe estimated liabilities, without attempting to define these costs using dollar amounts. If a pollution prevention option is being considered, a comparison highlighting the areas in which future liability would be reduced by implementing the pollution prevention option should be included. For example, this approach could be used to describe the future benefit of switching from lead-based paint to water based paint. Most likely, the best option may be to fully describe the potential liability if the change is not made and, if possible, document the remediation cost that could result if a liability event occurred today.

Refer to references 4 through 8 for more extensive discussions of the topics presented in this chapter.

Glossary of Important Terms

Discount Rate: The interest rate (sometimes called the Present Value Factor) used to discount future cash flows to their present values. This represents the rate of return that could be earned by investing in a project with risks comparable to the project being considered.

Hurdle Rate: In Internal Rate of Return calculations, the minimum rate of return that a project must generate in order to be considered worthy of investment. Although not common practice, we have used a U.S. government discount rate as the hurdle rate. Projects that provide a rate of return below this rate will not be pursued.

Internal Rate of Return (IRR): The discount rate at which the net savings (or NPV) on a project are equal to zero. The IRR of a project can be compared to the hurdle rate to determine economic attractiveness. The General IRR rule is: *If IRR > or = hurdle rate then accept project; If IRR < hurdle rate then reject project.*

Life Cycle Assessment: A method to evaluate the environmental effects of a product or process throughout its entire life cycle, from raw material acquisition to disposal. This includes identifying and quantifying energy and materials used and wastes released to the environment, assessing their environmental impact, and evaluating opportunities for improvement.

Life Cycle Costing: A method in which all costs are identified with a product, process, or activity throughout its lifetime, from raw material acquisition to disposal, regardless of whether these costs are borne by the organization making the investment, other organizations, or society as a whole.

Net Present Value (NPV): The present value of the future net revenues of an investment less the investment's current and future cost. An investment is profitable if the NPV of the net revenues it generates in the future exceeds its cost, that is, if the NPV is positive.

Payback Period: The amount of time required for an investment to generate enough net revenues or savings to cover the initial capital outlay for the investment.

Total Cost Assessment (TCA): A long-term comprehensive financial analysis of the full range of costs and savings of an investment that are or would be experienced directly by the organization making or contemplating the investment.

REFERENCES

1. Pearce, A., and J. Turner, *Economics of Natural Resources and the Environment*, Mc Graw-Hill Book Publishers, New York, 1977.
2. McHugh, R. T., *The Economics of Waste Minimization*, McGraw-Hill Book Publishers, New York, 1990.
3. Stephen, D. G., *A Primer for Financial Analysis of Pollution Prevention Projects*, USEPA, Cooperative Agreement No. CR-815932, April, 1993.
4. Joe K. Shim, J. K. and J. G. Siegel, *Managerial Finance*, Schaum's Outline Series (067306-9), New York, 1988.
5. Joe K. Shim, J. K. and J. G. Siegel, *Managerial Accounting*,Schaum's Outline Series (067303-0), New York, 1988.
6. Purcell, A. H., *Hazardous and Solid Waste Minimization,* Government Institutes, Inc., ISBN/ISSN: 0865871361.
7. *Waste Minimization Manual,* Government Institutes, Inc., ISBN/ISSN: 0865877319 - This document discusses waste minimization, economic imperatives, legal and regulatory incentives, and how to conduct waste

minimization audits. It also contains waste minimization case histories for Dow, DuPont, Chevron, Hewlett Packard, and the Navy.

8. Wittman, M. R., *Costing & Financial Analysis of Pollution Prevention Projects,* Massachusetts Office of Technical Assistance - This text provides a curriculum which is intended to familiarize environmental professionals with basic business terms, and to increase their awareness of the factors that influence an investment in pollution prevention options.

Chapter 4
The Pollution Prevention Audit

INTRODUCTION

There are several terms used by environmental professionals that apply to the subject matter in this chapter. The most common term used is *Pollution Prevention Audit*, which is a term that evolved from the so-called environmental audit nearly two decades ago; however, these two terms should not be confused. The environmental audit has a very different objective than a P2 audit. Environmental audits can be thought of as due-diligence exercises, i.e., their principal objectives are to define the extent to which a facility operation or its assets are out of compliance with environmental standards, and to develop recommendations that will bring the facility, its operations and even assets into compliance. In this regard, the term environmental audit can have a negative connotation for industry - especially with small to medium size operations. The reason for this being that in the past, environmental audits quite often focused on what was needed to correct out-of-compliance, and/or reduce environmental liabilities, but without taking into consideration the most economical technologies. The environmental audit in many ways was a golden opportunity for environmental consultants during the late 1980s and even through the 1990s, especially in the United States where so-called property transaction laws became popular and strict regulations on underground storage tanks forced many small business operations into costly site remediations and facility upgrades. Even today, lending institutions will insist that an independent environmental audit be conducted on commercial and industrial properties before considering loans. In the private sector - mergers, acquisitions and purchases of industrial properties simply do not take place, at least in technologically advanced countries, without conducting an initial environmental audit.

A P2 project really does not focus on due diligence, because to meet an environmental standard, one can always apply end-of-pipe treatment technologies. Instead, P2 focuses on those practices and technologies that either eliminate the pollution (and hence eliminate the cost of compliance), or it focuses on a more cost effective approach compared to conventional pollution control practices. Hence, the

P2 audit focuses on financial benefits or more specifically - it focuses on those recommendations which reduce the operating costs (present and future).

Although the distinction between a P2 audit and an environmental audit is for the most part clear in the West, this is not the case in Central and Eastern Europe, the Far East, and even parts of South America. In countries with economies in transition, confusion between the environmental audit and so-called P2 audit exists, particularly where privatization processes are taking place and some form of minimal due diligence requires the environmental audit or a variation thereof. As a result, other terms invented to distinguish the P2 audit from the environmental audit, and in general, to make it more "friendly" are *In-Plant Assessments*, *Waste Minimization Assessment*, *Cleaner Technologies Substitute Assessment* (CTSA - this term is in fact somewhat different than the P2 audit as discussed later), and others.

The objective of this chapter is to describe the approach and general protocol for conducting P2 audits. Although objectives differ, the P2 audit is not unlike an environmental audit - there are similar protocols and steps in both types of activities. A key difference though is that we now have both financial and LCA tools to help in the assessment phase of the audit. The assessment or analysis phase is the "meat and potatoes" of the entire affair, because this is where we either justify the economic benefits for a P2 opportunity identified, or we move on. In this chapter, a step-by-step approach to conducting a P2 audit is presented. The approach presented is not necessarily unique nor the best, but it is concise and easy to apply.

OVERVIEW OF POLLUTION PREVENTION

Let's not loose sight of the fact that pollution prevention focuses attention away from the treatment and disposal of wastes, and focuses towards the elimination or reduction of undesired by-products, within the production process itself. In the long run, pollution prevention through waste minimization and cleaner production is more cost- effective and environmentally sound than traditional pollution control methods. Pollution prevention techniques apply to any manufacturing process, and range from relatively easy operational changes and good housekeeping practices to more extensive changes such as the substitution of toxic substances, the implementation of clean technology, and the installation of state-of-the-art recovery equipment. Pollution prevention can improve plant efficiency, enhance the quality and quantity of natural resources for production, and make it possible to invest more financial resources in economic development.

Taking a simplistic view, all outputs from a manufacturing facility fall into two categories - they are either product or waste. Anything that the customer pays for is product; all else that leaves the facility is simply waste - whether it is regulated by environmental legislation or not. Ideally, manufacturing activities should produce zero waste. In reality, industry must strive to reduce the waste from manufacturing since this represents an inefficient use of scarce resources. We can argue that all waste can be indirectly associated with pollution since the management of waste consumes resources that would not otherwise be used, and pollution is often generated in these waste management activities.

Prevention is the act of taking advance measures against something possible or probable. Prevention is generally contrasted with control or cure. If we build quality into our products, then we can prevent defects - but that in itself does not necessarily eliminate waste or pollution. Inspection on the other hand enables us to control the defects. Hence quality, inspection and prevention are key terms and concepts in implementing P2 practices. In general, the effort, time, and money associated with prevention is less than that of control or cure. This idea is captured in the maxim "An ounce of prevention is worth a pound of cure." Thus, in many cases it is worthwhile for industry to prevent pollution rather than control it. In terms of a more formalized definition pollution prevention is any practice that:

- reduces the amount of any hazardous substance, pollutant, or contaminant reentering any waste stream or otherwise released into the environment prior to recycling, treatment, and disposal; or
- reduces the hazards to public health and the environment associated with the release of such substances, pollutants, or contaminants; or
- reduces or eliminates the creation of pollutants through (1) increased efficiency in the use of raw materials; or (2) protection of natural resources by conservation.

Another way of saying this - pollution prevention is *any action which reduces or eliminates the creation of pollutants or wastes at the source, achieved through activities which promote, encourage or require changes in the basic behavioral patterns of industry.* As noted in Chapter 1, some terms that are often used in connection and even sometimes interchangeably with P2 are *Cleaner Production*, *Clean Technology*, *Waste Reduction*, *Waste Prevention*, *Eco-efficiency*, and *Waste Minimization*.

Examples of Pollution Prevention

A few practical examples will help to illustrate how and why P2 is practiced. As noted in Chapter 3, it is important to define the objectives of a P2 project. We have only stated the most obvious and general reasons, but the reality is that P2 does not necessarily pay in all situations. The audit process itself is a tool that helps us to identify and quantify a P2 opportunity. When we combine the results from the audit with the financial tools for project assessment and with LCA tools, then we can make a critical judgement as to whether or not project opportunities are justified from a business standpoint. A very critical point to always keep in mind is that P2 is not a replacement for modernization investments or industry rationalization issues. There are some heavy polluting industry sectors that must face serious issues in terms of major investments into not only cleaner but more efficient technologies simply to sustain their operations. The iron and steel industry in Russia and Ukraine is one example. P2 on the whole can be thought of as a means to developing incremental savings. In other words - there are very few industry examples where a single P2 project has resulted in "mega-bucks" savings as opposed to small or incremental operational cost savings. It is only when you add up small incremental savings from a large number of P2 projects that one begins to see a long-term positive impact on business operations and profitability. As noted in the *Preface*, pollution prevention needs to be thought of as a form of corporate religion - where the true believer practices the philosophy on a daily basis. If on the other hand we are just occasional church goers, then we are less likely to see a significant benefit to our lifestyle. The same can be said in the manufacturing environment. Macro-economic improvements to manufacturing simply will not be achieved by occasional P2 successes.

Pollution Prevention in Auto Painting

Let us start off by examining a simple example of a pollution prevention practice. It is common for auto companies to change paint color with each car that goes through the paint process. As a result, old paint must be purged from the lines before painting each car. This results in excess paint sludge waste and fugitive emissions of toluene and xylene. Additionally, the purging and refilling qualifies as a setup activity that adds time to the process. Block painting, the process of painting batches of like colored cars, is a manufacturing process change that reduces the purged paint sludge and solvent emissions. Block painting not only decreases the waste, but also the setup time involved in the process. But this is not the only technological alternative. Cars can be painted without toxic toluene and xylene solvents. Electrostatic painting can adhere paint to treated metal. While the scrubber represents treatment, and block painting represents waste reduction, shifting to the electrostatic painting process represents pollution prevention by design. Both are examples of potential pollution prevention solutions that displace treatment, however a comparison of the economics between the two options, as well as with the status quo are needed in order to justify the investment into changing the process.

Pollution Prevention in the Polish Chemical Industry

The following is an analysis of a series of case studies for waste minimization projects funded by the U.S. Agency for International Development, and largely implemented by the *World Environmental Center* (WEC). The Chemical industry in Poland is diverse, providing a multitude of chemical products to different domestic and export consumer markets. Many plant operations can be characterized as old, lacking modern, automated controls and instrumentation, and in general, have enormous spare capacity. The types of chemical companies included in the analysis are listed in Table 1.

To examine the collective benefits of a number of P2 projects, we introduce the use of a *Pollution Prevention Matrix* (P2M). A P2M is a convenient way of tracking the performance of many pollution prevention projects within an organization. It can also serve as a useful management tool in developing environmental action plans for a facility. In this case, we apply the P2M to assessing the collective contributions of a number of pollution prevention projects from different chemical companies. Table 2 provides the P2M constructed for this discussion. Company-specific P2 case studies are organized into the matrix. The matrix provides the following information for each company-specific P2 practice:

- ☞ The pollution emissions reductions achieved to air, water and land, on a normalized tons per year basis;
- ☞ Raw materials savings in tons per year;
- ☞ Dollar investments for each P2 measure;
- ☞ The total dollar savings achieved by each P2 practice. Savings are derived from reductions in pollution fees, waste processing costs, reduced raw materials consumption, improved product quality and yields, and in some examples, energy savings;
- ☞ The payback period for the investment in months.

The matrix provides the reader not only with a sense of the overall cumulative benefits derived, but also identifies technology areas where P2 opportunities may exist in similar industry operations. In other words - *why reinvent the wheel*? If someone else has applied a cost effective solution to a pollution problem, why not use the same approach? That's the value of reviewing case studies.

Table 1. *Polish Chemical Companies Included in the Analysis.*

SPECIALTY CHEMICALS

Kedzierzyn Nitrogen Works, Kedzierzyn - nitrogen fertilizers, adhesive resins, technical gas components for plastics and intermediate products, specialty chemicals, high quality OXY-alcohols.

Boruta SA Dyestuff Industry Works, Zgierz - dyes and intermediate chemical byproducts.

Organika Azot Chemical Works, Jaworzno - pesticides, herbicides, and fungicides. Principle chemicals synthesized are organochlorophosphates and copper oxychloride.

POLYMERS

Oswiecim Chemical Works, Oswiecim - synthetic rubbers and latex products including styrene-butadiene, nitril-butadiene, and various monomers, vinyl plastics, styrene and related by-products, chlorinated polymers, chloroparaffins, esters, alkylbenzene, and detergents

Boryszew SA Chemical and Plastics Works, Sochaszew - compounding ingredients used in the production of polyvinyl chloride, polypropylene, anti-freezing fluids, polyvinyl acetate, and pyrotechnic products.

Organika-Zachem Chemical Works, Bydgoszcz - plastics, dyes, and foam products that supply the following consumer market sectors: auto parts, furniture, electrical cable insulation, textiles, and chemical intermediates. Major products are chlorine, epichlorohydrin (EPI), toluenediisocyanate, polyesters, polyurethane foams, polyvinyl chloride insulation compounds, synthetic dyes, soda lye, sodium phosphate, hydrochloric acid, and dinitrotoluene.

COKE CHEMICALS

Blachownia Chemical Works, Kedzierzyn Kozle - chemical products derived from coal and coke.

INORGANIC CHEMICALS

POLCHEM Chemical Works, Torun - chlorosulfonic acids and sulfites

Bonarka Chemical Works, Krakow - animal feed products - dicalcium fodder phosphate, which is a granular phosphorous-based animal feed.

HEALTH CARE AND PHARMACEUTICALS

VISCOPLAST SA Chemical Works, Wroclaw - medical plasters, technical tapes and glues, medical adhesive bandages, and polypropylene fibers

POLFA Pharmaceutical Works, Tarchomin - pharmaceutical dosage units of different drug forms; main drug groups are antibiotics, insulin, and psychotropic drugs.

Table 2. *Pollution Prevention Matrix of Company-specific Case Studies.*

	Emissions Reductions (t/yr)			Mtls. Savings, t/yr	Invest-ment U.S. $	1st Yr Savings U.S. $	Pay-back months
	Air	Water	Solid				
COMPANY SPECIFIC CASE STUDIES IN SPECIALTY CHEMICALS **Kedzierzyn Nitrogen Works Plant**							
Pollution Prevention Measure: Operational change to absorption towers resulting in collecting formaldehyde and methanol vapors that were normally released as air emissions. These wastes could be recycled, hence reducing raw materials purchase.							
	27.5			27.5	0	19,600	-
Pollution Prevention Measure: Direct venting of methane was stopped and the gases sent to a fuel gas collector during plant turnarounds. Methane losses of 177,000 cu. m were stopped and reused as fuel.							
	31.2			31.2	0	19,800	-
Pollution Prevention Measure: Condensate discharges from cooler segments normally sent to a sewer were recycled, also reducing fresh water feed purchases.							
		216,891		216,891	1,400	30,500	0.6
Pollution Prevention Measure: Replaced pressurized steam at 200° C with steam condensate at 135°C in venting system applications. By decreasing the venting system temperature, vapor pressures and hence, emissions of VOCs (volatile organic compounds) decreased.							
	35.2				2,000	55,700	0.4
Pollution Prevention Measure: Evaluation of the post-reaction cooling process showed that by installation of a liquid seal type degasifier, condensate could be safely discharged to recirculation waters, and post-reaction gases did not have to be discharged to air. This also reduced fresh water feed to recirculation waters.							
		7,817		7,817	2,260	2,880	9.4
Pollution Prevention Measure: A new procedure was developed in the recovery of formalin from a urea adhesive plant, eliminating the 350 cu. m of hazardous wastes.							
			733		2,700	71,400	0.5
Pollution Prevention Measure: Waste waters from the urea synthesis plant were found to be acceptable for recycling for process applications, thereby reducing wastes and raw water use.							
		500		500	18,900	106,000	2.1
Pollution Prevention Measure: Operating procedures were changed in the urea production plant which reduced ammonia losses to the atmosphere. This was accomplished by installing a pump to transfer condensate under pressure from a cooler directly to the urea absorption tower. This enabled condensate to be used in place of ammonia water, which previously had to be made.							
	1,000			1,831	40,100	89,000	5.4
Pollution Prevention Measure: A study of the heating system showed that combustion efficiency could be improved by installation of a carbon monoxide/oxygen monitoring instrument. This resulted in a savings of 250,000 cu. m/yr of fuel gas and reductions in emissions of NO_x and CO.							
				44.1	16,300	25,100	7.8
Subtotals	1,094	225,208	733	227,142	83,660	419,980	

	Emissions Reductions (t/yr)			Mtls. Savings, t/yr	Invest-ment U.S. $	1st Yr Savings U.S. $	Pay-back months
	Air	Water	Solid				
Pollution Prevention Measure: Application of a spectrophotometer enabled COD analysis on waste waters to be made rapidly and reliably. This information was used by process personnel to adjust process conditions to maximize sodium sulfate recovery from wastewater. This reduced nonmineral salts in wastewater discharges.							
		150			1,793	27,500	0.8
Pollution Prevention Measure: Application of a spectrophotometer enabled COD analysis on waste waters to be made rapidly and reliably. This information was used by process personnel to adjust process conditions and run the oxidation process in wastewater treatment, thereby reducing COD discharges.							
		40			1,794	31,200	0.7
Pollution Prevention Measure: Enhancements to a spectrophotometer enabled fast and reliable measurements of sulfites and sulfates in the post absorption process. This enabled better process control and recovery of sodium dioxide, which is raw material in the plant.							
		10			1,833	31,800	0.7
Subtotals		200			5,420	90,500	
Organika Azot Chemical Works							
Pollution Prevention Measure: Product recovery was done by redistillation of a part of organic waste during the raw production of DCAP. This also reduced solid waste.							
			1.7	5	800	11,900	0.8
Pollution Prevention Measure: Recycling of wastewaters in various parts of the plant resulted in reduced emissions and fresh water feeds.							
		165,146		165,145	1,050	21,600	0.6
Pollution Prevention Measure: Wastewaters from the second and third stages of product washing (DCAP) were reused in the hydrolysis process and replaced fresh water feed.							
		2			1,100	130,000	1.0
Pollution Prevention Measure: Water monitoring equipment was used to monitor and reduce plant consumption of hot and process waters at different locations in a plant							
				7,707	5,965	19,500	3.7
Subtotals		165,148	1.7	172,858	8,915	66,000	
COMPANY SPECIFIC CASE STUDIES IN THE POLYMERS SEGMENT							
Oswiecim Chemical Works							
Pollution Prevention Measure: A new latex degassing system and interstage latex heating system by steam improved the efficiency of the process and reduced the content of VCM (vinyl chloride monomer) in air emissions.							
	54			54	31,400	2,100,000	0.2
Pollution Prevention Measure: Problems with latex density measurements were eliminated by the use of a nuclear density gauge in the reactor. This improved PVC (polyvinyl chloride) product yield and quality, and reduced VCM emissions to air.							

<table>
<tr><th rowspan="2"></th><th colspan="3">Emissions Reductions (t/yr)</th><th rowspan="2">Mtls. Savings, t/yr</th><th rowspan="2">Invest- ment U.S. $</th><th rowspan="2">1st Yr Savings U.S. $</th><th rowspan="2">Pay- back months</th></tr>
<tr><th>Air</th><th>Water</th><th>Solid</th></tr>
<tr><td>Subtotals</td><td>54.01</td><td></td><td></td><td>54</td><td>41,100</td><td>2,134,800</td><td></td></tr>
<tr><td colspan="8">Boryszew Chemical and Plastics Works</td></tr>
<tr><td colspan="8">Pollution Prevention Measure: By replacing acetic acid used in the production of vegetable oil with formic acid, wastewater discharges became more biodegradable, resulting in a reduction in pollution fees.</td></tr>
<tr><td></td><td></td><td>116</td><td></td><td></td><td>8,600</td><td>165,000</td><td>0.6</td></tr>
<tr><td colspan="8">Pollution Prevention Measure: A series of low cost measures were undertaken to improve process yields and distillation efficiency, resulting in reduced wastewater discharges.</td></tr>
<tr><td></td><td></td><td>11</td><td></td><td></td><td>20,000</td><td>48,100</td><td>5.0</td></tr>
<tr><td>Subtotals</td><td></td><td>127</td><td></td><td></td><td>28,600</td><td>213,100</td><td></td></tr>
<tr><td colspan="8">Organika-Zachem Chemical Works</td></tr>
<tr><td colspan="8">Pollution Prevention Measure: Allyl chloride present in chloroorganic wastes were found to be recoverable in an existing distillation column, and reused in the production of epichlorohydrin (EPI)</td></tr>
<tr><td></td><td>24</td><td></td><td></td><td></td><td>1,250</td><td>13,750</td><td>1.1</td></tr>
<tr><td colspan="8">Pollution Prevention Measure: Introduction of a surface active agent to a dye production process, along with process parameter changes, resulted in reduction in raw materials consumption, improved product quality, and reduced chemical oxygen demand (COD) discharges to waste waters.</td></tr>
<tr><td></td><td></td><td>2.1</td><td></td><td></td><td>2,500</td><td>15,000</td><td>2.0</td></tr>
<tr><td colspan="8">Pollution Prevention Measure: A cost effective method for the recovery of calcium salts from process waste waters and reprocessing into calcium hydroxide (lime) was found. Lime used as a raw material in hydrolysis of EPI.</td></tr>
<tr><td></td><td></td><td>208</td><td></td><td>83,349</td><td>7,200</td><td>109,200</td><td>0.8</td></tr>
<tr><td colspan="8">Pollution Prevention Measure: This project was aimed at reducing steam and water use in toluenediisocyanate (TDI) production. By replacing a steam-water ejection system with a mechanical vacuum pump, a process solvent used as the pump's sealant, absorbs product and reactants from the distillation columns, and is recycled back to the TDI synthesis reactors. Results were reductions in discharges to sewers and the recovery of valuable products.</td></tr>
<tr><td></td><td></td><td>135</td><td></td><td>341,542</td><td>99,800</td><td>258,000</td><td>4.6</td></tr>
<tr><td colspan="8">Pollution Prevention Measure: The application of a plate-type heat exchanger at the EPI plant enabled recovery of heat energy from hot process waste waters. The heat is now used to preheat raw materials sent to distillation units. Savings in steam and energy were achieved.</td></tr>
<tr><td></td><td>21.5</td><td></td><td></td><td>12,000</td><td>120,000</td><td>252,000</td><td>5.7</td></tr>
<tr><td colspan="8">Pollution Prevention Measure: The application of a special evaporator enabled more efficient TDI product recovery from after-distillation tars. This resulted in product yield improvements and decreases in waste tar generation.</td></tr>
<tr><td></td><td></td><td></td><td>0.02</td><td></td><td>622,000</td><td>1,055,000</td><td>7.1</td></tr>
</table>

	Emissions Reductions (t/yr)			Mtls. Savings, t/yr	Invest-ment U.S. $	1st Yr Savings U.S. $	Pay-back months
	Air	Water	Solid				
Pollution Prevention Measure: Application of a spectrophotometer enabled more rapid measurements of COD (chemical oxygen demand) in waste waters. The more rapid and reliable measurements enabled process changes to be made in TDI product production. This action not only reduced wastewater discharges, but improved product yield substantially.							
		70		165,146	4,600	92,000	0.6
Pollution Prevention Measure: The production of dye is a pH sensitive process. By automating pH measurement and control, reductions in wastewater discharges and raw materials savings were achieved.							
		30		8,200	7,000	180,000	0.5
Pollution Prevention Measure: The application of steam monitoring devices improved the control of steam consumption and accountability of users in different parts of the plant. This resulted in steam consumption.							
	7			3,900	12,100	84,000	1.7
Subtotals	52.5	441	0.02	614,137	876,450	2,058,950	
COMPANY SPECIFIC CASE STUDIES IN COKE CHEMICALS **Blachownia Chemical Works**							
Pollution Prevention Measure: Compressed wheat substituted for coke as a filter media in the coke tar recovery process, reducing operating costs and eliminating over-fired coke waste.							
			130		0	25,700	-
Pollution Prevention Measure: Reuse of wash water in the cooling water loop of the carbochemical plant reduced water discharges and raw water consumption.							
		12,551		12,551	400	53,000	0.1
Pollution Prevention Measure: Equipment modifications were made which decreased naphthalene losses from a naphthalene oil distillation unit.							
	1.32			800	900	18,500	0.6
Pollution Prevention Measure: The installation of an aeration system in wastewater treatment enabled trace levels of organics to be removed from wastewater and be combusted in an existing catalytic burner.							
	600				11,500	255,000	0.5
Pollution Prevention Measure: Application of a gas chromatograph improved yields in the production of benzene, toluene, and light resin, and reduced the consumption of sulfuric acid.							
				0.01	25,200	157,000	1.9
Subtotals	601	12,551	130	126,311	38,000	509,200	
COMPANY SPECIFIC CASE STUDIES IN INORGANIC CHEMICALS **POLYCHEM Chemical Works**							
Pollution Prevention Measure: A solid waste containing high levels of lime was found. The waste was suitable as a raw material in a liquid waste neutralization process. To implement the recommendation, and automatic pH meter was purchased.							

	Emissions Reductions (t/yr)			Mtls. Savings, t/yr	Invest-ment U.S. $	1st Yr Savings U.S. $	Pay-back months
	Air	Water	Solid				
			204	10	2,700	3,600	9.0
Pollution Prevention Measure: Process modifications were made in which selected liquid wastes could be substituted for fresh water in a lime milk preparation step for wastewater treatment. This reduced both direct wastewater discharges and fresh water feed consumption.							
		50,094		50,094	3,850	9,500	4.9
Pollution Prevention Measure: A redesign of the decarbonation system used to recirculate a portion of decarbonated water back into a reactor was done. This allowed the facility to reuse decarbonated water in place of raw water for process purposes.							
			2.5	220,234	10,350	14,500	8.6
Pollution Prevention Measure: Installation of automatic process controls for temperature, pressure and flow improved production yields, reduced raw materials, and minimized wastewater flows.							
		44,500		40,800	250,500	1,100,000	2.7
Subtotals		50,094	44,707	311,139	267,400	1,127,600	
Bonarka Chemical Works							
Pollution Prevention Measure: Process changes could be better controlled through the use of a spectrophotometer. The faster control capability provided with a rapid measurement resulted in less phosphoric acid use, and a reduction in emissions from unreacted acid.							
	1,008			1,044	9,600	247,000	0.5
COMPANY SPECIFIC CASE STUDIES IN COKE CHEMICALS							
VISCOPLAST SA Chemical Works							
Pollution Prevention Measure: Continuous monitoring of naphtha and steam during absorption and desoprtion cycles in the production of acrylics and glues, enabled optimization of solvent recovery operations and steam. The program also reduced fugitive air emissions.							
	88			88	18,500	19,360	11.5
POLFA Pharmaceutical Works							
Pollution Prevention Measure: The project improved operations in a butyl acid regeneration process. By changing operating procedures during the washing operations, washing steps could continue without process shut downs. The new procedure minimized plugging of heat exchange equipment, and decreased steam and caustic consumption normally used for equipment cleanout purposes.							
		120		120	100	12,000	0.1
Pollution Prevention Measure: Repairs to leaking vacuum distillation equipment and cooling systems decreased solvent losses and improved process control. Additionally, major overhauls to vacuum pump equipment greatly improved yield and minimized losses. The use of an organic vapor analyzer was critical in identifying losses and leaks in the process operations.							
	248	26,203		26,203	51,000	198,000	3.1
Pollution Prevention Measure: A plant-wide leak detection and monitoring program was implemented to minimize solvent losses in the manufacture of various pharmaceuticals.							

	Emissions Reductions (t/yr)			Mtls. Savings, t/yr	Invest-ment U.S. $	1st Yr Savings U.S. $	Pay-back months
	Air	Water	Solid				
			204	10	2,700	3,600	9.0
Subtotals	383	26,323		26,458	61,100	298,000	
TOTALS	3,283	480,094	45,571	1,479,230	1,438,745	7,184,490	

The results achieved by the companies sampled in this analysis are significant. Overall, the total reductions in emissions to air, water and land were 528,949 tons per yr, with raw materials savings amounting to 1,479,230 tons per yr, both which translate into annual savings of $7,184,490. The total cost for these savings (i.e., total capital investment) was $1,479,230, which is a 400 percent simple return on investment (ROI) during the first year. Many of the P2 practices had immediate to less than 3 month payback periods.

A closer look at the achievements of these projects reveals some interesting trends. We arrive at these trends by defining three levels or *thresholds of investment* (i.e., the cost for a P2 practice). These levels are:

Level Designation	Investment Threshold	Dollar Range
A	Low-Cost/No-Cost	≤ 10,000
B	Moderate	> 10,000 to 50,000
C	Significant	> 50,000

The benefits derived for each investment threshold are summarized by the pie-charts in Figure 1. The comparisons provided in Figure 1 show:

- About 85% of the emissions reductions were achieved by low-cost/no-cost measures;
- Nearly 56% of the materials savings were derived from no-cost/low-cost measures; and
- Up to 60% of the total dollar savings were derived from low-cost to moderate threshold levels of investments.

In addition, the simple returns on investment within the first year of implementation were the greatest for low-cost/no-cost investments, as shown in

Figure 2. There are many companies where senior management believe that the only way their businesses can achieve more profitable operations, and to compete in international markets, are by making significant levels of investments. While this may be true in some industry segments and specific companies, this analysis suggests that a number of small-scale investments in pollution prevention practices cannot only achieve significant reductions in emissions, but collectively add up to sizable savings, which can in turn be reinvested into modernizing a company's operations. If we take the iron and steel industry in most of Eastern Europe and throughout Russia and Ukraine as examples, the technologies in these countries are old and highly inefficient. The levels of investments at a single steel mill operation can be hundreds of millions of dollars in order to modernize operations to the point where they would be comparable to Western technologies and capable of providing quality products. Such investments are unlikely to happen, and hence it makes sense for these operations to invest incrementally, with heavy emphasis on P2 programs that can generate savings that can be applied towards more critical upgrades in production.

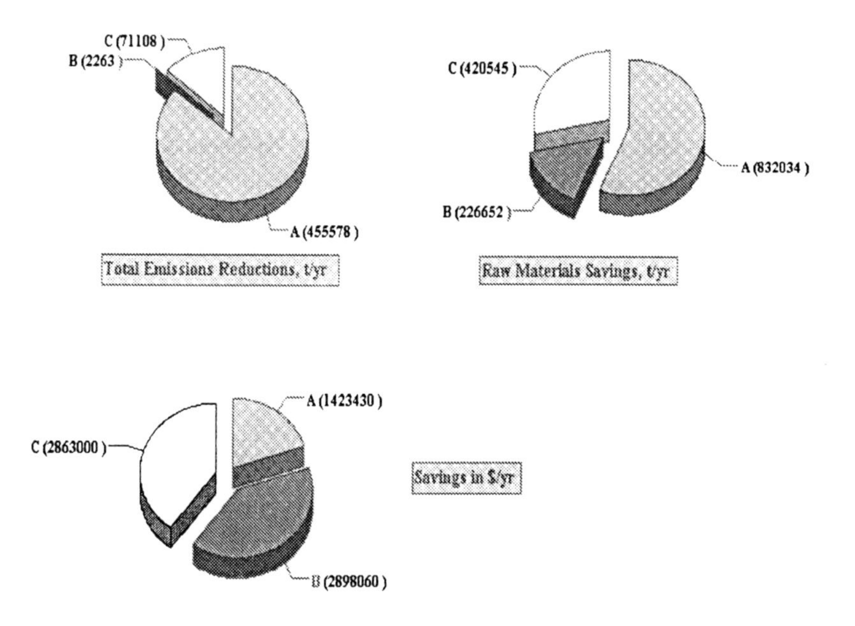

Figure 1. *Compares pollution reductions, raw materials savings and dollar savings by investment.*

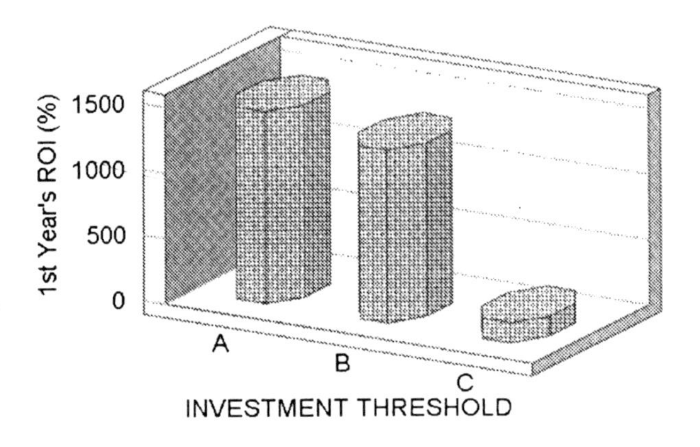

Figure 2. *Low-cost/no-cost P2 practices gave the highest returns.*

Elimination of a Toxic Solvent

Many industries use the solvent trichloroethylene (TCE) in their operations. This highly toxic chemical must be contained in a closed system, as releases of TCE can be fatal. Such releases often require the evacuation of the facility. Plant workers are the unwilling internal customers of TCE fumes. The external environment is also an unwilling customer. Rivers downstream can be effected by effluents. Aquatic life in the river and people dependent on the river for drinking water are unwilling customers.

It is important to recognize that quality can be built into, and not inspected into a product. This requires the producer to continuously identify and eliminate the *root cause* of the impediments to quality. Continuous improve is also the key to reducing the environmental impacts of the production process. The traditional approach to industrial waste has been to view it as a necessary by-product of manufacturing.

While production generates the waste, the responsibility to dispose of the waste in a safe and legal manner usually falls on the environmental engineering department. Because environmental engineers receive the waste after it has been created, they are not intimately familiar with the processes that created it. Further, because waste reduction is not a component of their performance review, environmental engineers do not have the institutional motivation to reduce the waste.

In the above example we introduced the term *Total Quality Environmental Management*, or TQEM. TQEM is the logical method for producing the results of pollution prevention. Pollution prevention calls for industry to prevent pollution wherever possible. Employing a customer focus, and classifying the waste itself and the activities required to control it as non-value added, TQEM calls for waste generation to be brought to a minimurn. Operators and process engineers, not environmental engineers, are responsible for identifying and eliminating the root causes of process waste. Employing the continuous improvement approach, *"zero waste" is as much a goal as "zero defects."*

As a result of TQEM projects, product quality often improves while waste is reduced. TQEM efforts empower employees to become more familiar with all aspects of the process, and not just those associated with production. When forced to question wastes from the process, improvements to quality characteristics can result. A team approach allows all factors of the environmental issue to be considered. Accountants are familiar with cost considerations; product engineers are familiar quality considerations; process and chemical engineers are familiar with feasibility considerations; and environmental engineers are familiar with environmental impacts. Because environmental engineers are trained to deal with waste as it has been generated, and not in methods of preventing it from being created in the first place, engineers with knowledge of the process characteristics must be involved.

The TCE story is an example of TQEM principles applied by the Ford Motor Company. Degreasing certain aluminum components with TCE has required extensive safety mechanisms and procedures. Building better containment systems reduces the risk of exposure, but does not get to the root cause of the problem; namely, the use of TCE.

With the above factors in mind, the Ford Motor Company team searched for a TCE-free solution to degreasing radiator coils. They formed a team that included a chemical engineer, an environmental engineer, a process engineer, an accountant, and a product engineer. The variety of backgrounds on the team ensured that the pertinent issues of cost, product quality, process feasibility and environmental impact were all addressed. The team designed an aqueous degreasing system (i.e., soap and water) to replace the TCE. Not only is the toxic chemical removed from the plant operations, but the water in the new system is recycled as well. The aqueous degreaser even exhibits better quality characteristics than the TCE degreaser.

The above project is an example of: improved quality, reduced cost, and reduced environmental impact. Certainly not all projects will prove so fruitful.

Some clean alternatives may cost more than their polluting rivals, but that cost must be balanced with the benefits of the environmental improvement.

To justify this viewpoint, one needs only to look to the increasing expectations of external customers for environmentally friendly products. In technologically advanced societies the consumer is more cognizant of environmentally friendly products and there are greater levels of confidence and investments into publicly held companies that have strong environmental tract records and management systems in place. The last two decades have simply had far too much negative press against companies that operated for decades without properly managing their environmental affairs. Despite this, we may still find many companies in the United States today that consider their environmental responsibilities as not going much beyond a policy statement and bare-bones staff to manage the environmental issues of an operation.

Environmental Management Hierarchy

Environmental management involves several strategies for dealing with wastes. A hierarchy has been developed to prioritize these strategies. Strategies that reduce or eliminate wastes before they are created are preferable to those that deal with treating or disposing wastes that are already generated. This hierarchy is:

- **Prevention** - The best waste reduction strategy is one that keeps waste from being formed in the first place. Waste prevention may in some cases require significant changes to process, but it can provide the greatest environmental and economic rewards.
- **Recycling** - If waste generation is unavoidable in a process, then strategies that minimize the waste to the greatest extent possible should be pursued, such as recycling and reuse.
- **Treatment** - When wastes cannot be prevented or minimized through reuse or recycling, strategies to reduce their volume or toxicity through treatment can be pursued. While end-of-pipe strategies can sometimes reduce the amount of waste, they are not as effective as preventing the waste in the first place.
- **Disposal** - The last strategy to consider is alternative disposal methods. Proper waste disposal is an essential component of an overall environmental management program; however, it is the least effective technique.

Table 3 provides examples of practices in the hierarchy.

Table 3. *Environmental Management Hierarchy Examples.*

Priority	Method	Example	Applications
1	Prevention (Source Reduction)	✓ Process Changes ✓ Design of Products that Minimize Environmental Impacts ✓ Source Elimination	✓ Modify Process to Avoid/Reduce Solvent Use ✓ Modify Product to Extend Coating Life
2	Recycling	✓ Reuse ✓ Reclamation	✓ Solvent Recycling ✓ Metal Recovery from Spent Bath ✓ Volatile Organic Recovery
3	Treatment	✓ Stabilization ✓ Neutralization ✓ Precipitation ✓ Evaporation ✓ Incineration ✓ Scrubbing	✓ Thermal Destruction of Organic Solvents ✓ Precipitation of Heavy Metals from Spent Plating Bath
4	Disposal	✓ Disposal at a Facility	✓ Land Disposal

Summary of Concepts

In most countries there is a need to balance economic growth with environmental protection. It is increasingly being recognized that economic development and the health and welfare of a society are closely linked to proper management of a country's natural resources and environment. In these countries pollution prevention offers the government and industry a way to manage the impacts of industrial growth on the environment while enabling economic development. Pollution prevention addresses three important components of the environmental protection/economic development issue:

- **Environment** - Offers a better solution for environmental management than "end-of pipe" pollution solutions;
- **Quality** - Encourages evaluation of production processes and product quality;
- **Cost** - Improves a facility's bottom-line by reducing treatment costs, saving on material and resource inputs, and reducing risk and liability insurance

Dealing with environmental wastes through "end-of-pipe" measures (such as wastewater treatment systems, hazardous waste incinerators and other treatment technologies, secure landfills, monitoring equipment, solid waste hauling equipment, air pollution control equipment, and catalytic converters) has proven to be very costly and does not address all environmental problems. Pollution prevention offers industry the advantages of:

1. Less need for costly pollution control equipment;
2. Getting ahead of environmental regulations;
3. Reduced reporting and permitting requirements;
4. Less operation and maintenance of pollution control equipment.

TQEM Principles

The process of identifying pollution prevention opportunities also provides the opportunity to identify measures to improve product quality. The P2 audit requires the enterprise to examine its production process in-depth. Finding ways to reduce wastes also requires a study of the root causes for generating wastes and finding improvements to processes. *Total Quality Management* (TQM) is the management system developed to achieve the goal of high product and service quality. The management elements of TQM include:

- Customer focus;
- Continuous improvement;
- Teamwork; and
- Strong management commitment

At first glance, TQM seems unrelated to these environmental concerns. Yet the inherent strengths of the TQM methodology can effectively address some of these issues. Professionals who apply TQM concepts to environmental issues have coined the term Total Quality Environmental Management (TQEM). TQEM is a logical method for achieving pollution prevention. First, through *Customer Focus* - In the context of quality, the customer is defined as the person who employs the "product and service characteristics." Customers fall into two categories, internal and external. The internal customer is the next person in the production chain, while the external customer is the end-user of the product. If the definition of the customer is expanded to include those people and environments that are affected by the production process waste, total quality management requires us to understand the impact of this waste on those customers, and take steps to reduce it.

Three of the elements of TQM - customer focus, continuous improvement, team approach - readily apply to environmental issues. As in traditional TQM settings, the last - strong management commitment - is perhaps the most important. No TQEM program will succeed without the commitment of senior management. Senior management, those who have built their careers when waste was seen as a necessary by-product, must come to understand that both internal and external customer expectations include environmentally conscious products and processes. They must learn to see the value of applying TQEM to get to the root causes of waste, and call on the cross-disciplinary teams to employ continuous improvement to implement ever "cleaner" solutions.

In many cases, pollution prevention measures can have clear environmental benefits in terms of pollution that is not generated, reductions in the toxic materials used in the production process, savings in energy use and other raw materials. Savings can accrue in five primary areas:

1. a company can save on raw materials and energy;
2. a company can save on labor costs;
3. disposal costs can be reduced or eliminated;
4. a facility can save on waste handling/treatment costs both in its own use of labor to collect, store, and process wastes and incur costs to transport wastes off-site;
5. decreasing the amount of toxic materials used, handled, and transported at a facility can reduce its future liability costs.

METHODOLOGY FOR P2 AUDITS

This Section describes a step-by-step approach for carrying out the pollution prevention audit, flowing P2 principles. It is designed to be generic and to apply to a broad spectrum of industries. The approach consists of three phases that are implemented in succession. The phases are:

1. a preassessment phase for assessment preparation;
2. a data collection phase to derive a material balance; and
3. a synthesis phase where the findings from material balances are translated into a waste reduction action plan.

It is possible that not all of the assessment steps will be relevant to every situation. Similarly, in some situations additional steps may be required. To augment the audit procedure, there are several key references that are cited later in this discussion that the reader should refer to.

Phase 1: The Pre-assessment

Step 1: Assessment Focus and Preparation

A thorough preparation for a pollution prevention audit is a prerequisite for an efficient and cost-effective study. Of particular importance is to gain support for the assessment from top-level management, and for the implementation of results; otherwise there will be no real action on recommendations. The pollution prevention auditing team should be identified. The number of people required on a team will depend on the size and complexity of the processes to be investigated. A pollution prevention audit of a small factory may be undertaken by one person with contributions from the employees. A more complicated process may require at least 3 or 4 people: technical staff, production employees and an environmental specialist. Involving personnel from each stage of the manufacturing operations will increase employee awareness of waste reduction and promote input and support for the program.

The audit may require external resources, such as laboratory and possibly equipment for sampling and flow measurement. You should attempt to identify external resource requirements at the outset. Analytical services and equipment may not be available to a small factory. If this is the case, investigate the possibility of forming a pollution prevention association with other factories or industries; under this umbrella the external resource costs can be shared.

It is important to select the focus of your assessment at the preparation stage. You may wish the audit to cover a complete process or you may want to concentrate on a selection of unit operations within a process. The focus will depend on the objectives of the assessment. You may wish to look at waste minimization as a whole or you may wish to concentrate on particular wastes, for example: raw material losses, wastes that cause processing problems, wastes considered to be hazardous or for which regulations exist, and/or wastes for which disposal costs are high. A good starting point for designing a pollution prevention assessment is to determine the major problems/wastes associated with your particular process or industrial sector.

All existing documentation and information regarding the process, the plant or the regional industrial sector should be collated and reviewed as a preliminary step. Regional or plant surveys may have been undertaken; these could yield useful information indicating the areas for concern and will also show gaps where no data are available. The following prompts give some guidelines on useful documentation.

- Is a site plan available?
- Are any process flow diagrams available?

- Have the process wastes ever been monitored -- do you have access to the records?
- Do you have a map of the surrounding area - indicating watercourses, hydrology and human settlements?
- Are there any other factories/plants in the area which may have similar processes?
- What are the obvious wastes associated with your process?
- Where is water used in greatest volume?
- Do you use chemicals that have special instructions for their use and handling?
- Do you have waste treatment and disposal costs -- what are they?
- Where are your discharge points for liquid, solid and gaseous emissions?

The plant employees should be informed that the assessment will be taking place, and they should be encouraged to take part. The support of the staff is imperative for this type of interactive study. It is important to undertake the assessment during normal working hours so that the employees and operators can be consulted, the equipment can be observed in operation and, most importantly, wastes can be quantified.

Step 2: Listing Unit Operations

Your process will comprise a number of unit operations. A unit operation may be defined as an area of the process or a piece of equipment where materials are input, a function occurs and materials are output, possibly in a different form, state or composition. For example, a process may comprise the following unit operations: raw material storage, surface treatment, rinsing, painting, drying, product storage and waste treatment.

Any initial site survey should include a walk around the entire manufacturing plant in order to gain a sound understanding of all the processing operations and their interrelationships. This will help the assessment team decide how to describe a process in terms of unit operations. During this initial overview, it is useful to record visual observations and discussions and to make sketches of process layout, drainage systems, vents, plumbing and other material transfer areas. These help to ensure that important factors are not overlooked.

The assessment team should consult the production staff regarding normal operating conditions. The production or plant staff are likely to know about waste discharge points, unplanned waste generating operations such as spills and washouts, and give the assessors a good indication of actual operating procedures. Investigations may reveal that night-shift procedures are different from day-shift procedures; also, a plant may disclose that actual material handling practices are different from those set out in written procedures. A long-standing employee could give some insight into recurring process problems. In the absence of any historical monitoring this information can be very useful. Such employee participation must however be a non-blaming process; otherwise it will not be as useful as it could be. During the initial survey, note imminent problems that need to be addressed before the assessment is complete.

The audit team needs to understand the function and process variables associated with each unit operation. Similarly, all the available information on the unit operations and the process in general should be collated, possibly in separate files. It is useful to tabulate this information, as illustrated by the following example:

Unit Operation	Function	File
(A) Surface Treatment	Surface treatment of glass bottles: 100 m^3 spray chamber, 6 jets, 100 gal/min pumps	1
(B) Rinsing	Washing glass bottles before labeling	2

Identification of materials handling operations (manual, automatic, bulk, drums, etc.) covering raw materials, transfer practices and products is also an important aspect which could usefully be included in the above tabulation as a prelude to development of a materials balance (Phase 2).

Step 3: Constructing Process Flow Diagrams

By connecting the individual unit operations in the form of a block diagram you can prepare a process flow diagram. Intermittent operations such as cleaning, make-up or tank dumping may be distinguished by using broken lines to link the boxes. Figure 3 is an example of a simplified process flow diagram for a pattern etch process for a printed wiring board process.

For complex processes prepare a general flow diagram illustrating the main process areas and, on separate sheets of paper, prepare detailed flow diagrams for each main processing area.

Now decide on the level of detail that you require to achieve your objectives.

It is important to realize that the less detailed or larger scale the assessment becomes, the more information is likely to be lost or masked by oversimplification. Establishing the correct level of detail and focusing in on specific areas is very important at an early stage.

Pay particular attention to correcting any obvious waste that can be reduced or prevented easily, before proceeding to development of a material balance (Phase 2).

By making simple changes at this early stage, the resultant benefits will help enlist the participation and stimulate the enthusiasm of employees for the total pollution prevention assessment/reduction program.

Phase 1: Summary

At the end of the P2 audit preassessment stage the team should be organized and be aware of the objectives of the pollution prevention assessment. Plant personnel should have been informed of the audit's purpose in order to maximize cooperation between all parties concerned. Any required financial resources should have been secured and external facilities checked out for availability and capability. The team should be aware of the overall history and local surroundings of the plant. The scope and focus of the audit should have been established, and a rough timetable worked out to fit in with production patterns. The audit team should be familiar with the

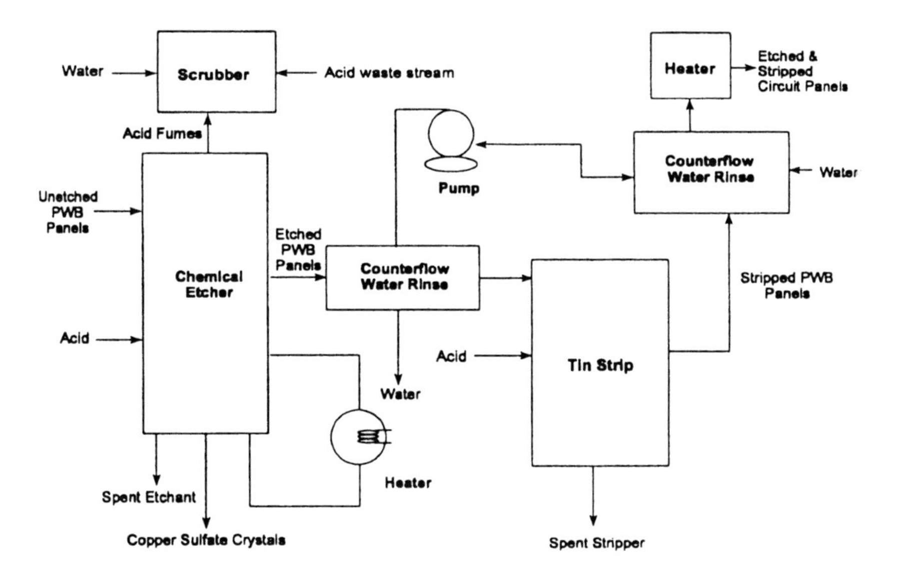

layout of the processes within the plant and should have listed the unit operations associated with each process. Sources of wastes and their causes should also have been identified. It should be possible to draw process flow diagrams highlighting areas to be covered in the pollution prevention assessment. Obvious waste saving measures that can be introduced easily should be implemented immediately. The findings of the Phase 1 investigations can be presented to management as a brief pre-assessment report to reaffirm their commitment into the next phase.

Phase 2: Material Balances: Process Inputs and Outputs

A material balance is defined as a precise account of the inputs and outputs of an operation. This phase describes a procedure for the collection and arrangement of input and output data. The procedure can be applied to derive the material balance of a plant, a process, or a unit operation. Note that infrequent outputs (e.g., the occasional dumping of an electroplating bath) may be as significant as continuous daily discharges and should therefore be accounted for in the material balance.

Step 4: Determining Inputs

Inputs to a process or a unit operation may include raw materials, chemicals, water, air and power. The inputs to the process and to each unit operation need to be quantified. As a first step towards quantifying raw material usage, examine purchasing records, this rapidly gives us an idea of the sort of quantities involved.

In many situations the unit operations where raw material losses are greatest are raw material storage and transfer. Look at these operations in conjunction with the purchasing records to determine the actual net input to the process.

Make notes regarding raw material storage and handling practices. Consider evaporation losses, spillages, leaks from underground storage tanks, vapor losses through storage tank pressure-relief vents and contamination of raw materials. Often these can be rectified very simply. Record raw material purchases and storage and handling losses in a table in order to derive the net input to the process. Table 4 provides an example.

Once the net input of raw materials to the process has been determined we can proceed with quantifying the raw material input to each unit operation. If accurate information about raw material consumption rates for individual unit operations is not available then we will need to take measurements to determine average figures. Measurements should be taken for an appropriate length of time. For example, if a batch takes one week to run, then measurements should be taken over a period of at least three weeks; these figures can be extrapolated for monthly, quarterly or annual figures. Note that some quantification is possible by observation and some simple accounting procedures.

Table 4. *Example of Recording Raw Materials Storage and Handling Losses.*

Raw Material	Quantity of Material	Yrly. Quantity Raw Mtl. Purchased	Type Storage Used in Production	Average Length of Storage (months)	Est. Annual Raw Mtl. Losses
Reagent grade sulfuric acid	2,500 liters	2,800 liters	vessels	1	100 liters
Nickel catalyst	600 kg	2,000 kg	sacks	3	50 kg
Surface treatment chemical	1,000 kg	7,000 kg	closed	2	50 kg

For solid raw materials, ask the warehouse operator how many sacks are stored at the beginning of the week or prior to using then in a unit operation; then ask him or her again at the end of the week or unit operation. Weigh a selection of sacks to check compliance with specifications.

For liquid raw materials such as water or solvents, check storage tank capacities and ask operators when a tank was last filled. Tank volumes can be estimated from the tank diameter and tank depth if automatic gauging is not used. Monitor the tank levels and the number of tankers arriving on site. While investigating the inputs, talking to staff and observing the unit operations in action, the auditing team should be thinking about how to improve the efficiency of the unit operations.

The audit relies on information gathered in the field - by interviewing operators and various shop personnel. These interviews help the team to identify possible ways to save raw materials, reduce pollution, conserve energy. These discussions should not be extemporaneous, but rather thought out in advance, and initially formulated during the walkthrough. It is advisable to have a list of questions and a checklist of issues for such meetings. Table 5 provides a sample list of questions which the reader can expand on and modify to make them more specific to the plant assessment.

The energy input to a unit operation should also be considered at this stage; however, energy use deserves a full assessment in its own right. For our discussions, we will focus on energy only as it relates to evaluating a P2 opportunity. In other words, our primary focus is on waste and pollution reduction opportunities, however these could reduce energy costs as well. If energy usage is a particularly prominent factor, then you should recommend that a separate energy audit be undertaken.

Table 5. *Sample Questions for Identifying Raw Materials Savings Opportunities.*

	Lead Question	Follow-on Question
1	*Is the size of the raw material inventory appropriate to ensure that material-handling losses can be minimized?*	*How often is inventory checked?*
2	*Could transfer distances between storage and process or between unit operations be reduced to minimize potential wastage?*	*Are bins and silos a source of product losses?*
3	*Do the same tanks store different raw materials depending on the batch product?*	*Is there a risk of cross contamination?*
4	*Are sacks of materials emptied or is some material wasted?*	
5	*Are viscous raw material used on site?*	*Is it possible to reduce residual wastage in drums?*
6	*Is the raw material storage area secure?*	*Could a building be locked at night, or could an area be fenced off to restrict access?*
7	*How could the raw materials be protected from direct sunlight or from heavy downpours?*	
8	*Is dust from stockpiles a problem?*	
9	*Is the equipment used to pump or transfer materials working efficiently?*	*Is it maintained regularly?*
10	*Could spillages be avoided?*	*Is there a formal spill handling procedure?*
11	*Is the process adequately manned?*	*What is the experience level of operators?*
12	*How could the input of raw materials be monitored?*	
13	*Are there any obvious equipment items in need of repair?*	*Is there a regularly scheduled maintenance program? Describe.*
14	*Are pipelines self-draining?*	*Where does the residue go?*
15	*Is vacuum pump water recirculated?*	

The input data collected for the material balance can be recorded on the process flow diagram or in tabular form on a spreadsheet. Table 6 provides a sample spreadsheet to follow.

Water is frequently used in the production process, for cooling, gas scrubbing, washouts, product rinsing and steam cleaning. This water usage needs to be quantified as an input. Some unit operations may receive recycled wastes from other unit operations. These also represent an input. Steps 5 and 6 describe how these two factors should be included in the audit.

Table 6. *Sample Spreadsheet for Input Materials.*

Unit Operation	Raw Material 1 (mm^3/yr)	Raw Material 2 (tons/yr)	Water (mm^3/yr)	Energy Source
Surface Treatment (A)	______	______	______	______
Rinse (B)	______	______	______	______
Painting (C)	______	______	______	______
Drying Stages (D)	______	______	______	______
Trimming Stages (E)	______	______	______	______
Totals	______	______	______	______

Step 5: Recording Water Usage

The use of water, other than for a process reaction, is a factor that should be covered in all pollution prevention assessments. The use of water to wash, rinse and cool is often overlooked, although it represents an area where waste reductions can frequently be achieved simply and cheaply. In fact, referring back to the pollution prevention example for the chemical industry in Poland, much of the raw materials savings reported in Table 2 are related to water.

Consider these general points about the site water supply before assessing the water usage for individual units.

- Identify the water sources within the plant operations.
- Is water extracted directly from a borehole, river or reservoir; is water stored on site in tanks or in a lagoon?
- What is the storage capacity for water on site?
- How is water transferred -- by pump, by gravity, manually?
- Is rainfall a significant factor on site?

For each unit operation consider the following,

- What is water used for in each operation - cooling, gas scrubbing, washing, product rinsing, dampening stockpiles, general maintenance, safety quench, etc.?
- How often does each action place?
- How much water is used for each action?

It is unlikely that the answers to these questions will be readily available -- you may need to undertake a monitoring program to assess the use of water in each unit operation. Again, the measurements must cover a sufficient period of time to ensure that all actions are monitored. Pay particular attention to intermittent actions such as steam cleaning and tank washout - water use is often indiscriminate during these operations. Find out when these actions will be undertaken so that detailed measurements can be made. Record water usage information in a tabular form -- ensure that the units used to describe intermittent actions indicate a time period. Table 7 provides a sample spreadsheet to follow.

Table 7. *Sample Spreadsheet for Water Balance.*

Unit Operation	Cleaning	Steam	Cooling	Other
Latex batch mix	________	________	________	________
Reactor washout	________	________	________	________
Reactor feed	________	________	________	________

Make sure that all measurements used for recording data in Table 7 are standard (e.g., m^3/day or m^3/yr, etc.). Using less water can be a cost-saving exercise. In many older plant operations water conservation programs are often overlooked. Even something as straightforward as a valve maintenance program can result in significant reductions in water consumption. Consider the following points while investigating water use:

- Tighter control of water use can reduce the volume of wastewater requiring treatment and result in cost savings - in the extreme, it can sometimes reduce volumes and increase concentrations to the point of providing economic material recovery in place of costly wastewater treatment.
- Attention to good house-keeping practices often reduces water usage and, in turn, the amount of wastewater passing to drains.

- The cost of storing wastewater for subsequent reuse may be far less than the treatment and disposal costs.
- Counter-current rinsing and rinse water reuse are useful tips for reducing water usage.

Step 6: Measuring Current Levels of Waste Reuse/Recycling

Some wastes lend themselves to direct reuse in production and may be transferred from one unit to another; others require some modifications before they are suitable for reuse in a process. These reused waste streams should be quantified. If reused wastes are not properly documented double-counting may occur in the material balance particularly at the process or complete plant level; that is, a waste will be quantified as an output from one process and as an input to another. The reuse or recycling of wastes can reduce the amount of fresh water and raw materials required for a process. While looking at the inputs to unit operations think about the opportunities for reusing and recycling outputs from other operations.

Steps 4, 5, and 6 Summary

By the end of Step 6 you should have quantified all your process inputs. The net input of raw materials and water to the process should be established having taken into account any losses incurred at the storage and transfer stages. Any reused or recycled inputs should be documented. All notes regarding raw material handling, process layout, water losses, obvious areas where problems exist should all be documented for consideration in Phase 3.

Step 7: Quantifying Process Outputs

To calculate the second half of the material balance - the outputs from unit operations and the process as a whole need to be quantified. Outputs include primary product, by-products, wastewater, gaseous wastes (emissions to atmosphere), liquid and solid wastes which need to be stored and/or sent off-site for disposal and reusable or recyclable wastes. You may find that a spreadsheet along the lines of that shown in Table 8 will help organize the input information. It is important to identify units of measurement.

The assessment of the amount of primary product or useful product is a key factor in process or unit operation efficiency. If the product is sent off-site for sale, then the amount produced is likely to be documented in company records. However, if the product is an intermediate to be input to another process or unit operation then the output may not be so easy to quantify. Production rates will have to be measured over a period of time. Similarly, the quantification of any by-products may require measurement.

Table 8. *Sample Spreadsheet for Tabulating Process Outputs.*

Unit Opera-tion	Product		Wastes/Pollutants				
	Main	By-	Recycle	Water	Gas	Stored	Off-Site
A	____	____	____	____	____	____	____
B	____	____	____	____	____	____	____
C	____	____	____	____	____	____	____
Totals	____	____	____	____	____	____	____

Step 8: Accounting for Wastewater

On many sites significant quantities of both clean and contaminated water are discharged to sewer or to a watercourse. In many cases, this wastewater has environmental implications and incurs treatment costs. In addition, wastewater may wash out valuable unused raw materials from the process areas. It is extremely important to know how much wastewater is going down the drain and what the wastewater contains. The wastewater flows, from each unit operation as well as from the process as a whole, need to be quantified, sampled and analyzed.

Identify the effluent discharge points; that is, where does wastewater leave the site? Wastewater may go to an effluent treatment plant or directly to a public sewer or watercourse. One factor that is often overlooked is the use of several discharge points - *it is important to identify the location, type and size of all discharge flows*. Identify where flows from different unit operations or process areas contribute to the overall flow. In this way, it is possible to piece together the drainage network for your site. This can lead to startling discoveries of what goes where! Once the drainage system is understood it is possible to design an appropriate sampling and flow measurement program to monitor the wastewater flows and strengths from each unit operation.

Plan your monitoring program thoroughly and try to take samples over a range of operating conditions such as full production, start up, shut down and washing out. In the case of combined storm water and wastewater drainage systems, ensure that sampling and flow measurements are carried out in dry weather. For small or batch wastewater flows it may be physically possible to collect all the flow for measurement using a pail and wristwatch. Larger or continuous wastewater flows can be assessed using flow measurement techniques.

The sum of the wastewater generated from each unit operation should be approximately the same as that input to the process. As indicated in Step 6, note that double-counting can occur where wastewater is reused. This emphasizes the importance of understanding your unit operation and their interrelationships.

The wastewater should be analyzed to determine the concentration of contaminants. You should include wastewater analyses such as:

- pH,
- chemical oxygen demand (COD),
- biochemical oxygen demand (BOD_5),
- suspended solids,
- grease and oil.

Other parameters that should be measured depend on the raw material inputs. For example, an electroplating process is likely to use nickel and chromium. The metal concentrations of the wastewater should be measured to ensure that the concentrations do not exceed discharge regulations, but also to ensure that raw materials are not being lost to drain. Any toxic substances used in the process should be measured.

Take samples for laboratory analysis. Composite samples should be taken for continuously-running wastewater. For example, a small volume, 100 ml, may be collected every hour through a production period of ten hours to gain a 1 liter composite sample. The composite sample represents the average wastewater conditions over that time. Where significant flow variations occur during the discharge period, consideration should be given to varying the size of individual samples in proportion to flow rate in order to ensure that a representative composite sample is obtained. For batch tanks and periodic drain down, a single spot sample may be adequate (check for variations between batches before deciding on the appropriate sampling method). Wastewater flows and concentrations should be tabulated. Table 9 provides a sample spreadsheet to follow.

Table 9. *Example of Spreadsheet for Tabulating Wastewater Flows.*

	Discharge Receptacle				
Source	**Sewer**	**Storm Water**	**Reuse**	**Storage**	**Total**
Unit Ops. A	______	______	______	______	______
Unit Ops. B	______	______	______	______	______
Unit Ops. C	______	______	______	______	______
Total	______	______	______	______	______

Step 9: Accounting for Gaseous Emissions

To arrive at an accurate material balance some quantification of gaseous emissions associated with the process is necessary. It is important to consider the actual and potential gaseous emissions associated with each unit operation from raw material storage through to product storage. Gaseous emissions are not always obvious and can be difficult to measure. Where quantification is impossible, estimations can be made using stoichiometric information. The following example illustrates the use of indirect estimation.

Consider coal burning in a boiler house. The assessor may not be able to measure the mass of sulphur dioxide leaving the boiler stack due to problems of access and lack of suitable sampling ports on the stack. The only information available is that the coal is of soft quality containing 3% sulphur by weight and, on average, 1000 kg of coal is burned each day.

First calculate the amount of sulphur burned:

1000 kg coal x 0.03 kg sulphur/kg coal
= 30 kg sulphur/day.

The combustion reaction is approximately:

$$S + O_2 = SO_2$$

The number of moles of sulphur burned equals the number of moles of sulphur dioxide produced. The atomic weight of sulphur is 32 and molecular weight of sulphur dioxide is 64. Therefore:

kg-moles S = 30 kg/32 kg per kg-mole = kg-mole Of SO_2 formed

kg SO_2 formed = (64 kg SO_2/kg-mole) x kg-molesSO_2
= 64 x 30/32 = 60 kg

Thus, it may be estimated that an emission of 60 kg sulphur dioxide will take place each day from the boiler stack.

These types of stoichiometric calculations are commonplace and can provide reliable estimates for the material balance. As with an calculation methods, one should list the assumptions in order to qualify the accuracy of the estimate. Limited field measurements can always be done later on to verify the estimated emissions. Record the quantified emission data in tabular form and indicate which figures are estimates and which are actual measurements. The assessor should consider qualitative characteristics at the same time as quantifying gaseous wastes. The following are some typical questions to address when developing the material balance around the gaseous emissions components.

- Are odors associated with a unit operation?
- Are there certain times when gaseous emissions are more prominent -- are they linked to temperature?
- Is any pollution control equipment in place?
- Are gaseous emissions from confined spaces (including fugitive emissions) vented to the outside?
- If gas scrubbing is practiced, what is done with the spent scrubber solution? Could it be converted to a useful product?
- Do employees wear protective clothing, such as masks?

Step 10: Accounting for Off-Site Wastes

Your process may produce wastes which cannot be treated on-site. These need to be transported off-site for treatment and disposal. Wastes of this type are usually nonaqueous liquids, sludges or solids. Often, wastes for off-site disposal are costly to transport and to treat, and the represent a third-party liability. Therefore, minimization of these wastes yields a direct cost benefits, both present and future.

Measure the quantity and note the composition of any wastes associated with your process which need to be sent for off-site disposal. Record your results in a table. Table 10 provides an example of a spreadsheet to follow.

Table 10. *Example of Spreadsheet for Tabulating Off-Site Disposal.*

Unit Ops.	Liquids		Sludge		Solid Waste	
	Qty[a]	Conc[b]	Qty	Conc	Qty	Conc
A	_____	_____	_____	_____	_____	_____
B	_____	_____	_____	_____	_____	_____
C	_____	_____	_____	_____	_____	_____
D	_____	_____	_____	_____	_____	_____

a - Quantities in m3/yr or t/yr; b - concentration in mass per unit volume.

It is useful to ask the following questions during the data collection stage:

- Where does the waste originate from within our process operations?
- Could the manufacturing operations be optimized to produce less waste?
- Could alternative raw materials be used which would produce less waste?
- Is there a particular component that renders the whole waste hazardous -- could this component be isolated? *This can be a key question. Under RCRA for example, if we have a waste with only 1% of a carcinogenic material as a component - then the entire waste is classified as carcinogenic. By eliminating the hazardous, regulated component, we potentially eliminate a much larger waste problem.*
- Does the waste contain valuable materials that could be recovered or possibly sold offsite?
- Wastes for off-site disposal need to be stored on-site prior to dispatch. Does storage of these wastes cause additional emission problems? For example, are solvent wastes stored in closed tanks?
- How long are wastes stored on-site, and are we in compliance with storage requirements under RCRA?
- Are stockpiles of solid waste secure or are dust storms a regular occurrence? Also, do waste piles result in stormwater runnoff issues?

Steps 7, 8, 9, and 10 Summary

At the end of Step 10 the pollution prevention assessment team should have collated all the information required for evaluating a material balance for each unit operation and for a whole process. All actual and potential wastes should be quantified. Where direct measurement is not possible, estimates based on stoichiometric information should be made. The data should be arranged in clear tables with standardized units. Throughout the data collection phase the assessors should make notes regarding actions, procedures and operations that could be improved.

Step 11: Assembling Input and Output Information for Unit Operations

From the *Law of Conservation of Mass* - the total of what goes into a process must equal the total of what comes out. Prepare a material balance at a scale appropriate for the level of detail required in your study. For example, you may require a material balance for each unit operation or one for a whole process may sufficient. Preparing a material balance is designed to gain a better understanding of the inputs and outputs, especially waste, of a unit operation such that areas where information is inaccurate or lacking can be identified. Imbalances require further investigation. Do not expect a perfect balance -- your initial balance should be considered as a rough assessment to be refined and improved. Assemble the input and output information for each unit operation and then decide whether all the inputs and outputs need to be included in the material balance. For example, this is not essential where the cooling water input to a unit operation equals the cooling water output.

Make sure to standardize units of measurement (liters, tons or kilograms) on a per day, per year or per batch basis. Finally, summarize the measured values in standard units by reference to your process flow diagram. It may have been necessary to modify your process flow diagram following the in-depth study of the plant.

Step 12: Deriving a Preliminary Material Balance for Unit Operations

Now it is possible to complete a preliminary material balance. For each unit operation utilize the data developed in Steps 1 through 10 and construct the material balance. Display your information clearly. Table 11 offers the reader one way of presenting the material balance information. This spreadsheet, of course, can be modified as necessary, but it does provide a general approach to tabulating the information in a concise manner.

Table 11. *Spreadsheet for Developing a Preliminary Material Balance.*

INPUTS (AMOUNTS IN STANDARD UNITS PER ANNUM)	
Raw Material 1	
Raw Material 2	
Raw Material 3	
Waste Reuse	
Water	
Total	

▼

Unit Process A

▼

OUTPUTS (AMOUNTS IN STANDARD UNITS PER ANNUM)	
Product	
By-Product	
Raw Material Storage and Handling Losses	
Reused Wastes	
Wastewater	
Gaseous Emissions	
Stored Wastes	
Hazardous Liquid Waste Transported Off-Site	
Hazardous Solid Waste Transported Off-Site	
Nonhazardous Liquid Waste Transported Off-Site	
Nonhazardous Solid Waste Transported Off-Site	
Total	

Note that a material balance will often need to be carried out in weight units since volumes are not always conserved. Where volume measurements have to be converted to weight units, take account of the density of the liquid, gas or solids concerned. Once the material balance for each unit operation has been completed for raw material inputs and waste outputs it might be worthwhile repeating the procedure with respect to each contaminant of concern. It is highly desirable to carry out a water balance for all water inputs and outputs to and from unit operations because water imbalances may indicate serious underlying process problems such as leaks or spills. The individual material balances may be summed to give a balance for the whole process, a production area or factory.

Step 13: Evaluating the Material Balance

The individual and sum totals making up the material balance should be reviewed to determine information gaps and inaccuracies. If you do have a significant material imbalance then further investigation is needed. For example, if outputs are less than inputs look for potential losses or waste discharges (such as evaporation, or fugitive emissions not accounted for such as significant valve, pump and reactor seals leakage, etc.). Outputs may appear to be greater than inputs if large measurement or estimating errors are made or some inputs have been overlooked.

At this stage you should take time to re-examine the unit operations to attempt to identify where unnoticed losses may be occurring. It may be necessary to repeat some data collection activities. Remember that you need to be thorough and consistent to obtain a satisfactory material balance. The material balance not only reflects the adequacy of your data collection, but by its very nature, ensures that you have a sound understanding of the processes involved.

Step 14: Refining the Material Balance

Now you can reconsider the material balance equation by adding those additional factors identified in the previous step. If necessary, estimates of unaccountable losses will have to be calculated. Note that, in the case of relatively simple manufacturing plants, preparation of a preliminary material balance and its refinement (Steps 13 and 14) can usefully be combined. For more complex pollution prevention assessments however, two separate steps are likely to be more appropriate. An important rules to remember is that the inputs should ideally equal the outputs but in practice this will rarely be the case and some judgement will be required to determine what level of accuracy is acceptable.

In the case of high concentrations or hazardous wastes, accurate measurements are needed to develop cost-effective waste reduction options. It is possible that the material balance for a number of unit operations will need to be repeated. Again, continue to review, refine and, where necessary, expand your database. The compilation of accurate and comprehensive data is essential for a successful pollution prevention assessment and subsequent waste reduction action plan. But remember - you cannot reduce what you do not know is there!

Steps 11, 12, 13, and 14 Summary

By the end of Step 14, you should have assembled information covering process inputs and process outputs. These data should be organized and presented clearly in the form of material balances for each unit operation. These data form the basis for the development of an action plan for waste minimization. We now are ready to move onto the next phase in the process - which is to develop P2/waste minimization options, and to evaluate their financial merits.

Phase 3: Synthesis

Phases 1 and 2 have covered planning and undertaking a pollution prevention audit, resulting in the preparation of a material balance for each unit operation. Phase 3 represents the interpretation of the material balance to identify process areas or components of concern. The material balance focuses on the attention of the assessor. The arrangement of the input and output data in the form of a material balance facilitates your understanding of how materials flow through a production process. To interpret a material balance it is necessary to have an understanding of normal operating performance. *How can you assess whether a unit operation is working efficiently if you do not know what is normal?* Therefore, it is essential that a member of your team must have a good working knowledge of the process. To an experienced process engineer, with the aid of the team members - the material balance will indicate areas for concern and help to prioritize problem wastes.

You should use the material balance to identify the major sources of waste, to look for deviations from the norm in terms of waste production, to identify areas of unexplained losses and to pinpoint operations which contribute to flows that exceed national, local or site discharge regulations. Also, a good thing to remember is that from a practical standpoint, *process efficiency* is synonymous with *waste minimization*.

Different waste reduction measures require varying degrees of effort, time and financial resources. They can be categorized as two groups:

Group 1 - Obvious waste reduction measures, including improvements in management techniques and house-keeping procedures that can be implemented cheaply and quickly. These can be referred to as **Low Cost/No Cost**.

Group 2 - Long-term reduction measures involving process modifications or process substitutions to eliminate problem wastes.

Increased reuse/recycling to reduce waste falls between the immediate and the more substantial waste reduction measures. Steps 15, 16, and 17 describe how to identify waste reduction measures.

Step 15: Examining Obvious Waste Reduction Measures

It may have been possible to implement very obvious waste reduction measures already, before embarking on obtaining a material balance (refer back to Step 3). Now consider the material balance information in conjunction with visual observations made during the whole of the data collection period in order to pinpoint areas or operations where simple adjustments in procedure could greatly improve the efficiency of the process by reducing unnecessary losses.

Use the information gathered for each unit operation to develop better operating practices for all units. Significant waste reductions can often be achieved by improved operation, better handling and generally taking more care in performing operations and handling materials. Table 12 provides a list of waste reduction hints that can be implemented immediately with no or only small extra costs.

Table 12. *Common Waste Reduction Hints.*

Specifying and Ordering Materials

- ❑ **Do not over-order materials especially if the raw materials or components can spoil or are difficult to store.**
- ❑ **Try to purchase raw materials in a form which is easy to handle, for example, pellets instead of powders.**
- ❑ **It is often more efficient and certainly cheaper to buy in bulk.**

Receiving Materials

- ❑ **Demand quality control from suppliers by refusing damaged, leaking or unlabeled containers.**
- ❑ **Undertake a visual inspection of all materials coming on to the site.**
- ❑ **Check that a sack weighs what it should weigh and that the volume ordered is the volume supplied.**
- ❑ **Check that composition and quality are correct.**

Material Storage

- ❑ **Install high-level control on bulk tanks to avoid overflows.**
- ❑ **Bund tanks to contain spillages.**
- ❑ **Use tanks that can be pitched and elevated, with rounded edges for ease of draining and rinsing.**
- ❑ **Dedicated tanks, receiving only one type of material, do not need to be washed out as often as tanks receiving a range of materials.**
- ❑ **Make sure that drums are stored in a stable arrangement to avoid damaging drums while in storage.**

- ❑ **Implement a tank checking procedure -- dip tanks regularly and document to avoid discharging a material into the wrong tank.**
- ❑ **Evaporation losses are reduced by using covered or closed tanks.**

Material and Water Transfer and Handling

- ❑ **Minimize the number of times materials are moved on site.**
- ❑ **Check transfer lines for spills and leaks.**
- ❑ **Is flexible pipework too long?**
- ❑ **Catch drainings from transfer hoses.**
- ❑ **Plug leaks and fit flow restrictions to reduce excess water consumption.**

Table 12. continued

Process Control
❑ Design a monitoring program to check the emissions and wastes from each unit operation.
❑ Regular maintenance of all equipment will help to reduce fugitive process losses.
❑ Feedback on how waste reduction is improving the process motivates the operators - it is vital that employees are informed of why actions are taken and what it is hoped to achieve.
Cleaning Procedures
❑ Minimize the amount of water used to wash out and rinse vessels -- on many sites indiscriminate water use contributes a large amount to wastewater flows. Ensure that hoses are not left running by fitting self-sealing valves.
❑ Investigate how washing water can be contained and used again before discharge to drains. The same applies to solvents used to clean; these can often be used more than once.

Tightening up house-keeping procedures can reduce waste considerably. Simple, quick adjustments could be made to your process to achieve a rapid improvement in process efficiency. Where such obvious reduction measures do not however solve the entire waste disposal problem, more detailed consideration of waste reduction options will be needed. These are addressed in Steps 16-18.

Step 16: Targeting and Characterizing Problem Wastes

Use the material balance for each unit operation to pinpoint the problem areas associated with your process. The material balance exercise may have brought to light the origin of wastes with high treatment costs or may indicate which wastes are causing process problems in which operations. The material balance should be used for your priorities for long-term waste reduction.

At this stage, it may be worthwhile considering the underlying causes as to why wastes are generated and the factors which lead to these; for examples, poor or outdated technology, lack of maintenance and non-compliance with company procedures may be contributing or even underlying factors. Additional sampling and characterization of your wastes might be necessary involving more in-depth analysis to ascertain the exact concentrations of contaminants. A worthwhile exercise is to list the wastes in order of priority for reduction actions. This will alert both the team and management to the most costly waste issues and also help to better define what resources may be needed to address them.

Step 17: Segregation

Segregation of wastes can offer enhanced opportunities for recycling and reuse with resultant savings in raw material costs. Concentrated simple wastes are more likely to be of value than dilute or complex wastes. In contrast, the practice of mixing wastes can aggravate pollution problems. If a highly-concentrated waste is mixed with a large quantity of weak, relatively uncontaminated effluent the result is a larger volume of waste requiring treatment. Isolating the concentrated waste from the weaker waste can reduce treatment costs. The concentrated waste could be recycled/reused or may require physical, chemical and biological treatment to comply with discharge consent levels whereas the weaker effluent could be reused or may only require settlement before discharge. Therefore, waste segregation can provide more scope for recycling and reuse while at the same time reducing treatment costs. Step 17 then is to review your waste collection and storage facilities to determine if waste segregation is possible. If so, then adjust your list of priority wastes accordingly.

Step 18: Developing Long-Term Waste Reduction Options

Waste problems that cannot be solved by simple procedural adjustments or improvements in house-keeping practices will require more substantial long-term changes. It is necessary to develop possible prevention options for the waste problems. Process or production changes which may increase production efficiency and reduce waste generation include:

- Changes in production process - continuous versus batch
- Equipment and installation changes
- Changes in process control - automation
- Changes in process conditions such as retention times, temperatures, agitation, pressure, catalysts
- Use of dispersants in place of organic solvents where appropriate
- Reduction in the quantity or type of raw materials used in production
- Raw material substitution through the use of wastes as raw materials or the use of different raw materials that produce less waste or less hazardous waste
- Process substitution with cleaner technology

Waste reuse can often be implemented if materials of sufficient purity can be concentrated or purified. Technologies such as reverse osmosis, ultrafiltration, electrodialysis, distillation, electrolysis, and ion exchange may enable materials to be reused and reduce or eliminate the need for waste treatment. Where waste treatment is necessary, a variety of technologies should be considered. These include physical, chemical and biological treatment processes. In some cases the treatment method can also recover valuable materials for reuse. Another industry or factory may be able to use or treat a waste that you cannot treat on-site. It may be worth investigating the possibility of setting up a waste exchange bureau as a structure for sharing treatment and reuse facilities. A simple example of this kind of practice is the following. Figure 4 shows photographs of a wire and cable manufacturing operation that the author visited in Russia. The operation is comprised of a series of wire drawing and cold extrusion machines that draw copper wire down to various diameters and then spool the product. The spools (closeup shown in the lower photograph of Figure 4) are an intermediate product in the plant operation. These spools are sent to another unit operation where the wire is coated with an insulator and multi-strand cables of various specifications are made. Figure 5 shows a close-up photograph of the cable being drawn through a set of close tolerance rollers that are a part of a cold extrusion machine. From this stage in the operation, the wire is fed to a final stage of drawing and washing. The washing is comprised of drawing the wire through a bath containing a solvent cleaning solution. In this manufacturing operation, a considerable amount of raw material (copper) is lost since the abrasion of the soft wire surface occurs as it is drawn by the rollers. The washing stage is where much of this loss accumulates. Figure 6 shows a photograph of one of the baths. The cleaning solution is recirculated, but because of the significant sludge build-up, the bath solution must be changed every ten days or so. In the past, the sludge was bagged and stored on-site. Because this not a huge accumulation of solid waste (up to 3 tons per year), the facility chose to stockpile the waste on site, and every quarter, ship the waste, along with other solid wastes to a local landfill. Hence, the normal practice is to mix the solid waste streams at the time of transport and to dispose of them in a common landfill operation off-site. Figure 7 shows some of the wastes. Obviously this would be totally unacceptable in the United States, but Russia has no solid waste laws that are comparable to RCRA. The mixed waste comprised not only of copper contaminated sludge (note that copper is a heavy metal and is toxic), but HDPE and other wire coating insulating materials used in the downstream process. It never occurred to management to measure the concentration of copper in the waste sludge. Measurements showed the sludge to have copper levels as high as 60%, which made it attractive as a feedstock to a local smelter. Although the economics were not well defined at the time of the very brief audit conducted by the author, conceptually, the plant could capture benefits from the sale of the waste and reductions in transport and landfill tipping fees by segregating this waste stream.

Figure 4. *Cable producing plant in Samara, Russia.*

The pollution prevention opportunity in this example is relatively straightforward and seemingly very obvious - but as is often the situation incremental cost savings are likely achievable from obvious housekeeping and better management of small scale waste problems that are overlooked.

Figure 5. *Copper wire being drawn through rollers.*

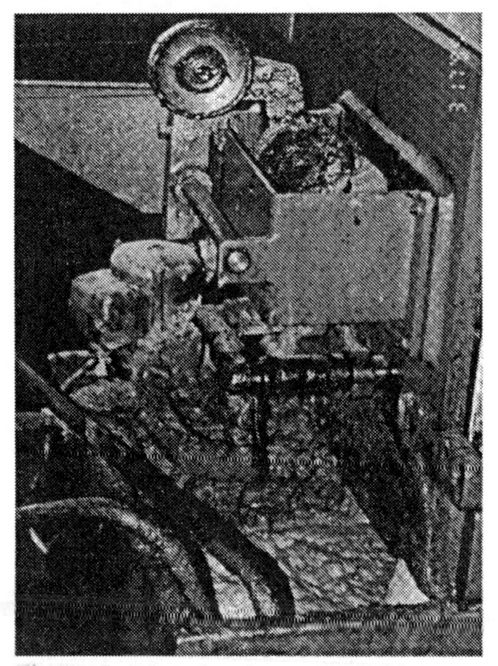

Figure 6. *Wire cleaning operation produces waste sludge.*

Figure 7. *The final waste form.*

Consider also the possibilities for product improvements or changes yielding cleaner, more environmentally-friendly products, both for existing products and in the development of new products.

Steps 15, 16, 17, and 18 Summary

At the end of Step 18 you should have identified all the waste reduction options which could be implemented. It is now time to begin assessing whether options identified have economic incentives worth going after.

Step 19: Environmental and Economic Evaluation of Waste Reduction Options

In order to decide which options should be developed to formulate a waste reduction action plan each option should be considered in terms of environmental and economic benefits using the theory and principles discussed in Chapter 3.

Environmental Evaluation: It is often taken for granted that reduction of a waste will have environmental benefits. This is generally true; however, there are exceptions to the rule. For example, reducing one waste may give rise to pH imbalances or may produce another waste that is more difficult to treat, resulting in a net environmental disadvantage. Hence, there may be environmental trade-offs between status quo and the alternatives identified. In many cases, the benefits may be obvious such as the result of the removal of a toxic element from an aqueous effluent by segregating the polluted waste or by changing the process in such a way that the waste is prevented. In other cases the environmental benefits may be less tangible. Creating a cleaner, healthier workplace will increase production efficiency but this may be difficult to quantify. For each option a series of questions or issues should be considered:

- Consider the effect of each option on the volume and degree of contamination of process wastes.
- Does a waste reduction option have cross-media effects? For example, does the reduction of a gaseous waste produce a liquid waste?
- Does the option change the toxicity, degradability, or treatability of the wastes?
- Does the option use more or less nonrenewable resources?
- Does the option use less energy?

The LCA principles outlined in Chapter 3 should be applied at this stage.

Economic Evaluation: A comparative economic analysis of the waste reduction options and the existing situation should be undertaken. Where benefits or changes cannot be quantified (e.g., reduction in future liability, worker health and safety costs) some form of qualitative assessment should be made.

Economic evaluations of waste reduction options should involve a comparison of operating costs to illustrate where cost savings would be made. For example, a waste reduction measure that reduces the amount of raw material lost to drain during the process results in reduced raw material costs. Raw material substitution or process changes may reduce the amount of solid waste that has to be transported off-site. Therefore, the transport costs for waste disposal would be reduced.

In many cases, it is appropriate to compare the waste treatment costs under existing conditions with those associated with the waste reduction option. The size of a treatment plant and the treatment processes required may be altered significantly by the implementation of waste reduction options. This should be considered in an economic evaluation.

Calculate the annual operating costs for the existing process indicating waste treatment and estimate how these would be altered with the introduction of waste reduction options. Tabulate and compare the process and waste treatment operating costs for both the existing and proposed future waste management options. The example given in Table 13 provides some of the typical cost components. In addition, if there are any monetary benefits (e.g., recycled or reused materials or wastes), then these should be subtracted from the total process or waste treatment costs as appropriate. The expanded cost analysis scheme discussed in Chapter 3 is appropriate to include at this point.

Now that you have determined the likely savings in terms of annual process and waste treatment operating costs associated with each option, consider the necessary investment required to implement each option. Investment can be assessed by looking at the payback period for each option. Recall that payback period is the time taken for a project to recover its financial outlay. A more detailed investment analysis may involve an assessment of the internal rate of return (IRR) and net present value (NPV) of the investment based on discounted cash flows. An analysis of investment risk allows you to rank the options identified. Consider the environmental benefits and the savings in process and waste treatment operating costs along with the payback period for an investment, to decide which options are viable candidates. Once this is done, the audit team can develop draft recommendations to be included in the final report and presentation to management.

Table 13. *Summarizing Annual Process and Waste Treatment Operating Costs.*

Process Operating Costs	**Annual Cost**
Raw Material 1	
Raw Material 2	
Water	
Energy	
Labor	
Maintenance	
Administration	
Other	
Total	
Waste Treatment Operating Costs	**Annual Cost**
Raw Material, e.g., Lime	
Raw Material, e.g., Flocculent	
Water	
Energy	
Trade Effluent Discharge Costs	
Transportation	
Off-Site Disposal	
Labor	
Maintenance	
Administration	
Other, e.g., violation, fines	
Total	

Step 20: Developing and Implementing An Action Plan: Reducing Wastes and Increasing Production Efficiency

Consider the immediate reduction measures identified in Step 15 along with the long-term waste reduction measures that have been evaluated in Steps 18 and 19. These measures should form the basis of the waste reduction action plan. Discuss your findings with members of staff and develop a workable action plan. Prepare the ground for the waste reduction action plan. Its implementation should be preceded by an explanation of the objectives behind undertaking a pollution prevention assessment. It is necessary to convince those who must work to new procedures that the change in philosophy from end-of-pipe treatment to waste prevention makes sense and serves to improve efficiency. Use posters around the site to emphasize the importance of waste reduction to minimize production and waste treatment/disposal costs and, where appropriate, for improving the health and safety of company personnel. Set out the intended action plan within an appropriate schedule. Remember it may take time for the staff to feel comfortable with a new way of thinking. Therefore, it is a good idea to implement waste reduction measures slowly but consistently to allow everyone time to adapt to these changes.

Set up a monitoring program to run alongside the waste reduction action plan so that actual improvements in process efficiency can be measured. For multiple pollution prevention projects, the P2 matrix presented in Table 2 is a good way to track and report overall performance. Relay these results back to the workforce as evidence of the benefits of waste reduction. Adopt an internal record-keeping system for maintaining and managing data to support material balances and waste reduction assessments.

It is likely that you will have highlighted significant information gaps or inconsistencies during the pollution prevention assessment investigations. You should concentrate on these gaps and explore ways of developing the additional data. Ask yourself repeatedly - is outside help required?

A good way of providing waste reduction incentives is to set up an internal waste charging system, those processes that create waste in great volume or that are difficult and expensive to handle having to contribute to the treatment costs on a proportional basis. Another method of motivating staff is to offer financial reward for individual waste-saving efforts, drawing on the savings gained from implementing waste reduction measures.

Pollution prevention assessments/audits should be a regular event -- attempt to develop a specific pollution prevention assessment approach for your own situation, keeping abreast of technological advances that could lead to waste reduction and the development of cleaner products. Train process employees to undertake material balance exercises. Training people who work on the process to undertake a pollution prevention assessment will help to raise awareness in the workforce. Without the support of the operators waste reduction actions will be ineffectual - these are the people who can really make a difference to process performance.

Step 20: Summary

The key elements in the final step of the auditing process are:

1. Prepare the ground for the waste reduction action plan, ensuring that support for the assessment, and implementation of the results, is gained from senior management.
2. Implement the plan slowly to allow the workforce to adjust.
3. Monitor process efficiency. Relay results back to the workforce to show them the direct benefits.
4. Train personnel to undertake your own pollution prevention assessment for waste reduction.

Although there are variations of the three-phase auditing process described in this chapter, the general philosophy and protocol are similar. A key concept to bear in mind is that the audit itself is only a tool. In order for pollution prevention to work, there must be strong management support, and there must be a team approach to implementing the audit. Without this mix, the process simply will not yield positive results.

The references listed below are key resources that the reader should refer to. The United Nations EP3 project (1) provides a similar step-by-step methodology for implementing a pollution prevention audit. Bernstein (2) has provided a good overview of waste management issues and approaches to waste minimization and pollution prevention. The USEPA study (3) on industry motivation for pollution prevention also provides some general background reading material. EPA's cleaner technologies substitute assessment guide (4) is a very different approach than pollution prevention, but still offers some useful information that can be incorporated into a P2 audit. Finally, Chopey and Hicks (6) provide examples of mass and energy balances that are needed in quantitative evaluations in the audit.

REFERENCES

1. United Nations Environmental Program, *Environmental Pollution Prevention Project (EP3) Training Manual*, New York, March 1995.

2. Bernstein, J. D., *Alternative Approaches to Pollution Control and Waste Management*, Urban Management Programme, World Bank, Washington, D.C., 1993.
3. *Prototype Study of Industry Motivation for Pollution Prevention*, USEPA, EPA Document Number 100-R-96-001, June 1996.
4. *Cleaner Technologies Substitute Assessments: A Methodology and Resource Guide*, USEPA, EPA Document Number EPA 744-R-95-002, December 1996.
5. Chopey, N. P. and T. G. Hicks, *Handbook of Chemical Engineering Calculations*, McGraw-Hill Book Company, New York, 1984.

Chapter 5
Pollution Prevention Practices in the Chemical Process Industries

INTRODUCTION

Chemicals can be described as the foundation of a modern, progressive society. They are an integral and ever-increasing part of our complex technological world, making it possible for us to enjoy a high standard of living. Yet, as the 1984 catastrophe in Bhopal, India dramatically and tragically demonstrated, those same chemicals are the source of danger to those in the workplace and surrounding locales who are regularly exposed to them. The city of Bhopal (1991 pop. 1,063,662), central India, capital of Madhya Pradesh state, was founded in 1728. Bhopal is a railway junction and industrial center, producing electrical equipment, textiles, and jewelry. Landmarks include the old fort (built 1728) and the Taj-ul-Masajid mosque, the largest in India. On Dec. 3, 1984, the worst industrial accident in history occurred there when a toxic gas leak from a Union Carbide insecticide plant killed over 6,400 people and seriously injured 30,000 to 40,000. The Indian government sued on behalf of over 500,000 victims and in 1989 settled for $470 million in damages and exempted company employees from criminal prosecution. The Indian judiciary rejected that exemption in 1991, and the company's Indian assets were seized (1992) after its officials failed to appear to face charges.

We have seen how their improper use and handling impact and exact unacceptable human and economic costs on families, industries, communities, and even nations. As a result, we have learned that correcting situations that could lead to disasters and catastrophes is more responsible and less expensive than hoping accidents will not occur and responding only when they inevitably do. Pollution prevention practices in the Chemical Process Industries (CPI) offer not only reduced costs in manufacturing operations, but they are essential in developing more efficient and safe means of manufacturing many of the chemical products we have grown to depend on.

This chapter provides an overview of P2 practices and opportunities within the CPI and allied industries. Not all industry subsectors are covered since the CPI is so diverse and fragmented; however, the author has tried to highlight some of the manufacturing operations that deal with bulk and commodity chemical products. In addition, the chapter provides some small case studies for the reader to gain a flavor for P2 practices, and to possibly draw ideas from for similar applications. Finally, the chapter concludes with discussions on the USEPA methodology developed dealing with Cleaner Technology Substitute Assessments (CTSA). This methodology and protocol is distinct from pollution prevention - but incorporates P2 as an important module. The CTSA is a methodology for evaluating the comparative risk, performance, cost, and resource conservation of alternatives to chemicals currently used by specific industry sectors.

PETROCHEMICALS MANUFACTURING

Industry Description and Practices

Natural gas and crude distillates such as naphtha from petroleum refining are used as feedstocks to manufacture a wide range of petrochemicals that are in turn used in the manufacture of consumer goods. The following discussions on petrochemical processes and products are for illustrative purposes only. The basic petrochemicals are manufactured by cracking, reforming, and other processes, and include olefins (such as ethylene, propylene, butylenes, and butadiene) and aromatics (such as benzene, toluene, and xylenes). The capacity of naphtha crackers is generally of the order of 250,000-750,000 metric tons per year (tpy) of ethylene production. Some petrochemical plants also have alcohol and oxo-compound manufacturing units on site. The base petrochemicals or products derived from them, along with other raw materials, are converted to a wide range of products. Among them are:

- Resins and plastics such as low-density polyethylene (LDPE), high-density polyethylene (HDPE), linear low-density polyethylene (LLDPE), polypropylene, polystyrene, and polyvinyl chloride (PVC);
- Synthetic fibers such as polyester and acrylic engineering polymers such as acrylonitrile butadiene styrene (ABS);
- Rubbers, including styrene butadiene rubber (SBR) and polybutadiene rubber (PBR);

- Solvents;
- Industrial chemicals, including those used for the manufacture of detergents such as linear alkyl benzene (LAB) and of coatings, dyestuffs, agrochemicals, pharmaceuticals, and explosives.

Chemical compounds manufactured at petrochemical plants include methanol, formaldehyde, and halogenated hydrocarbons. Formaldehyde is used in the manufacture of plastic resins, including phenolic, urea, and melamine resins. Halogenated hydrocarbons are used in the manufacture of silicone, solvents, refrigerants, and degreasing agents.

Olefins (organics having at least one double bond for carbon atoms) are typically manufactured from the steam cracking of hydrocarbons such as naphtha. Major olefins manufactured include ethylene, propylene, butadiene, and acetylene. The olefins manufactured are used in the manufacture of polyethylene, including low-density polyethylene (LDPE) and high-density polyethylene (HDPE), and for polystyrene, polyvinyl chloride, ethylene glycol (used along with dimethyl terephthalate, DMT, as feedstock to the polyester manufacturing process), ethanol amines (used as solvents), polyvinyl acetate (used in plastics), polyisoprene (used for synthetic rubber manufacture), polypropylene, acetone (used as a solvent and in cosmetics), isopropanol (used as a solvent and in pharmaceuticals manufacturing), acrylonitrile (used in the manufacture of acrylic fibers and nitrile rubber), propylene glycol (used in pharmaceuticals manufacturing), and polyurethane. Butadiene is used in the manufacture of polybutadiene rubber (PBR) and styrene butadiene rubber (SBR). Other C_4 compounds manufactured include butanol, which is used in the manufacture of solvents such as methyl ethyl ketone.

The major aromatics (organics having at least one ring structure with six carbon atoms) manufactured include benzene, toluene, xylene, and naphthalene. Other aromatics manufactured include phenol, chlorobenzene, styrene, phthalic and maleic anhydride, nitrobenzene, and aniline. Benzene is generally recovered from cracker streams at petrochemical plants and is used for the manufacture of phenol, styrene, aniline, nitrobenzene, sulfonated detergents, pesticides such as hexachlorobenzene, cyclohexane (an important intermediate in synthetic fiber manufacture), and caprolactam, used in the manufacture of nylon. Benzene is also used as a solvent.

The main uses of toluene are as a solvent in paints, rubber, and plastic cements and as a feedstock in the manufacture of organic chemicals, explosives, detergents, and polyurethane foams. Xylenes (which exist as three isomers) are used in the manufacture of DMT, alkyd resins, and plasticizers. Naphthalene is mainly used in the manufacture of dyes, pharmaceuticals, insect repellents, and phthalic anhydride (used in the manufacture of alkyd resins, plasticizers, and polyester).

The largest user of phenol in the form of thermosetting resins is the plastics industry. Phenol is also used as a solvent and in the manufacture of intermediates for pesticides, pharmaceuticals, and dyestuffs. Styrene is used in the manufacture of synthetic rubber and polystyrene resins. Phthalic anhydride is used in the manufacture of DMT, alkyd resins, and plasticizers such as phthalates. Maleic anhydride is used in the manufacture of polyesters and, to some extent, for alkyd resins. Minor uses include the manufacture of malathion and soil conditioners. Nitrobenzene is used in the manufacture of aniline, benzidine, and dyestuffs and as a solvent in polishes. Aniline is used in the manufacture of dyes, including azo dyes, and rubber chemicals such as vulcanization accelerators and antioxidants.

A number of alternative methods for manufacturing the desired products are available; however, it is beyond the scope of this book to discuss these processes and products.

Overview of Pollution and Prevention

In technologically advanced countries, it is fair to say that fugitive air emissions from pumps, valves, flanges, storage tanks, loading and unloading operations, and wastewater treatment are of greatest concern. Some of the compounds released to air are carcinogenic or toxic. Ethylene and propylene emissions are of concern because their release can lead to the formation of extremely toxic oxides. Compounds considered carcinogenic that may be present in air emissions include benzene, butadiene, 1,2-dichloroethane, and vinyl chloride. A typical naphtha cracker at a petrochemical complex may release annually about 2,500 metric tons of alkenes, such as propylenes and ethylene, in producing 500,000 metric tons of ethylene. Boilers, process heaters, flares, and other process equipment (which in some cases may include catalyst regenerators) are responsible for the emission of PM (particulate matter), carbon monoxide, nitrogen oxides (200 tpy), based on 500,000 tpy of ethylene capacity, and sulfur oxides (600 tpy).

The release of volatile organic compounds (VOCs) into the air depends on the products handled at the plant. VOCs released may include acetaldehyde, acetone, benzene, toluene, trichloroethylene, trichlorotoluene, and xylene. VOC emissions are mostly fugitive and depend on the production processes, materials-handling and effluent-treatment procedures, equipment maintenance, and climatic conditions. VOC emissions from a naphtha cracker range from 0.6 to 10 kilograms per metric ton (kg/t) of ethylene produced. Of these emissions, 75% consists of alkanes, 20% of unsaturated hydrocarbons, about half of which is ethylene, and 5% of aromatics. For a vinyl chloride plant, VOC emissions are 0.02-2.5 kg/t of product; 45% is ethylene dichloride, 20% vinyl chloride, and 15% chlorinated organics; for an SBR plant, VOC emissions are typically 3 or more kg/t of product; for an ethyl benzene plant, 0.1-2 kg/t of product; for an ABS plant, 1.4-27 kg/t of product; for a styrene plant, 0.25-18 kg/t of product; and for a polystyrene plant, 0.2-5 kg/t of product. Petrochemical units generate waste waters from process operations such as vapor condensation, from cooling tower blowdown, and from stormwater runoff. Process waste waters are generated at a rate of about 15 cubic meters per hour (m^3/hr), based on 500,000 tpy ethylene production, and may contain biochemical oxygen demand (BOD) levels of 100 mg/l, as well as chemical oxygen demand (COD) of 1,500-6,000 mg/l, suspended solids of 100-400 mg/l, and oil and grease of 30-600 mg/l. Phenol levels of up to 200 mg/l and benzene levels of up to 100 mg/l may also be present.

Petrochemical plants also generate significant amounts of solid wastes and sludges, some of which may be considered hazardous because of the presence of toxic organics and heavy metals. Spent caustic and other hazardous wastes may be generated in significant quantities; examples are distillation residues associated with units handling acetaldehyde, acetonitrile, benzyl chloride, carbon tetrachloride, cumene, phthallic anhydride, nitrobenzene, methyl ethyl pyridine, toluene diisocyanate, trichloroethane, trichloroethylene, perchloroethylene, aniline, chlorobenzenes, dimethyl hydrazine, ethylene dibromide, toluenediamine, epichlorohydrin, ethyl chloride, ethylene dichloride, and vinyl chloride.

Of great concern in this industry is accidental discharges as a result of abnormal operation, especially from polyethylene and ethylene-oxide-glycol plants in a petrochemical complex. These can be a major environmental hazard, releasing large quantities of pollutants and products into the environment. The universe of chemical accidents within the United States cannot now be accurately tallied. No comprehensive, reliable historical records exist. Further, the EPA acknowledges that many accidents occurring today at fixed facilities and during transport are not reported to the federal government. This underreporting is documented by several studies (National Environmental Law Center et al. 1994). What is known, however, is that in 1991 the National Response Center received over 16,300 calls reporting the release or potential release of a hazardous material (USEPA 1993). Also, NTSB's statistics indicate that, in 1992, chemicals were involved in 3,500 fatal highway accidents and 6,500 railroad accidents (NTSB 1992). One study analyzed information contained in the EPA's Emergency Response Notification System (ERNS) database. ERNS is acknowledged to be the largest and most comprehensive United States database of chemical accident notifications, covering both transportation and fixed facility accidents. The study found that from 1988 through 1992 an average of 19 accidents occurred each day . . . 6,900 per year, with more than 34,500 accidents involving toxic chemicals occurring over the five-year period. The study's report emphasized that the findings gravely understated the severity of the United States' chemical accident picture (National Environmental Law Center et al. 1994).

Although the absolute numbers vary depending on the source of statistics and period of time examined, there is no doubt about the effects of chemical accidents on human life . . . year after year, large numbers of people are killed and injured. Added to these imprecise numbers must be those long-term consequences of exposure that are not immediately discernable and may not be reflected in studied databases . . . low-level exposure to some chemicals may result in debilitating diseases that appear only years later. During the years 1988 through 1992, six percent, or 2,070, of the 34,500 accidents that occurred resulted in immediate death, injury and/or evacuation; an average of two chemical-related injuries occurred every day during those five years (National Environmental Law Center et al. 1994). Between 1982 and 1986, the EPA's Acute Hazard Events (AHE) database, which contains information only for chemical accidents having acute hazard potential, recorded 11,048 events involving releases of extremely hazardous substances; these events resulted in 309 deaths, 11,341 injuries and, based on evacuation information for the one-half of the recorded events reporting whether such activity occurred, evacuation of 464,677 people from their homes and jobs (USEPA 1989). During the years 1987 through 1991, chemical accidents resulted in 453 deaths and 1,576 injuries at fixed facilities, while transportation accidents involving chemicals claimed 55 lives and injured 1,252 persons (USEPA 1993).

Pollution Prevention Practices and Opportunities

Petrochemical plants are typically large and complex, and the combination and sequence of products manufactured are often unique to the plant. Specific pollution prevention practices or source reduction measures are best determined by a dedicated technical staff. However, there are a number of broad areas where improvements are often possible, and site-specific emission reduction measures in these areas should be designed into the plant and targeted by plant management. General areas where efforts should be concentrated are listed in Table 1. A good practice target for a petrochemical complex is to reduce total organic emissions (including VOCs) from the process units to 0.6% of the throughput. Target maximum levels for air releases, per ton of product, are, for ethylene, 0.06 kg; for ethylene oxide, 0.02 kg; for vinyl chloride, 0.2 kg; and for 1,2-dichloroethane, 0.4 kg.

Table 1. *Recommended P2 Practices in the CPI.*

Area of Opportunity	RECOMMENDED POLLUTION PREVENTION PRACTICE
Reduction of Air Emissions	1. Minimize leakages of volatile organics, including benzene, vinyl chloride, and ethylene oxide, from valves, pump glands (through use of mechanical seals), flanges, and other process equipment by following good design practices and equipment maintenance procedures. 2. Use mechanical seals where appropriate. 3. Minimize losses from storage tanks, product transfer areas, and other process areas by adopting methods such as vapor recovery systems and double seals (for floating roof tanks). 4. Recover catalysts and reduce particulate missions. 5. Reduce nitrogen oxide NO_x emissions by using low-NO_x burners. Optimize fuel usage. 6. In some cases, organics that cannot be recovered are effectively destroyed by routing them to flares and other combustion devices.
Elimination or Reduction of Pollutants	7. Use nonchrome-based additives in cooling water. 8. Use long-life catalysts and regeneration to extend the cycle.
Recycling and Reuse	9. Recycle cooling water and treated wastewater to the extent feasible. 10. Recover and reuse spent solvents and other chemicals to the extent feasible.
Improved Operating Procedures	11. Segregate process waste waters from stormwater systems. 12. Optimize the frequency of tank and equipment cleaning. 13. Prevent solids and oily wastes from entering the drainage system. 14. Establish and maintain an emergency preparedness and response plan.

Vapor recovery systems to control losses of VOCs from storage tanks and loading areas should achieve close to 100% recovery. In addition, a wastewater generation rate of 15 cubic meters per 100 tons of ethylene produced is achievable with good design and operation; and new petrochemical complexes should strive to achieve this.

The following summarizes the acceptable treatment technologies that are relied upon in the CPI. Control of air emissions normally includes the capturing and recycling or combustion of emissions from vents, product transfer points, storage tanks, and other handling equipment. Catalytic cracking units should be provided with particulate removal devices. Particulate removal technologies include fabric filters, ceramic filters, wet scrubbers, and electrostatic precipitators. Gaseous releases are minimized by condensation, absorption, adsorption (using activated carbon, silica gel, activated alumina, and zeolites), and, in some cases, biofiltration and bioscrubbing (using peat or heather, bark, composts, and bioflora to treat biodegradable organics), and thermal decomposition.

Petrochemical waste waters often require a combination of treatment methods to remove oil and other contaminants before discharge. Separation of different streams (such as stormwater) is essential to minimize treatment requirements. Oil is recovered using separation techniques. For heavy metals, a combination of oxidation/reduction, precipitation, and filtration is used. For organics, a combination of air or steam stripping, granular activated carbon, wet oxidation, ion exchange, reverse osmosis, and electrodialysis is used. A typical system may include neutralization, coagulation/flocculation, flotation/sedimentation/filtration, biodegradation (trickling filter, anaerobic, aerated lagoon, rotating biological contactor, and activated sludge), and clarification. A final polishing step using filtration, ozonation, activated carbon, or chemical treatment may also be required. Examples of pollutant loads that can be achieved are: COD, less than 1 kg per 100 tons of ethylene produced; suspended solids, less than 0.4 kg/100 t; and dichloroethane less than 0.001 kg/100 t.

For solid and hazardous wastes, combustion (preceded in some cases by solvent extraction) of toxic organics is considered an effective treatment technology for petrochemical organic wastes. Steam stripping and oxidation are also used for treating organic waste streams. Spent catalysts are generally sent back to the suppliers. In some cases, the solid wastes may require stabilization to reduce the leachability of toxic metals before disposal of in an approved, secure landfill.

For emissions guidelines, we will follow those recommended by the World

Bank Organization (WBO). The reader should bear in mind that the guidelines cited present emissions levels normally acceptable to the World Bank Group in making decisions regarding provision of World Bank Group financial assistance, and do not necessarily conform to environmental standards in the United States. Since this handbook is intended to assist in industry P2 practices on a global basis, we have chosen a more internationally recognized set of guidelines as opposed to the U.S. standards - even though in general, the U.S. standards are likely the most stringent in many cases.

The emissions levels cited in this section and others can be consistently achieved by well-designed, well-operated, and well-maintained pollution control systems. Note that guidelines are typically expressed as concentrations to facilitate monitoring.

Dilution of air emissions or effluents to achieve these guidelines is unacceptable. All of the maximum levels should be achieved for at least 95% of the time that the plant or unit is operating, to be calculated as a proportion of annual operating hours. WBO recommended air emissions levels are presented in Table 2. Standards for liquid effluents are reported in Table 3.

Table 2. *Target Ambient Levels to Achieve.*

PARAMETER	MAXIMUM VALUE
PM	20 mg/m^3
Nitrogen oxides	300 mg/m^3
Hydrogen chloride	10 mg/m^3
Sulfur oxides	500 mg/m^3
Benzene	5 mg/m^3 for emissions; 0.1 ppb at the plant fence
1,2-dichloroethane	5 mg/m^3 for emissions;
	1.0 ppb at the plant fence
Vinyl chloride	5 mg/m^3 for emissions;
	0.4 ppb at the plant fence
Ammonia	15 mg/m^3

Note: Maximum ambient levels for ethylene oxide are 0.3 ppb at the plant fence. Maximum total emissions of the VOCs acetaldehyde, acrylic acid, benzyl chloride, carbon tetrachloride, chlorofluorocarbons, ethyl acrylate, halons, maleic anhydride, 1, 1, 1-trichlorethane, trichloroethylene, and trichlorotoluene are 20 mg/Nm^3. Maximum total heavy metal emissions are 1.5 mg/Nm^3.

Table 3. *Effluents from Petrochemicals Manufacturing (milligrams per liter, except for pH and temperature).*

PARAMETER	MAXIMUM VALUE	PARAMETER	MAXIMUM VALUE
pH	6-9	Copper	0.5
BOD	30	Phenol	0.5
COD	150	Benzene	0.05
TSS	30	Vinyl chloride	0.05
Oil and grease	10	Sulfide	1
Cadmium	0.1	Nitrogen (total)	10
Chromium (hexavalent)	0.1	Temperature increase	3 °C [a]

Note: Effluent requirements are for direct discharge to surface waters. a. The effluent should result in a temperature increase of no more than 3° C at the edge of the zone where initial mixing and dilution take place. Where the zone is not defined, use 100 meters from the point of discharge.

Note that wherever possible, generation of sludges should be minimized. Sludges must be treated to reduce toxic organics to nondetectable levels. Wastes containing toxic metals should be stabilized before disposal.

Air emissions from stacks should be visually monitored for opacity at least once every eight hours. Annual emissions monitoring of combustion sources should be carried out for sulfur oxides, nitrogen oxides, and the organics listed above, with fuel sulfur content and excess oxygen maintained at acceptable levels during normal operations. Leakages should be visually checked every eight hours and at least once a week using leak detection equipment.

Liquid effluents should be monitored at least once every eight hours for all the parameters cited above except metals, which should be monitored at least monthly. Each shipment of solid waste going for disposal should be monitored for toxics. All monitoring data should be analyzed and reviewed at regular intervals and compared with the operating standards so that any necessary corrective actions can be taken. Records of monitoring results should be kept in an acceptable format. From a compliance standpoint, the results should be reported to the responsible authorities and relevant parties, as required.

To summarize, in addition to using properly designed pollution control equipment, the following P2 practices are likely to assist in positive environmental benefits:

- Implement an equipment maintenance program that minimizes releases of volatile organics, including ethylene oxide, benzene, vinyl chloride, and 1,2-dichloroethane.
- Install vapor recovery systems to reduce VOC emissions.
- Use low-NO_x burners.
- Optimize fuel usage.
- Regenerate and reuse spent catalysts, solvents, and other solutions to the extent feasible.
- Recycle cooling water and reuse waste waters.
- Segregate stormwater from process wastewater.
- Use nonchrome-based additives in cooling water.
- Design and practice emergency preparedness and prevention measures.

CHLOR-ALKALI PLANTS

Industry Description and Practices

There are three basic processes for the manufacture of chlorine and caustic soda from brine: the mercury cell, the diaphragm cell, and the membrane cell. Among these technologies, the membrane cell is the most modern and has both economic and environmental advantages. The other two processes generate hazardous wastes containing mercury or asbestos. Mercury cell technology is being phased out in worldwide production.

In the membrane process, the chlorine (at the anode) and the hydrogen (at the cathode) are kept apart by a selective polymer membrane that allows the sodium ions to pass into the cathodic compartment and react with the hydroxyl ions to form caustic soda. The depleted brine is dechlorinated and recycled to the input stage. As noted already, the membrane cell process is the preferred process for new plants. Diaphragm processes may be acceptable, in some circumstances, but only if nonasbestos diaphragms are used. The energy consumption in a membrane cell process is of the order of 2,200-2,500 kilowatt-hours per metric ton (kWh/t), as compared with 2,400-2,700 kWh/t of chlorine for a diaphragm cell process.

Pollution Prevention Practices and Opportunities

The major waste stream from the process consists of brine muds - the sludges from the brine purification step. The sludge is likely to contain magnesium, calcium, iron, and other metal hydroxides, depending on the source and purity of the brines. The muds are normally filtered or settled, the supernatant is recycled, and the mud is dried and then landfilled. Chlorine is a highly toxic gas, and strict precautions are necessary to minimize risk to workers and possible releases during its handling. Major sources of fugitive air emissions of chlorine and hydrogen are vents, seals, and transfer operations. Acid and caustic waste waters are generated in both the process and the materials recovery stages of the operation. The following pollution prevention measures should be considered in plant operations:

- Use metal rather than graphite anodes to reduce lead and chlorinated organic matter.
- Resaturate brine in closed vessels to reduce the generation of salt sprays.
- Use noncontact condensers to reduce the amount of process wastewater.
- Scrub chlorine tail gases to reduce chlorine discharges and to produce hypochlorite.
- Recycle condensates and waste process water to the brine system.
- Recycle brine wastes, if possible.

For the chlor-alkali industry, an emergency preparedness and response plan is mandatory for potential uncontrolled chlorine and other releases. Carbon tetrachloride is sometimes used to scrub nitrogen trichloride (formed in the process) and to maintain its levels below 4% to avoid fire and explosion. Substitutes for carbon tetrachloride may have to be used, as the use of carbon tetrachloride may be banned in the near future due to its carcinogenicity. Implementation of cleaner production processes and pollution prevention measures can yield both economic and environmental benefits. The primary treatment technologies afforded to this manufacturing include the following: Caustic scrubber systems should be installed to control chlorine emissions from condensers and at storage and transfer points for liquid chlorine. Sulfuric acid used for drying chlorine should be neutralized before discharge.

Brine muds should be discharged to lined settling ponds (or the equivalent) to prevent contamination of soil and groundwater. Effluents should be controlled for pH by neutralization. Settling and filtration are performed to control total suspended solids. Dechlorination of waste waters is performed using sulfur dioxide or bisulfite.

Daily monitoring for parameters other than pH (for effluents from the diaphragm process) is recommended. The pH in the liquid effluent should be monitored continuously. Chlorine monitors should be strategically located within the plant to detect chlorine releases or leaks on a continuous basis. Monitoring data should be analyzed and reviewed at regular intervals and compared with the operating standards so that any necessary corrective actions can be taken. Records of monitoring results should be kept in an acceptable format. The results should be reported to the responsible authorities and relevant parties, as required.

To summarize, preference should be given to the membrane process due to its less polluting characteristics over other technologies. In addition, the following pollution prevention measures should be considered for use with the membrane technology:

- Use metal instead of graphite anodes.
- Resaturate brine in closed vessels.
- Recycle brine wastes.
- Scrub chlorine from tail gases to produce hypochlorite.
- Provide lined settling ponds for brine muds.

AGRO-INDUSTRY CHEMICALS

Mixed Fertilizer Plants

Industry Description and Practices

Mixed fertilizers contain two or more of the elements nitrogen, phosphorus, and potassium (NPK), which are essential for good plant growth and high crop yields. This subsection briefly addresses the production of ammonium phosphates (monoammonium phosphate, or MAP, and diammonium phosphate, or DAP), nitrophosphates, potash, and compound fertilizers.

Ammonium phosphates are produced by mixing phosphoric acid and anhydrous ammonia in a reactor to produce a slurry. This is referred to as the mixed acid route for producing NPK fertilizers; potassium and other salts are added during the process. The slurry is sprayed onto a bed of recycled solids in a rotating granulator, and ammonia is sparged into the bed from underneath. Granules pass to a rotary dryer followed by a rotary cooler. Solids are screened and sent to storage for bagging or for bulk shipment.

Nitrophosphate fertilizer is made by digesting phosphate rock with nitric acid. This is the nitrophosphate route leading to NPK fertilizers; as in the mixed-acid route, potassium and other salts are added during the process. The resulting solution is cooled to precipitate calcium nitrate, which is removed by filtration methods. The filtrate is neutralized with ammonia, and the solution is evaporated to reduce the water content. The process of prilling may follow. The calcium nitrate filter cake can be further treated to produce a calcium nitrate fertilizer, pure calcium nitrate, or ammonium nitrate and calcium carbonate. Nitrophosphate fertilizers are also produced by the mixed-acid process, through digestion of the phosphate rock by a mixture of nitric and phosphoric acids.

Potash (potassium carbonate) and sylvine (potassium chloride) are solution-mined from deposits and are refined through crystallization processes to produce fertilizer. Potash may also be dry-mined and purified by flotation.

Compound fertilizers can be made by blending basic fertilizers such as ammonium nitrate, MAP, DAP, and granular potash; this route may involve a granulation process.

Pollution Prevention Practices and Opportunities

The principal pollutants from the production of MAP and DAP are ammonia and fluorides, which are given off in the steam from the reaction. Fluorides and dust are released from materials-handling operations. Ammonia in uncontrolled air emissions has been reported to range from 0.1 to 7.8 kilograms of nitrogen per metric ton (kg/t) of product, with phosphorus ranging from 0.02 to 2.5 kg/t product (as phosphorous pentoxide, P_2O_5).

In nitrophosphate production, dust will also contain fluorides. Nitrogen oxides NO_x are given off at the digester. In the evaporation stage, fluorine compounds and ammonia are released. Unabated emissions for nitrogen oxides from selected processes are less than 1,000 milligrams per cubic meter (mg/m^3) from digestion of phosphate rock with nitric acid, 50-200 (mg/m^3) from neutralization with ammonia, and 30-200 mg/m^3 from granulation and drying. Dust is the primary air pollutant from potash manufacturing.

The volumes of liquid effluents from mixed fertilizer plants are reported to range from 1.4 to 50 cubic meters per metric ton (m^3/t) of product. Where water is used in scrubbers, the scrubbing liquors can usually be returned to the process. Effluents can contain nitrogen, phosphorus, and fluorine; the respective ranges of concentrations can be 0.7-15.7 kg/t of product (as N), 0.1-7.8 kg/t of product (as P_2O_5), and 0.1-3.2 kg/t of product.

Generally, there is little solid waste from a fertilizer plant, since dust and fertilizer spillage can be returned to the process.

Materials handling and milling of phosphate rock should be carried out in closed buildings. Fugitive emissions can be controlled by, for example, hoods on conveying equipment, with capture of the dust in fabric filiters. In the ammonium phosphate plant, the gas streams from the reactor, granulator, dryer, and cooler should be passed through cyclones and scrubbers, using phosphoric acid as the scrubbing liquid, to recover particulates, ammonia, and other materials for recycling. In the nitrophosphate plant, nitrogen oxide (NO_x) emissions should be avoided by adding urea to the digestion stage. Fluoride emissions should be prevented by scrubbing the gases with water. Ammonia should be removed by scrubbing. Phosphoric acid may be used for scrubbing where the ammonia load is high. The process water system should be balanced, if necessary, by the use of holding tanks to avoid the discharge of an effluent.

Additional pollution control devices-beyond the scrubbers, cyclones, and baghouses that are an integral part of the plant design and operations are generally not required for mixed fertilizer plants. Good housekeeping practices are essential to minimize the amount of spilled material. Spills or leaks of solids and liquids should be returned to the process. Liquid effluents, if any, need to be controlled for TSS, fluorides, phosphorus, and ammonia. An effluent discharge of less than 1.5 m^3/t product as P_2O_5, is realistic, but use of holding ponds makes feasible a discharge approaching zero. In many countries outside of the United States, wastewater treatment discharges are often used for agricultural purposes and may contain heavy metals. Of particular concern is the cadmium content.

Air emissions at point of discharge should be monitored continuously for fluorides and particulates and annually for ammonia and nitrogen oxides. Liquid effluents should be continuously monitored for pH. Other parameters should be monitored at least weekly. Monitoring data should be analyzed and reviewed at regular intervals and compared with the operating standards so that any necessary corrective actions can be taken. Records of monitoring results should be kept in an acceptable format. The results should be reported to the responsible authorities and relevant parties, as required.

The key production and control practices that will lead to compliance with emissions requirements can be summarized as follows:

- Maximize product recovery and minimize air emissions by appropriate maintenance and operation of scrubbers and baghouses.
- Eliminate effluent discharges by operating a balanced process water system.
- Prepare and implement an emergency preparedness and response plan. Such a plan is required because of the large quantities of ammonia and other hazardous materials stored and handled on site.

The reader should consult references 1 through 5 for more detailed information and data.

Nitrogenous Fertilizer Plants

Industry Description and Practices

This subsection discusses the production of ammonia, urea, ammonium sulfate, ammonium nitrate (AN), calcium ammonium nitrate (CAN), and ammonium sulfate nitrate (ASN). The manufacture of nitric acid used to produce nitrogenous fertilizers typically occurs on site and is therefore included here.

Ammonia (NH_3) is produced from atmospheric nitrogen and hydrogen from a hydrocarbon source. Natural gas is the most commonly used hydrocarbon feedstock for new plants; other feedstocks that have been used include naphtha, oil, and gasified coal. Natural gas is favored over the other feedstocks from an environmental perspective.

Ammonia production from natural gas includes the following processes: desulfurization of the feedstock; primary and secondary reforming; carbon monoxide shift conversion and removal of carbon dioxide, which can be used for urea manufacture; methanation; and ammonia synthesis. Catalysts used in the process may include cobalt, molybdenum, nickel, iron oxide/chromium oxide, copper oxide/zinc oxide, and iron.

Urea fertilizers are produced by a reaction of liquid ammonia with carbon dioxide. The process steps include solution synthesis, where ammonia and carbon dioxide react to form ammonium carbamate, which is dehydrated to form urea; solution concentration by vacuum, crystallization, or evaporation to produce a melt; formation of solids by prilling (pelletizing liquid droplets) or granulating; cooling and screening of solids; coating of the solids; and bagging or bulk loading. The carbon dioxide for urea manufacture is produced as a by-product from the ammonia plant reformer.

Ammonium sulfate is produced as a caprolactam by-product from the petrochemical industry, as a coke by-product, and synthetically through reaction of ammonia with sulfuric acid. Only the third process is covered in our discussion. The reaction between ammonia and sulfuric acid produces an ammonium sulfate solution that is continuously circulated through an evaporator to thicken the solution and to produce ammonium sulfate crystals. The crystals are separated from the liquor in a centrifuge, and the liquor is returned to the evaporator. The crystals are fed either to a fluidized bed or to a rotary drum dryer and are screened before bagging or bulk loading.

Ammonium nitrate is made by neutralizing nitric acid with anhydrous ammonia. The resulting 80-90% solution of ammonium nitrate can be sold as is, or it may be further concentrated to a 95-99.5% solution (melt) and converted into prills or granules. The manufacturing steps include solution formation, solution concentration, solids formation, solids finishing, screening, coating, and bagging or bulk shipping. The processing steps depend on the desired finished product. Calcium ammonium nitrate is made by adding ammonia calcite or dolomite to the ammonium nitrate melt before prilling or granulating. Ammonium sulfate nitrate is made by granulating a solution of ammonium nitrate and ammonium sulfate. The production stages for nitric acid manufacture include vaporizing the ammonia; mixing the vapor with air and burning the mixture over a platinum/rhodium catalyst; cooling the resultant nitric oxide (NO) and oxidizing it to nitrogen dioxide (NO_2) with residual oxygen; and absorbing the nitrogen dioxide in water in an absorption column to produce nitric acid (HNO_3). Because of the large quantities of ammonia and other hazardous materials handled on site, an emergency preparedness and response plan is required.

Pollution Prevention Practices and Opportunities

Emissions to the atmosphere from ammonia plants include sulfur dioxide (SO_2), nitrogen oxides (NO_x), carbon monoxide (CO), carbon dioxide (CO_2), hydrogen sulfide (H_2S), volatile organic compounds (VOCs), particulate matter, methane, hydrogen cyanide, and ammonia. The two primary sources of pollutants, with typical reported values, in kilograms per ton (kg/t) for the important pollutants, are as follows:

- Flue gas from primary reformer: CO_2: 500 kg/t NH_3, NO_x: 0.6 -1.3 kg/t NH_3 as NO_2, SO_2: less than 0.1 kg/t; CO: less than 0.03 kg/t.
- Carbon dioxide removal: CO_2:1,200 kg/t.

Nitrogen oxide emissions depend on the process features. Nitrogen oxides are reduced, for example, when there is low excess oxygen, with steam injection; when post-combustion measures are in place; and when low-NO_x burners are in use. Other measures will also reduce the total amount of nitrogen oxides emitted. Concentrations of sulfur dioxide in the flue gas from the reformer can be expected to be significantly higher if a fuel other than natural gas is used. Energy consumption ranges from 29 to 36 gigajoules per metric ton (GJ/t) of ammonia. Process condensate discharged is about 1.5 cubic meters per metric ton (m^3/t) of ammonia. Ammonia tank farms can release upward of 10 kg of ammonia per ton of ammonia produced. Emissions of ammonia from the process have been reported in the range of less than 0.04 to 2 kg/t of ammonia produced.

In a urea plant, ammonia and particulate matter are the emissions of concern. Ammonia emissions are reported as recovery absorption vent (0.1-0.5 kg/t), concentration absorption vent (0.1-0.2 kg/t), urea prilling (0.5-2.2 kg/t), and granulation (0.2-0.7 kg/t). The prill tower is a source of urea dust (0.5-2.2 kg/t), as is the granulator (0.1-0.5 kg/t).

Particulate matter are the principal air pollutant emitted from ammonium sulfate plants. Most of the particulates are found in the gaseous exhaust of the dryers. Uncontrolled discharges of particulates may be of the order of 23 kg/t from rotary dryers and 109 kg/t from fluidized bed dryers. Ammonia storage tanks can release ammonia, and there may be fugitive losses of ammonia from process equipment.

The production of ammonium nitrate yields emissions of particulate matter (ammonium nitrate and coating materials), ammonia, and nitric acid. The emission sources of primary importance are the prilling tower and the granulator. Total quantities of nitrogen discharged are in the range of 0.01-18.4 kg/t of product. Values reported for calcium ammonium nitrate are in the range of 0.13-3 kg nitrogen per ton of product.

Nitric acid plants emit nitric oxide, nitrogen dioxide (the visible emissions), and trace amounts of nitric acid mist. Most of the nitrogen oxides are found in the tail gases of the absorption tower. Depending on the process, emissions in the tail gases can range from 215 to 4,300 milligrams per cubic meter (mg/m^3) for nitrogen oxides. Flow may be of the order of 3,200 m^3 per ton of 100% nitric acid. Nitrogen oxide values will be in the low range when high-pressure absorption is used; medium-pressure absorption yields nitrogen oxide emissions at the high end of the range. These values are prior to the addition of any abatement hardware.

Ammonia plant effluents may contain up to 1 kg of ammonia and up to 1 kg of methanol per cubic meter prior to stripping. Effluent from urea plants may discharge from less than 0.1 kg to 2.6 kg nitrogen per ton product. Effluents from ammonium nitrate plants have been reported to discharge 0.7-6.5 kg nitrogen per ton product. Comparable values for CAN plants are 0-10 kg nitrogen per ton of product. Nitric acid plants may have nitrogen in the effluent of the order of 0.1-1.7 kg nitrogen per ton of nitric acid.

Solid wastes are principally spent catalysts that originate in ammonia production and in the nitric acid plant. Other solid wastes are not normally of environmental concern. It is important to note that hot ammonium nitrate, whether in solid or in concentrated form, carries the risk of decomposition and is unstable and may even detonate under certain circumstances. Special precautions are therefore required in its manufacture. Implementation of cleaner production processes and pollution prevention measures can yield both economic and environmental benefits. The following describes production-related targets that can be achieved by measures such as those described above. The numbers relate to the production processes before the addition of pollution control measures.

Ammonia Plants - New ammonia plants should set as a target the achievement of nitrogen oxide emissions of not more than 0.5 kg/t of product (expressed as NO_2 at 3% 0). Ammonia releases in liquid effluents can be controlled to 0.1 kg/t of product. Condensates from ammonia production should be reused.

Nitric Acid Plant - Nitrogen oxide levels should be controlled to a maximum of 1.6 kg/t of 100% nitric acid. Extended absorption and technologies such as nonselective catalytic reduction (NSCR) and selective catalytic reduction (SCR) are used to control nitrogen oxides in tail gases.

To attain a level of 150 parts per million by volume (ppmv) of nitrogen oxides in the tail gases, the following approaches should be considered: High-pressure, single-pressure process with absorbing efficiency high enough to avoid additional abatement facilities; Dual-absorption process with an absorption efficiency high enough to avoid additional treatment facilities; Dual-pressure process with SCR; Medium-pressure, single-pressure process with SCR.

Urea Plants - In urea plants, wet scrubbers or fabric filters are used to control fugitive emissions from prilling towers; fabric filters are used to control dust emissions from bagging operations. These equipment are an integral part of the operations, to retain product. New urea plants should achieve levels of particulate matter in air emissions of less than 0.5 kg/t of product for both urea and ammonia.

Ammonium Sulfate Plants - In ammonium sulfate plants, use of fabric filters, with injection of absorbent as necessary, is the preferred means of control. Discharges of not more than 0.1 kg/t of product should be attainable for particulate matter.

Ammonium Nitrate Plants - In ammonium nitrate plants, wet scrubbers can be considered for prill towers and the granulation plant. Particulate emissions of 0.5 kg/t of product for the prill tower and 0.25 kg/t of product for granulation should be the target. Similar loads for ammonia are appropriate. Other effluents that originate in a nitrogenous fertilizer complex include boiler blowdown, water treatment plant backwash, and cooling tower blowdown from the ammonia and nitric acid plants. They may require pH adjustment and settling. These effluents should preferably be recycled or reused. Spent catalysts are usually sent for regeneration or disposed of in a secure landfill.

Air emissions should be monitored annually, except for nitrate acid plants, where nitrogen oxides should be monitored continuously. Effluents should be monitored continuously for pH and monthly for other parameters. Monitoring data should be analyzed and reviewed at regular intervals and compared with the operating standards so that any necessary corrective actions can be taken. Records of monitoring results should be kept in an acceptable format. The results should be reported to the responsible authorities and relevant parties, as required.

The key production and control practices that will lead to compliance with emissions requirements can be summarized as follows:

- Choose natural gas, where possible, as feedstock for the ammonia plant.
- Give preference to high-pressure processes or absorption process in combination with catalytic reduction units.
- Use low-dust-forming processes for solids formation.
- Reuse condensates and other waste waters.
- Maximize product recovery and minimize air emissions by appropriate maintenance and operation of scrubbers and baghouses.

Table 4 provides a summary of P2 recommended practices.

Table 4. *Recommended Pollution Prevention Practices in the Fertilizer Plants.*

Area of Opportunity	RECOMMENDED POLLUTION PREVENTION PRACTICE
Ammonia Plant	1. Where possible, use natural gas as the feedstock for the ammonia plant, to minimize air emissions. 2. Use hot process gas from the secondary reformer to heat the primary reformer tubes (the exchanger-reformer concept), thus reducing the need for natural gas. 3. Direct hydrogen cyanide (HCN) gas in a fuel oil gasification plant to a combustion unit to prevent its release. 4. Consider using purge gases from the synthesis process to fire the reformer; strip condensates to reduce ammonia and methanol. 5. Use carbon dioxide removal processes that do not release toxics to the environment. When monoethanolamine (MEA) or other processes, such as hot potassium carbonate, are used in carbon dioxide removal, proper operation and maintenance procedures should be followed to minimize releases to the environment.
Urea Plant	6. Use total recycle processes in the synthesis process; reduce microprill formation and carryover of fines in prilling towers.
Ammonium Nitrate Plant	7. *Prill tower*: reduce microprill formation and reduce carryover of fines through entrainment. 8. *Materials handling*: where feasible use covers and hoods on conveyors and transition points. Good cleanup practices must be in place to minimize contamination of stormwater runoff from the plant property. 9. *Granulators*: reduce dust emissions from the disintegration of granules.

Phosphate Fertilizer Plants

Industry Description and Practices

Phosphate fertilizers are produced by adding acid to ground or pulverized phosphate rock. If sulfuric acid is used, single or normal, phosphate (SSP) is produced, with a phosphorus content of 16-21% as phosphorous pentoxide (P_2O_5). If phosphoric acid is used to acidulate the phosphate rock, triple phosphate (TSP) is the result. TSP has a phosphorus content of 43-48% as P_2O_5.

SSP production involves mixing the sulfuric acid and the rock in a reactor.

The reaction mixture is discharged onto a slow-moving conveyor in a den. The mixture is cured for 4 to 6 weeks before bagging and shipping.

Two processes are used to produce TSP fertilizers: run-of-pile and granular. The run-of-pile process is similar to the SSP process. Granular TSP uses lower-strength phosphoric acid (40%, compared with 50% for run-of-pile). The reaction mixture, a slurry, is sprayed onto recycled fertilizer fines in a granulator. Granules grow and are then discharged to a dryer, screened, and sent to storage.

Phosphate fertilizer complexes often have sulfuric and phosphoric acid production facilities. Sulfuric acid is produced by burning molten sulfur in air to produce sulfur dioxide, which is then catalytically converted to sulfur trioxide for absorption in oleum. Sulfur dioxide can also be produced by roasting pyrite ore. Phosphoric acid is manufactured by adding sulfuric acid to phosphate rock. The reaction mixture is filtered to remove phosphogypsum, which is discharged to settling ponds or waste heaps.

Pollution Prevention Practices and Opportunities

Fluorides and dust are emitted to the air from the fertilizer plant. All aspects of phosphate rock processing and finished product handling generate dust, from grinders and pulverizers, pneumatic conveyors, and screens. The mixer/reactors and dens produce fumes that contain silicon tetrafluoride and hydrogen fluoride. Liquid effluents are not normally expected from the fertilizer plant, since it is feasible to operate the plant with a balanced process water system. The fertilizer plant should generate minimal solid wastes.

A sulfuric acid plant has two principal air emissions: sulfur dioxide and acid mist. If pyrites ore is roasted, there will also be particulates in air emissions that may contain heavy metals such as cadmium, mercury and lead. Sulfuric acid plants do not normally discharge liquid effluents except where appropriate water management measures are absent. Solid wastes from a sulfuric acid plant will normally be limited to spent vanadium catalyst. Where pyrite ore is roasted, there will be pyrite residue, which will require disposal. The residue may contain a wide range of heavy metals such as zinc, copper, lead, cadmium, mercury, and arsenic. The phosphoric acid plant generates dust and fumes, both of which contain hydrofluoric acid, silicon tetrafluoride, or both. Phosphogypsum generated in the process (at an approximate rate of about 5 tons per ton of phosphoric acid produced) is most often disposed of as a slurry to a storage/settling pond or waste heap. (Disposal to a marine environment is practiced at some existing phosphoric acid plants.) Process water used to transport the waste is returned to the plant after the solids have settled out. It is preferable to use a closed-loop operating system, where possible, to avoid a liquid effluent. In many climatic conditions, however, this is not possible, and an effluent is generated that contains phosphorus (as PO_4), fluorides, and suspended solids. The phosphogypsum contains trace metals, fluorides, and radionuclides (especially radon gas) that have been carried through from the phosphate rock.

In a fertilizer plant, the main source of potential pollution is solids from spills, operating upsets, and dust emissions. It is essential that tight operating procedures be in place and that close attention be paid to constant cleanup of spills and to other housecleaning measures. Product will be retained, the need for disposal of waste product will be controlled, and potential contamination of stormwater runoff from the property will be minimized. The discharge of sulfur dioxide from sulfuric acid plants should be minimized by using the double-contact, double-absorption process, with high efficiency mist eliminators. Spills and accidental discharges should be prevented by using well-bunded storage tanks, by installing spill catchment and containment facilities, and by practicing good housekeeping and maintenance. Residues from the roasting of pyrites may be used by the cement and steel manufacturing industries. In the phosphoric acid plant, emissions of fluorine compounds from the digester/reactor should be minimized by using well-designed, well-operated, and well-maintained scrubbers. Design for spill containment is essential for avoiding inadvertent liquid discharges. An operating water balance should be maintained to avoid an effluent discharge.

The management of phosphogypsum tailings is a major problem because of the large volumes and large area required and because of the potential for release of dust and radon gases and of fluorides and cadmium in seepage. The following measures will help to minimize the impacts:

- Maintain a water cover to reduce radon gas release and dust emissions.
- Where water cover cannot be maintained, keep the tailings wet or revegetate to reduce dust. (Note, however, that the revegetation process may increase the rate of radon emissions.)
- Line the tailings storage area to prevent contamination of groundwater by fluoride. Where contamination of groundwater is a concern, a management and monitoring plan should be implemented.
- Phosphogypsum may find a use in the production of gypsum board for the construction industry.

Implementation of cleaner production processes and pollution prevention measures can yield both economic and environmental benefits. The following production-related targets can be achieved by measures such as those described above. The numbers relate to the production processes before the addition of pollution control measures. In sulfuric acid plants that use the double-contact, double-absorption process, emissions levels of 2-4 kilograms of sulfur dioxide per metric ton (kg/t) of sulfuric acid can be achieved, and sulfur trioxide levels of the order of 0.15-0.2 kg/t of sulfuric acid are attainable. Scrubbers are used to remove fluorides and acid from air emissions. The effluent from the scrubbers is normally recycled to the process.

If it is not possible to maintain an operating water balance in the phosphoric acid plant, treatment to precipitate fluorine, phosphorus, and heavy metals may be necessary. Lime can be used for treatment. Spent vanadium catalyst is returned to the supplier for recovery, or, if that cannot be done, is locked in a solidification matrix and disposed of in a secure landfill. Opportunities to use gypsum wastes as a soil conditioner (for alkali soil and soils that are deficient in sulfur) should be explored to minimize the volume of the gypsum stack.

Pesticides Formulation

Industry Description and Practices

This subsection describes formulation of pesticides from active ingredients. Manufacture of pesticides merits an entirely separate discussion. The major chemical groups that are formulated include: Insecticides (organophosphates, carbamates, organochlorines, pyrethroids, biorationals, and botanicals); Fungicides (dithiocarbamates, triazoles, MBCs, morpholines, pyrimidines, phthalamides, and inorganics); Herbicides (triazines, carbamates, phenyl ureas, phenoxy acids, bipyridyls, glyphosates, sulfonyl ureas, amide xylenols, and imidazole inones); Rodenticides (coumarins).

The main purpose of pesticide formulation is to manufacture a product that has optimum biological efficiency, is convenient to use, and minimizes environmental impacts. The active ingredients are mixed with solvents, adjuvants (boosters), and fillers as necessary to achieve the desired formulation. The types of formulations include wettable powders, soluble concentrates, emulsion concentrates, oil-in-water emulsions, suspension concentrates, suspoemulsions, water-dispersible granules, dry granules, and controlled release, in which the active ingredient is released into the environment from a polymeric carrier, binder, absorbent, or encapsulant at a slow and effective rate. The formulation steps may generate air emissions, liquid effluents, and solid wastes.

Pollution Prevention Practices and Opportunities

The principal air pollutants are particulate matter (PM) and volatile organic compounds (VOCs). These are released from mixing and coating operations. Most liquid effluents result from spills, the cleaning of equipment, and process waste waters. The effluents may contain toxic organics, including pesticide residues. Major solid wastes of concern include contaminated discarded packaging and process residues. There will also be effluent treatment sludges. The solid wastes generated depend on the process. They can amount to about 3.3 grams per kilogram (g/kg) of product and may contain 40% active ingredient. The recommended pollution prevention measures are as follows:

- Use equipment washdown waters as makeup solutions for subsequent batches.
- Use dedicated dust collectors to recycle recovered materials.
- Use suction hoods to collect vapors and other fugitive emissions.
- Return toxic materials packaging to the supplier for reuse.
- Find productive uses for off-specification products to avoid disposal problems.
- Minimize raw material and product inventory to avoid degradation/wastage.
- Label and store toxic and hazardous materials in secure, bunded areas.

A pesticide formulation plant should prepare and implement an emergency preparedness and response plan that takes into account neighboring land uses and the potential consequences of an emergency or accidental release of harmful substances. Measures to avoid the release of harmful substances should be incorporated in the design, operation, maintenance, and management of the plant. Pollution control equipment employed in this sector include baghouses for removal of particulate matter and carbon adsorption for removal of VOCs. Reverse osmosis or ultrafiltration is used to recover process materials from wastewater. Effluent treatment may include carbon adsorption, detoxification of pesticides by oxidation (using ultraviolet systems or peroxide solutions), and biological treatment. Exhausted carbon from absorption processes may be sent for regeneration or combustion. Due to the relatively small volumes of solid wastes, it is difficult to find acceptable and affordable methods of disposal.

Ideally, solid wastes should be incinerated in a facility where combustion conditions such as 1,100° C and at least 0.5 second flame residence time are maintained, to ensure effective destruction of toxics. Toxic solid wastes should be treated to destroy toxic organics and bring them to levels below 0.05 milligrams per kilogram (mg/kg). It is recommended to continuously monitor air emissions exiting the air pollution control system where toxic organics are being emitted at rates greater than 0.5 kilograms per hour (kg/h). It is also advisable to analyze liquid effluents generated from the process before discharge (or at least once per shift). Where the effluents are suspected to be toxic, a bioassay test should be performed to assess their acceptability in the environment. The toxicity factor for fish should not be greater than 2; toxicity to Daphnia = 8; toxicity to algae = 16; and toxicity to bacteria = 8. Monitoring data should be analyzed and reviewed at regular intervals and compared with the operating standards so that any necessary corrective actions can be taken. Records of monitoring results should be kept in an acceptable format. The results should be reported to the responsible authorities and relevant parties, as required.

Pesticides Manufacturing

Industry Description and Practices

The following discussion deals with the synthesis of the active ingredients used in pesticide formulations. The major chemical groups manufactured include:

- Carbamates and dithiocarbamates (carbofuran, carbaryl, ziram, and benthiocarb)
- Chlorophenoxy compounds (2,4-D, 2,4,5-T, and silvex)
- Organochlorines (dicofol and endosulfan)
- Organophosphorus compounds (malathion, dimethoate, phorate, and parathion methyl)
- Nitro compounds (trifluralin)
- Miscellaneous compounds such as biopesticides (for example, *Bacillus thuringiensis* and pherhormones), heterocycles (for example, atrazine), pyrethroids (for example, cypermethrin), and urea derivatives (for example, diuron).

Special attention must be given to restricted substances. Production proposals for the following pesticides should be carefully evaluated: hexachlorobenzene, toxaphene, chlordane, aldrin, DDT, mirex, dieldrin, endrin, and heptachlor. The principal manufacturing steps are (a) preparation of process intermediates; (b) introduction of functional groups; (c) coupling and esterification; (d) separation processes, such as washing and stripping; and (e) purification of the final product. Each of these steps may generate air emissions, liquid effluents, and solid wastes.

Pollution Prevention Practices and Opportunities

The principal air pollutants are VOCs and PM. Liquid effluents resulting from equipment cleaning after batch operation contain toxic organics and pesticide residues. Cooling waters are normally recirculated. Typical wastewater concentrations are: chemical oxygen demand (COD), 13,000 milligrams per liter (mg/l), with a range of 0.4-73,000 mg/l; oil and grease, 800 mg/l, (with a range of 1-13,000 mg/l; total suspended solids, 2,800 mg/l, with a range of 4-43,000 mg/l. Major solid wastes of concern include process and effluent treatment sludges, spent catalysts, and container residues. Approximately 200 kilograms (kg) of waste is generated per metric ton of active ingredient manufactured. Every effort should be made to replace highly toxic and persistent ingredients with degradable and less toxic ones. Recommended pollution prevention measures include the following:

- Meter and control the quantities of active ingredients to minimize wastage.
- Reuse by-products from the process as raw materials or as raw material substitutes in other processes.
- Use automated filling to minimize spillage.
- Use "closed" feed systems for batch reactors.
- Use nitrogen blanketing where appropriate on pumps, storage tanks, and other equipment to minimize the release of toxic organics.
- Give preference to nonhalogenated and nonaromatic solvents where feasible.
- Use high-pressure hoses for equipment cleaning to reduce wastewater.
- Use equipment washdown waters and other process waters (such as leakages from pump seals) as makeup solutions for subsequent batches.
- Use dedicated dust collectors to recycle recovered materials.
- Vent equipment through a recovery system.

- Maintain losses from vacuum pumps (such as water ring and dry) at low levels.
- Return toxic materials packaging to the supplier for reuse or incinerate/ destroy in an environmentally acceptable manner.
- Minimize storage time of off-specification products through regular reprocessing.
- Find productive uses for off-specification products to avoid disposal problems.
- Minimize raw material and product inventory to avoid degradation and wastage that could lead to the formation of inactive but toxic by-products.
- Label and store toxic and hazardous materials in secure, bunded areas.

A pesticide manufacturing plant should prepare a hazard assessment and operability study and also prepare and implement an emergency preparedness and response plan that takes into account neighboring land use and the potential consequences of an emergency. Measures to avoid the release of harmful substances should be incorporated in the design, operation, maintenance, and management of the plant. Implementation of cleaner production processes and pollution prevention measures can yield both economic and environmental benefits. Specific reduction targets for the different processes are not well established internationally. In the absence of specific pollution reduction targets, new plants should always achieve better than the industry averages and should approach the load-based effluent levels.

Typical pollution control equipment employed by the industry include the following. For air emissions - stack gas scrubbing and/or carbon adsorption (for toxic organics) and baghouses (for particulate matter removal) are applicable and effective technologies for minimizing the release of significant pollutants to air. Combustion is used to destroy toxic organics.

Combustion devices should be operated at temperatures above 1,100° C with a flame residence time of at least 0.5 second to achieve acceptable destruction efficiency of toxics. However, temperatures of around 900° C are acceptable provided that at least 99.99% destruction/ removal efficiency of toxics is achieved.

For liquid effluents and solid wastes - reverse osmosis or ultrafiltration is used to recover and concentrate active ingredients. Effluent treatment normally includes flocculation, coagulation, settling, carbon adsorption, detoxification of pesticides by oxidation (using ultraviolet systems or peroxide solutions), and biological treatment. Exhausted carbon from absorption processes may be sent for regeneration or combustion. When the wastewater volumes are small and an onsite incinerator is appropriate, combustion of toxic waste waters may be feasible.

Contaminated solid wastes are generally incinerated, and the flue gases are scrubbed. The emissions levels cited in Table 5 are those recommended by the World bank Organization that should be achieved. Recommended effluent levels that should be achieved are reported in Table 6. Bioassay testing should be performed to ensure that the toxicity of the effluent is acceptable (toxicity to fish = 2; toxicity to *Daphnia* =8; toxicity to algae = 16; and toxicity to bacteria = 8). Effluent requirements are for direct discharge to surface waters. Also, a BOD test should be performed only in cases where the effluent does not contain any substance toxic to the microorganisms used in the test.

Table 5. *Emissions from Pesticides Manufacturing.*

PARAMETER	MAXIMUM LEVEL milligrams per normal cubic meter
PM	20; 5 where very toxic compounds are present[a]
VOCs	20
Chlorine (or chloride)	5

a. See the World Health Organization's list of extremely hazardous substances (WHO 1996).

Table 6. *Effluents from Pesticides Manufacturing.*

PARAMETER	MAXIMUM VALUE milligrams per liter, except for pH
pH	6-9
BOD	30
COD	150
AOX	1
TSS	10
Oil and grease	10
Phenol	0.5
Arsenic	0.1
Chromium (hexavalent)	0.1
Copper	0.5
Mercury	0.01
Active ingredient (each)	0.05

Contaminated solid wastes should be treated to achieve toxic organic levels of no more than 0.05 milligrams per kilogram.

Monitoring of air emissions should be done on a continuous basis when the mass flow of toxic substances exceeds 0.5 kg per hour. Otherwise, it can be done annually. Liquid effluents should be monitored for active ingredients at least once every shift. The remaining parameters should be monitored at least daily. Monitoring data should be analyzed and reviewed at regular intervals and compared with the operating standards so that any necessary corrective actions can be taken. Records of monitoring results should be kept in an acceptable format. The results should be reported to the responsible authorities and relevant parties, as required.

To summarize, the primary P2 practices that the industry should adhere to include:

- Replace highly toxic and persistent ingredients with less toxic, degradable ones.
- Control loss and wastage of active ingredients.
- Return packaging for refilling.
- Use equipment washdown waters as makeup solutions for subsequent batches.
- Minimize wastage by inventory control and find uses for off-specification products.

References 6 through 14 provide detailed information on both pesticide manufacturing operations, formulations and standard pollution control practices.

COKE MANUFACTURING

Industry Description and Practices

Coke and coke by-products, including coke oven gas, are produced by the pyrolysis (heating in the absence of air) of suitable grades of coal. The process also includes the processing of coke oven gas to remove tar, ammonia (usually recovered as ammonium sulfate), phenol, naphthalene, light oil, and sulfur before the gas is used as fuel for heating the ovens. This section provides an overview of the production of metallurgical coke and the associated by-products using intermittent horizontal retorts, as well as the pollution prevention practices.

The coking industry is a heavy polluting industry, which for the most part has been eliminated in the United States, but is still very much alive in many parts of the world. Figure 1 shows a photograph of a large operation in Ukraine. In countries like Ukraine and Russia, coke serves as a merchant industry to the iron and steel sector.

In the coke-making process, bituminous coal is fed (usually after processing operations to control the size and quality of the feed) into a series of ovens, which are sealed and heated at high temperatures in the absence of oxygen, typically in cycles lasting 14 to 36 hours. Volatile compounds that are driven off the coal are collected and processed to recover combustible gases and other by-products. The solid carbon remaining in the oven is coke.

The coke is taken to the quench tower, where it is cooled with a water spray or by circulating an inert gas (nitrogen), a process known as dry quenching. The coke is screened and sent to a blast furnace or to storage. Coke oven gas is cooled, and by-products are recovered. Flushing liquor, formed from the cooling of coke oven gas, and liquor from primary coolers contain tar and are sent to a tar decanter. Note that the coke oven gas has a heating value and can be used effectively in cogeneration type projects.

Figure 1. *Avdeyevka coke-chemical plant in Donetsk Ukraine.*

An electrostatic precipitator is used to remove more tar from coke oven gas. The tar is then sent to storage. Ammonia liquor is also separated from the tar decanter and sent to wastewater treatment after ammonia recovery. Coke oven gas is further cooled in a final cooler. Naphthalene is removed in the separator on the final cooler. Light oil is then removed from the coke oven gas and is fractionated to recover benzene, toluene, and xylene. Some facilities may include an onsite tar distillation unit. The Claus process is normally used to recover sulfur from coke oven gas. During the coke quenching, handling, and screening operation, coke breeze is produced. It is either reused on site (e.g., in the sinter plant) or sold off site as a by-product.

The coke oven is a major source of fugitive air emissions. The coking process emits particulate matter (PM); volatile organic compounds (VOCs); polynuclear aromatic hydrocarbons (PAHs); methane, at approximately 100 grams per metric ton (g/t) of coke; ammonia; carbon monoxide; hydrogen sulfide (50-80 g/t of coke from pushing operations); hydrogen cyanide; and sulfur oxides, SO_x, (releasing 30% of sulfur in the feed). Significant amount of VOCs may also be released in by-product recovery operations. For every ton of coke produced, approximately 0.7 to 7.4 kilograms (kg) of PM, 2.9 kg of SO_x, (ranging from 0.2 to 6.5 kg), 1.4 kg of nitrogen oxides NO_x, 0.1 kg of ammonia, and 3 kg of VOCs (including 2 kg of benzene) may be released into the atmosphere if there is no vapor recovery system. Coal-handling operations may account for about 10% of the particulate load. Coal charging, coke pushing, and quenching are major sources of dust emissions.

Wastewater is generated at an average rate ranging from 0.3 to 4 cubic meters (m^3) per ton of coke processed. Major wastewater streams are generated from the cooling of the coke oven gas and the processing of ammonia, tar, naphthalene, phenol, and light oil. Process wastewater may contain 10 milligrams per liter (mg/l) of benzene, 1,000 mg/l of biochemical oxygen demand (BOD) (4 kg/t of coke), 1,500-6,000 mg/l of chemical oxygen demand (COD), 200 mg/l of total suspended solids (TSS), and 150-2,000 mg/l of phenols (0.3-12 kg/t of coke). Wastewaters also contain PAHs at significant concentrations (up to 30 mg/l), ammonia (0.1-2 kg nitrogen/t of coke), and cyanides (0.1-0.6 kg/t of coke). Coke production facilities generate process solid wastes other than coke breeze (which averages 1 kg/t of product).

Most of the solid wastes contain hazardous components such as benzene and PAHs. Waste streams of concern include residues from coal tar recovery (typically 0.1 kg/t of coke), the tar decanter (0.2 kg/t of coke), tar storage (0.4 kg/t of coke), light oil processing (0.2 kg/t of coke), wastewater treatment (0.1 kg/t of coke), naphthalene collection and recovery (0.02 kg/t of coke), tar distillation (0.01 kg/t of coke), and sludges from biological treatment of wastewater.

Pollution Prevention Practices and Opportunities

Pollution prevention in coke making is focused on reducing coke oven emissions and developing cokeless iron- and steel-making techniques. Table 7 provides a list of pollution prevention and control measures that should be considered.

Implementation of cleaner production processes and pollution prevention measures can yield both economic and environmental benefits. By way of some general guidelines; the generation rate for wastewater should be less than 0.3 m^3/t of coke. New coke plants should not generate more than 1 kg of process solid waste (excluding coke breeze and biosludges) per ton of coke.

With regard to air emissions, the World Bank Organization recommends that benzene emissions should not be more than 5 milligrams per normal cubic meter (mg/Nm^3) in leaks from light oil processing, final cooler, tar decanter, tar storage, weak ammonia liquor storage, and the tar/water separator. VOC emissions should be less than 20 mg/Nm^3. Particulate matter emissions from the stacks should not exceed 50 mg/Nm^3. Sulfur recovery from coke oven gas should be at least 97% but preferably over 99%. Air emission control technologies include scrubbers (removal efficiency of 90%) and baghouses and electrostatic precipitators (ESPs), with removal efficiencies of 99.9%.

Baghouses are preferred over venturi scrubbers for controlling particulate matter emissions from loading and pushing operations because of the higher removal efficiencies. ESPs are effective for final tar removal from coke oven gas. Stack air emissions should be monitored continuously for particulate matter. Alternatively, opacity measurements of stack gases could suffice. Fugitive emissions should be monitored annually for VOCs.

Wastewater treatment systems include screens and settling tanks to remove total suspended solids, oil, and tar; steam stripping to remove ammonia, hydrogen sulfide, and hydrogen cyanide; biological treatment; and final polishing with filters. Wastewater discharges should be monitored daily for flow rate and for all parameters, except for dibenz(a,h)anthracene and benzo(a)pyrene. The latter should be monitored at least on a monthly basis or when there are process changes. Frequent sampling may be required during start-up and upset conditions.

Table 7. *Recommended Pollution Prevention Practices in the Coke Industry.*

Area of Opportunity	RECOMMENDED POLLUTION PREVENTION PRACTICE
General	1. Use cokeless iron- and steel-making processes, such as the direct reduction process, to eliminate the need to manufacture coke. 2. Use beneficiation (preferably at the coal mine) and blending processes that improve the quality of coal feed to produce coke of desired quality and reduce emissions of sulfur oxides and other pollutants. 3. Use enclosed conveyors and sieves for coal and coke handling. Use sprinklers and plastic emulsions to suppress dust formation. Provide windbreaks where feasible. Store materials in bunkers or warehouses. Reduce drop distances. 4. Use and preheat high-grade coal to reduce coking time, increase throughput, reduce fuel consumption, and minimize thermal shock to refractory bricks.
Coke Oven Emissions	5. *Charging:* dust particles from coal charging should be evacuated by the use of jumper-pipe systems and steam injection into the ascension pipe or controlled by fabric filters 6. *Coking:* use large ovens to increase batch size and reduce the number of chargings and pushings, thereby reducing the associated emissions. Reduce fluctuations in coking conditions, including temperature. Clean and seal coke oven openings to minimize emissions. Use mechanical cleaning devices (preferably automatic) for cleaning doors, door frames, and hole lids. Seal lids, using a slurry. Use lowleakage door construction, preferably with gas sealings. 7. *Pushing:* emissions from coke pushing can be reduced by maintaining a sufficient coking time, thus avoiding "green push." Use sheds and enclosed cars, or consider use of traveling hoods. The gases released should be removed and passed through fabric filters. 8. *Quenching:* where feasible, use dry instead of wet quenching. Filter all gases extracted from the dry quenching unit. If wet quenching, is used, provide interceptors (baffles) to remove coarse dust. When wastewater is used for quenching, the process transfers pollutants from the wastewater to the air, requiring subsequent removal. Reuse quench water. 9. *Conveying and sieving:* enclose potential dust sources, and filter evacuated gases.

Area of Opportunity	RECOMMENDED POLLUTION PREVENTION PRACTICE
By-Product Recovery	10. Use vapor recovery systems to prevent air emissions from light oil processing, tar processing, naphthalene processing, and phenol and ammonia recovery processes.
	11. Segregate process water from cooling water.
	12. Reduce fixed ammonia content in ammonia liquor by using caustic soda and steam stripping.
	13. Recycle all process solid wastes, including tar decanter sludge to coke oven.
	14. Recover sulfur from coke oven gas. Recycle Claus tail gas into coke oven gas system.

All process hazardous wastes except for coke fines should be recycled to coke ovens. Wastewater treatment sludges should be dewatered. If toxic organics are detectable, dewatered sludges are to be charged to coke ovens or disposed in a secure landfill or an appropriate combustion unit.

Solid hazardous wastes containing toxic organics should be recycled to a coke oven or treated in a combustion unit, with residues disposed of in a secure landfill. In summary, the key production and control practices that will lead to compliance with emissions guidelines can be summarized as follows:

- Use cokeless iron- and steel-making processes, such as the direct reduction process for iron-making, to eliminate the need for coke manufacturing.
- Where feasible, use dry quenching instead of wet quenching.
- Use vapor-recovery systems in light oil processing, tar processing and storage, naphthalene processing, and phenol and ammonia recovery operations.
- Recover sulfur from coke oven gas.
- Segregate process and cooling water.
- Recycle process solid wastes to the coke oven.

References 15 through 20 have been consulted in developing the above discussions and recommendations, and are recommended for further information.

DYE MANUFACTURING

Industry Description and Practices

This section provides an overview of the synthesis of dyes and pigments used in textiles and related industries. An understanding of these operations helps to identify pollution prevention opportunities within the industry sector. Dyes are soluble at some stage of the application process, whereas pigments, in general, retain essentially their particulate or crystalline form during application. A dye is used to impart color to materials of which it becomes an integral part. An aromatic ring structure coupled with a side chain is usually required for resonance and thus to impart color. Resonance structures cause displacement or appearance of absorption bands in the visible spectrum of light, and hence they are responsible for color. Correlation of chemical structure with color has been accomplished in the synthesis of dye using a chromogen-chromophore with auxochrome. Chromogen is the aromatic structure containing benzene, naphthalene, or anthracene rings. A chromophore group is a color giver or donor and is represented by the following radicals, which form a basis for the chemical classification of dyes when coupled with the chromogen: azo (-N=N-); carbonyl (=C=O); carbon (=C=C=); carbon-nitrogen (>C=NH or -CH=N-); nitroso (-NO or N-OH); nitro (-NO_2 or=NO-OH); and sulfur (>C=S, and other carbon-sulfur groups). The chromogen-chromophore structure is often not sufficient to impart solubility and cause adherence of dye to fiber. The auxochrome or bonding affinity groups are amine, hydroxyl, carboxyl, and sulfonic radicals, or their derivatives. These auxochromes are important in the use classification of dyes. A listing of dyes by use classification comprises the following:

- *Acetate rayon dyes:* developed for cellulose acetate and some synthetic fibers.
- *Acid dyes:* used for coloring animal fibers via acidified solution (containing sulfuric acid, acetic acid, sodium sulfate, and surfactants) in combination with amphoteric protein.
- *Azoic dyes:* contain the azo group (and formic acid, caustic soda, metallic compounds, and sodium nitrate); especially for application to cotton.
- *Basic dyes:* amino derivatives (and acetic acid and softening agents); used mainly for application on paper.
- *Direct dyes:* azo dyes, and sodium salts, fixing agents, and metallic (chrome

and copper) compounds; used generally on cotton-wool, or cotton-silk combinations.

- *Mordant or chrome dyes:* metallic salt or lake formed directly on the fiber by the use of aluminum, chromium, or iron salts that cause precipitation in situ.
- *Lake or pigment dyes:* form insoluble compounds with aluminum, barium, or chromium on molybdenum salts; the precipitates are ground to form pigments used in paint and inks.
- *Sulfur or sulfide dyes:* contain sulfur or are precipitated from sodium sulfide bath; furnish dull shades with good fastness to light, washing, and acids but susceptible to chlorine and light.
- *Vat dyes:* impregnated into fiber under reducing conditions and reoxidized to an insoluble color.

Chemical classification is based on chromogen. For example, nitro dyes have the chromophore $-NO_2$. The *Color Index (C.I.),* published by the Society of Dyers and Colourists (United Kingdom) in cooperation with the American Association of Textile Chemists and Colorists (AATC), provides a detailed classification of commercial dyes and pigments by generic name and chemical constitution.

Dyes are synthesized in a reactor, then filtered, dried, and blended with other additives to produce the final product. The synthesis step involves reactions such as sulfonation, halogenation, amination, diazotization, and coupling, followed by separation processes that may include distillation, precipitation, and crystallization.

In general, organic compounds such as naphthalene are reacted with an acid or an alkali along with an intermediate (such as a nitrating or a sulfonating compound) and a solvent to form a dye mixture. The dye is then separated from the mixture and purified. On completion of the manufacture of actual color, finishing operations, including drying, grinding, and standardization, are performed; these are important for maintaining consistent product quality.

Pollution Prevention Practices and Opportunities

The principal air pollutants from dye manufacturing are volatile organic compounds (VOCs), nitrogen oxides NO_x hydrogen chloride (HCl), and sulfur oxides SO_x. Liquid effluents resulting from equipment cleaning after batch operation can contain toxic organic residues. Cooling waters are normally recirculated. Wastewater generation rates are of the order of 1-700 liters per kg (l/kg) of product except for vat dyes. The wastewater generation rate for vat dyes can be of the order of 8,000 (ℓ/kg) of product. Biochemical oxygen demand (BOD) and chemical oxygen demand (COD) levels of reactive and azo dyes can be of the order of 25 kg/kg of product and 80 kg/kg of product, respectively. Values for other dyes are, for example, BOD, 6 kg/kg; COD, 25 kg/kg; suspending solids, 6 kg/kg; and oil and grease, 30 kg/kg of product.

Major solid wastes of concern include filtration sludges, process and effluent treatment sludges, and container residues. Examples of wastes considered toxic include wastewater treatment sludges, spent acids, and process residues from the manufacture of chrome yellow and orange pigments, molybdate orange pigments, zinc yellow pigments, chrome and chrome oxide green pigments, iron blue pigments, and azo dyes.

Dedicated effort should be made to substitute degradable and less toxic ingredients for highly toxic and persistent ingredients in this industry sector. Recommended pollution prevention measures include:

- Avoid the manufacture of toxic azo dyes and provide alternative dyestuffs to users such as textile manufacturers.
- Meter and control the quantities of toxic ingredients to minimize wastage.
- Reuse by-products from the process as raw materials or as raw material substitutes in other processes.
- Use automated filling to minimize spillage.
- Use equipment washdown waters as makeup solutions for subsequent batches.
- Return toxic materials packaging to suppliers for reuse, where feasible.
- Find productive uses for off-specification products to avoid disposal problems.
- Use high-pressure hoses for equipment cleaning to reduce the amount of wastewater generated.
- Label and store toxic and hazardous materials in secure areas.

A dye and pigment manufacturing plant should prepare and implement an emergency response plan that takes into account neighboring land uses and the potential consequences of an emergency such as a spill or fire. Measures to avoid the release of harmful substances should be incorporated in the design, operation, maintenance, and management of the plant.

Implementation of cleaner production processes and pollution prevention measures can yield both economic and environmental benefits. Specific reduction targets for the different processes have not been well established or even defined. In the absence of specific pollution reduction targets, new plants should always achieve better than the industry averages as cited in the general waste characteristics above. The following is a brief description of the standard treatment technologies applied in this industry sector:

Air Emissions: Stack gas scrubbing and/or carbon adsorption (for toxic organics) are applicable and effective technologies for minimizing the release of significant pollutants to air. Combustion is used to destroy toxic organic compounds. Combustion devices should be operated at temperatures above 1,100° C (when required for the effective destruction of toxic organics), with a residence time of at least 0.5 second.

Liquid Effluents: Effluent treatment normally includes neutralization, flocculation, coagulation, settling, carbon adsorption, detoxification of organics by oxidation (using ultraviolet systems or peroxide solutions), and biological treatment. Exhausted carbon from adsorption processes may be sent for regeneration or combustion. Reverse osmosis, ultrafiltration, and other filtration techniques are used to recover and concentrate process intermediates.

Solid Hazardous Wastes: Contaminated solid wastes are generally incinerated, and the flue gases, when acidic wastes are scrubbed. Contaminated solid wastes should be incinerated under controlled conditions to reduce toxic organics to nondetectable levels, in no case exceeding 0.05 mg/kg or the health-based level. Emissions levels for the design and operation of each project must be established based upon national and local emissions standards. Use Table 8 as a general.

The guidelines given present emissions levels normally acceptable to the World Bank Group and other international lending institutions (e.g., U.S. Export Import Bank) in making decisions regarding provisions or contingencies for loans. The emissions levels given here can be consistently achieved by well-designed, well-operated, and well-maintained pollution control systems. The guidelines are expressed as concentrations to facilitate monitoring. Of course the goal would be to try an eliminate the use of pollution controls (i.e., end-of-pipe treatment technologies). Pollution control devices can represent significant O&M costs, as well as sizable capital investments. These are the direct cost items discussed earlier in Chapter 3.

Table 8. *Emissions and Effluent Guidelines Established by the World Bank.*

Gas Emissions (Units in milligrams per normal cubic meter)	
Parameter	*Maximum value*
Chlorine (or chloride)	10
VOCs	20
Effluents (Units in milligrams per liter, except for pH)	
pH	*6-9*
BOD	30
COD	150
TSS	50
Oil and grease	10
Phenol	0.5
Chromium (hexavalent)	0.1
Copper	0.5
Zinc	2
AOX	1
Toxic organics such as benzidine (each)	0.05

Note: Effluent requirements are for direct discharge to surface waters.

Dilution of air emissions or effluents to achieve these guidelines is considered unacceptable by the World Bank, the U.S. Export-Import Bank, and other international lending organizations. All of the maximum levels should be achieved for at least 95% of the time that the plant or unit is operating, to be calculated as a proportion of annual operating hours.

Because of the toxicity of many of the pollutants from this industry sector, it is recommended that monitoring of air emissions be done on a continuous basis. Liquid effluents should be monitored for toxic ingredients at least once every shift. The remaining parameters should be monitored at least daily. Monitoring data should be analyzed and reviewed at regular intervals and compared with the operating standards so that any necessary corrective actions can be taken. Records of monitoring results should be kept in an acceptable format. The results should be reported to the responsible authorities and relevant parties, as required. The above discussions were summarized from references 21 and 22, which the reader can consult for additional information.

PHARMACEUTICALS MANUFACTURING

Industry Description and Practices

The pharmaceutical industry includes the manufacture, extraction, processing, purification, and packaging of chemical materials to be used as medications for humans or animals. Pharmaceutical manufacturing is divided into two major stages: the production of the active ingredient or drug (primary processing, or manufacture) and secondary processing, the conversion of the active drugs into products suitable for administration. This section briefly deals with the synthesis of the active ingredients and their usage in drug formulations to deliver the prescribed dosage. Formulation is also referred to as galenical production. The main pharmaceutical groups manufactured include:

- Proprietary ethical products or prescription only medicines (POM), which are usually patented products.
- General ethical products, which are basically standard prescription-only medicines made to a recognized formula that may be specified in standard industry reference books.
- Over-the counter (OTC), or nonprescription, products.

The products are available as tablets, capsules, liquids (in the form of solutions, suspensions, emulsions, gels, or injectables), creams (usually oil-in-water emulsions), ointments (usually water-in-oil emulsions), and aerosols, which contain inhalable products or products suitable for external use. Propellants used in aerosols include chlorofluorocarbons (CFCs), which are being phased out. Recently, butane has been used as a propellant in externally applied products. The major manufactured groups include:

- Antibiotics such as penicillin, streptomycin, tetracyclines, chloramphenicol, and antifungals;
- Other synthetic drugs, including sulfa drugs, antituberculosis drugs, antileprotic drugs, analgesics, anesthetics, and antimalarials;
- Vitamins;
- Synthetic hormones;

- Glandular products drugs of vegetable origin, such as quinine, strychnine and brucine, emetine, and digitalis glycosides;
- Vaccines and sera;
- Other pharmaceutical chemicals such as calcium gluconate, ferrous salts, nikethamide, glycerophosphates, chloral hydrate, saccharin, antihistamines (including meclozine, and buclozine), tranquilizers (including meprobarnate and chloropromoazine), antifilarials, diethyl carbamazine citrate, and oral antidiabetics, including tolbutamide and chloropropamide;
- Surgical sutures and dressings.

The principal manufacturing steps are

(a) preparation of process intermediates;

(b) introduction of functional groups;

(c) coupling and esterification;

(d) separation processes such as washing and stripping; and

(e) purification of the final product.

Additional product preparation steps include granulation; drying; tablet pressing, printing, and coating; filling; and packaging. Each of these steps may generate air emissions, liquid effluents, and solid wastes.

The manufacture of penicillin, for example, involves batch fermentation-using 100-200 m^3 batches-of maize steep liquor or a similar base, with organic precursors added to control the yield. Specific mold culture such as *Penicillium chrysogenum* for Type 11 is inoculated into the fermentation medium. Penicillin is separated from the fermentation broth by solvent extraction.

The product is further purified using acidic extraction. This is followed by treatment with a pyrogen-free distilled water solution containing the alkaline salt of the desired element. The purified aqueous concentrate is separated from the solvent in a supercentrifuge and pressurized through a biological filter to remove the final traces of bacteria and pyrogens.

The solution can be concentrated by freeze drying or vacuum spray drying. Oil-soluble procaine penicillin is made by reacting a penicillin concentrate (20-30%) with a 50% aqueous solution of procaine hydrochloride. Procaine penicillin crystallizes from this mixture.

Pollution Prevention Practices and Opportunities

In some countries, the manufacture of pharmaceuticals is controlled by Good Management Practices (GMP). Some countries require an environmental assessment (EA) report addressing the fate and toxicity of drugs and their metabolized by-products. The EA data relate to the parent drug, not to all metabolites, and include:

(a) physical and chemical properties;

(b) biodegradability;

(c) photolysis propensity;

(d) aqueous toxicity to fish;

(e) prediction of existing or planned treatment plant to treat wastes and wastewaters; and

(f) treatment sequences that are capable of treating wastes and waste waters. The principal air pollutants are volatile organic compounds (VOCs) and particulate matter (PM). Liquid effluents resulting from equipment cleaning after batch operation contain toxic organic residues. Their composition varies, depending on the product manufactured, the materials used in the process, and other process details. Cooling waters are normally recirculated. Some wastewaters may contain mercury, in a range of 0.1-4 milligrams per liter (mg/l), cadmium (10-600 mg/l), isomers of hexachlorocyclohexane, 1,2-dichloroethane, and solvents. Typical amounts released with the wastewater are 25 kilograms of biochemical oxygen demand (BOD) per metric ton of product (kg/t), or 2,000 mg/l; 50 kg/t chemical oxygen demand (COD), or 4,000 mg/l; 3 kg/t of suspended solids; and up to 0.8 kg/t of phenol. The principal solid wastes of concern include process and effluent treatment sludges, spent catalysts, and container residues. Approximately 200 kg wastes per ton of product of waste are generated. Some solid wastes contain significant concentrations of spent solvents and other toxic organics. Every effort should be made to replace highly toxic and persistent ingredients with degradable and less toxic ones.

Recommended pollution prevention measures are outlined in Table 9. Where appropriate, a pharmaceutical manufacturing plant should prepare a hazard assessment and operability study and also prepare and implement an emergency plan that takes into account neighboring land uses and the potential consequences of an emergency. Measures to avoid the release of harmful substances should be incorporated in the design, operation, maintenance, and management of the plant.

Table 9. *Recommended P2 Practices.*

1.	Meter and control the quantities of active ingredients to minimize wastage.
2.	Reuse by-products from the process as raw materials or as raw material substitutes in other processes.
3.	Recover solvents used in the process by distillation or other methods.
4.	Give preference to the use of nonhalogenated solvents.
5.	Use automated filling to minimize spillage.
6.	Use "closed" feed systems into batch reactors. Use equipment washdown waters and other process waters (such as leakages from pump seals) as makeup solutions for subsequent batches.
7.	Recirculate cooling water.
8.	Use dedicated dust collectors to recycle recovered materials and prevent cross-contamination.
9.	Vent equipment through a vapor recovery system.
10.	Use loss-free vacuum pumps.
11.	Return toxic materials packaging to the supplier for reuse, or incinerate/destroy it in an environmentally acceptable manner.
12.	Minimize storage time of off-specification products through regular reprocessing.
13.	Find productive uses for off-specification products to avoid disposal problems.
14.	Use high-pressure hoses for equipment cleaning to reduce wastewater.
15.	Provide stormwater drainage and avoid contamination of stormwater from process areas.
16.	Label and store toxic and hazardous materials in secure, bunded areas. Spillage should be collected and reused.
17.	Minimize raw material and product inventory to avoid degradation and wastage.

This now leads us to some pollution prevention case studies. P2 case studies are worthwhile reviewing because they can oftentimes present common industry solutions, not only within the same industry sector, but between different types of operations. It is important to recognize that many unit operations and intermediate steps in manufacturing are common.

P2 CASE STUDIES

Low-Cost Pollution Prevention at a PVC Plant

OHIS is the abbreviated name for Organsko Hemiska Industrija Skopje, which is a joint stock company in the Republic of Macedonia. The plant was placed into operation in 1964. OHIS manufactures basic chemicals which include caustic soda and chlorine, herbicides, detergents, various synthetic fibers, PVA, PVC. Vinyl chloride monomer (VCM) is the precursor to PVC, one of the most widely used commodity plastics in the world. PVC resin is manufactured by a suspension process. The suspension polymerization is performed in reactors with conversion rates typically between 89 to 91% of the monomer (VCM). The polymerization reaction is terminated by injecting an inhibitor, whereupon unconverted VCM is removed from the polymerization reactors by heating up the polymer slurry and applying vacuum technology to remove unreacted VCM vapors. The polymer slurry is removed from the reactors and sent to slurry tanks, and from there, the slurry resin product goes on to polymer drying and finishing stages to produce a final resin product. The PVC plant is comprised of four single agitator reactors (CSTRs). Residual or unconverted VCM is also removed from the polymer slurry prior to sending the resin to the finishing stages of the operation. This is traditionally accomplished by the use of a series of stripper columns.

There are significant quantities of unreacted VCM that are in the form of fugitive emissions. Figure 2 shows a photograph of the monomer feed tank and vacuum system to the reactors, which in itself is a significant source of fugitive emissions. VCM is a highly toxic chemical that is a suspect human carcinogen (i.e., exposure has been linked to liver cancer - angiosarcoma). In addition, long-term exposure to low levels of VCM (i.e., chronic exposure) has been linked to several other occupational diseases including a bone condition (acro-osteolysisli), a condition leading to a narrowing of the blood vessels (Raynaud's phenomenon), and hardening of the skin (scleroderma). The losses or fugitive emissions are not always at a steady rate, meaning that under transient process operating conditions, there are periods where there are greater losses than at others. This poses an additional problem of a serious fire potential because VCM is a highly flammable gas. Under the proper set of conditions, the VCM emissions pose a serious threat of fire and explosion. A catastrophic fire at this facility would represent a potential loss of lives for plant personnel, an extremely high health risk to the surrounding community from exposure to noxious vapors resulting from a fire incident, and likely hundreds of millions of dollars in property damages. A pollution prevention audit was performed, in which it was determined that about 5% of the monomer was lost through agitator seals, various draw-off points, through pump seals in the vacuum system. The audit revealed that recycling the VCM vapors that are being lost to the atmosphere from the stripper column operations was feasible.

Figure 2. *Shows monomer feed tank and vacuum system to reactors.*

In addition, it was shown that product quality demands in the international marketplace call for lower levels of residual VCM in the PVC resin product. Existing fugitive VCM monitoring practices involve slow and inaccurate methods, based on random grab samples of product and air samples surrounding equipment. These methods are incapable of accurately determining the sources and quantities of fugitive VCM emissions. Based on the P2 audit, OHIS invested in reliable portable monitoring instrumentation, which is now applied as a diagnostic tool to identify the sources of fugitive emissions, which are being corrected. An investment into a thermoparamagnetic oxygen analyzer and sample conditioning system, which cost $15,000 identified approximately 240 tons/year of VCM emissions that can be recycled through the process (to the polymerization reactors) with only minor modifications to existing process piping. This resulted in a projected annual material savings of $144,000, and more consistent and higher quality product.

Ammonium Nitrate Production

This case study provides an analysis of a pollution prevention program conducted at the Stirol Chemical Plant in the Donetsk region of Ukraine. The pollution prevention demonstration was conducted as a part of USAID's Environmental Policy & Technology Project (EPT), and was implemented in cooperation with the World Environmental Center (WEC). The EPT Project provided technical assistance in strengthening environmental management at industrial enterprises, and was a collaborative effort among the governments of Ukraine and the United States The first plant at this location was started in 1933, producing ammonia from coke gas. Over the years, plant operations were expanded to produce nitric and sulfuric acids, sulfates and ammonium nitrate (AN). In 1975, the company was reorganized as "Stirol." Today, the plant is a world class nitrogen fertilizer plant that can produce 1,350,000 tons/year ammonia in three plants, 330,000 tons/year prilled urea using two plants, 660,000 tons/year of prilled AN using 4 AN reactors and 3 prilling towers, and nitric acid production which is required for AN manufacturing. Liquid fertilizer from urea and AN is also produced for export to the U.S. In addition, the complex, now a "Joint Stock Company," produces sulfuric acid and oleum, sodium nitrate, nitrous oxide and many forms of styrene and polystyrene.

To illustrate the benefits of low-cost pollution prevention measures, a rapid in-plant assessment was made. An in-plant assessment consists of a team of engineers performing a walk-through of the plant operations, with the intent of identifying one or more opportunities for reducing wastes, pollution, saving raw materials and energy, and improving yields and efficiencies. These opportunities are then quantified by performing material and energy balances, from which dollar savings and returns on investments can be calculated for a proposed technical solution. During the walk through of the granulated ammonium nitrate plant, it was observed that liquor vapor condensate from the AN reactors was being discharged directly to the sewers. Plant personnel understood that there were product losses in this stream, but did not have fast and reliable measurements to minimize the losses.

The economic incentives for this case study are based on 1996 data, which is the time-frame of the project. Based on analytical sampling, it was determined that the liquid vapor condensate from the AN reactors had an average free ammonia content 384 mg/l, while the average content of ammonium nitrate was 1,111 mg/l. The condensate quality was typically determined by grab samples that were brought back to a remote on-site laboratory. A review of historical condensate quality data and process conditions at the AN reactors indicated that real time monitoring of the quality of condensate could be used as a basis to control process conditions in the AN reactors. In other words, production yields could be improved by monitoring the condensate quality and taking corrective actions on pressure, temperature, and flow rates into the reactors. An important design parameter in this plant is the capacity of the condensers on the AN reactors. These were designed and operated for the normal conditions of 0.8 cubic meters of condensate for every ton of ammonium nitrate produced. Of this amount, 0.73 cubic meters of condensate per ton of ammonium nitrate production are sent to a biological treatment plant. For a normal plant production of 360,000 tons of ammonium nitrate (NH_4NO_3) there are about 339,000 tons each of free ammonia and ammonium nitrate lost in the condensate stream per year. The value of ammonia is $135.61 per ton, and that of ammonium nitrate is $123.61 per ton. Hence, the plant loses enormous specialty products that are in the league of tens of millions of dollars on a yearly basis. Unfortunately, there are technology limitations to recovering low concentrations of these products, but the magnitude of losses clearly indicates that even incremental savings would be worth pursuing.

The P2 audit team recommended a continuous in-line conductivity meter, as well as upgrades to an existing pH meter, both to be used in parallel to provide real time condensate quality data (i.e., product concentrations in the condensate were correlated against pH and conductivity). Conductivity and pH data could be used as a basis to define optimum process conditions for the AN reactors that would maximize yields and minimize the concentrations of NH_3 and NH_4NO_3 losses to the condensate. This effort took several months of continuous testing, however, the instrumentation was nonintrusive, and therefore did not impact on normal production runs. The configuration for the installed instrumentation are illustrated by the schematic diagram in Figure 3.

By developing a set of process operating expressions for the AV rectors, whereby production yields for free-ammonia and ammonium nitrate where correlated to pH and conductivity in the overhead vapor condensate, the plant was able to optimize the process and reduce the product losses to the condensate stream. Under the optimum operating conditions experimentally identified, the ammonia content in the condensate was reduced from 384 to 260 mg/l, and that of the ammonium nitrate from 1,111 to 480 mg/l. The higher production yield, as verified by the lower condensate concentrations, resulted in product savings of 42 and 214 tons of NH_3 and NH_4NO_3, respectively. Combined, these product savings are worth a modest $32,070.

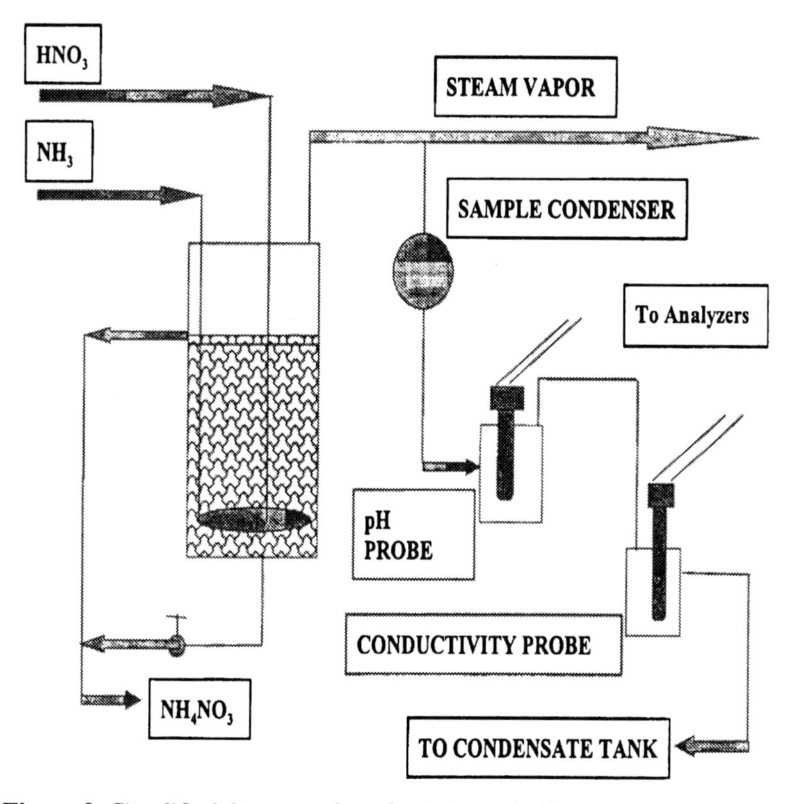

Figure 3. *Simplified diagram of conductivity and pH analyzers installed.*

In addition to the savings due to improved yields, a second source of savings were derived from reduced wastewater pollution fees. The 256 tons per year of reduced contaminants in the condensate stream (NH_3 + NH_4NO_3) was worth \$104,050 in pollution fees.

Finally, a third savings was identified from energy credits. Use of the conductivity meter to control the condensate quality made it possible to reduce the consumption of heat energy normally used for evaporating purposes. The historical operation require heating 10 tons of condensate per hour in order to recover some lost product. However, this practice was no longer needed since higher production yields were now possible. The vaporization step consumed 0.1 Giga-cal per hour (or roughly 72 Giga-cal per month). The cost for this energy in 1996 was about \$17.11 per Giga-cal. On a yearly basis, this amounted to a savings of \$14,690 in energy.

The project resulted in a yearly savings of \$150,810, with a one time investment for the conductivity meter and pH meter upgrades of only \$6000. The payback period was well under one month. Table 10 summarizes the savings.

This P2 case study shows significant savings from a small investment. The savings were in the form of reduced pollution fees, energy savings, and improved production yields. By implementing a number of small scale projects with relatively low levels of investments, plant operations can become safer for the employees and surrounding communities, as well as achieve significant savings to improve the profitability or help sustain operations during difficult economic times.

The information for this case study is based upon publications by the World Environmental Center (WEC). WEC is located at 419 Park Avenue South, Suite 1800, New York, New York 10016. More information about this project can be obtained from the WEC report titled "Waste Minimization Demonstration Project at Stirol Chemical Plant in Gorlovka," Final Report, September 1997.

Table 10. *P2 Matrix Summary of Savings.*

Recovered Product (tons/yr)	Economic Benefits (Savings, \$/yr)				Invest-ment, \$	Payback Period (months)
	Reduced Pollution Fees	Energy Savings	Recovered Products	Total		
250	104,050	14,690	32,070	150,810	6,000	< 1

CLEAN TECHNOLOGIES SUBSTITUTE ASSESSMENTS

This section provides an overview of the methods and resources needed to conduct a Cleaner Technologies Substitutes Assessment (CTSA), a methodology for evaluating the comparative risk, performance, cost, and resource conservation of alternatives to chemicals currently used by specific industry sectors. The CTSA methodology was developed by the USEPA (23). The CTSA *methodology* is a means of systematically evaluating the comparative human health and environmental risk, competitiveness (e.g., performance, cost, etc.) and resource conservation of traditional and alternative chemicals manufacturing methods and technologies. The CTSA differs from a conventional P2 audit in that it not only focuses on evaluating a particular group of traditional, but also nontraditional (i.e., unusual, new, or novel) substitutes or alternatives that can be used to perform a key function within a given industrial process. In contrast, P2 programs, because of their very nature and often uphill battle needed to obtain management support, focus on conventional technologies. The two approaches, however utilize many of the same elements, but the CTSA was developed for specific industry sectors like printed circuit board manufacturing. That does not mean that it cannot be applied to other industry sectors, once the methodology is understood.

In CTSA terminology, the term *use cluster* is crucial. A use cluster is a product- or process-specific application in which a set of chemical products, technologies, or processes can substitute for one another to perform a particular function. Unfortunately a CTSA does not recommend alternatives. Instead, the goal is to promote informed business decisions that integrate risk, performance, and cost concerns by providing businesses with easily accessible information. A project team implementing a CTSA uses data from the assessment to develop fact sheets and summary reports designed to reach individual users and suppliers who may not have the resources to develop the information on their own.

Project partners in a CTSA perform a number of preliminary steps prior to embarking on the detailed analyses of a CTSA. These include recruiting partners, preparing scoping documents, selecting a use cluster for evaluation, and setting the boundaries of the evaluation. These preliminary steps not only ensure the selection of a productive project focus, they also help build relationships among the potential team members and lay the foundation for the culture of collaboration essential to project success.

The following are the basic steps leading up to and following a CTSA. First, the project organizers recruit partners from various stakeholder communities to create a project team. Team members then develop an Industry and Use Cluster Profile document and a Regulatory Profile document to help define the project focus. An Industry and Use Cluster Profile gives market data for the industry, describes technological trends, and presents a summary of key industry processes, individual steps within processes, chemicals typically used in each step, and a preliminary list of substitutes for each step. These sets of substitutes make up the use clusters for the industry. A Regulatory Profile identifies the principal federal environmental regulations that may affect the industry under study and the factors that determine which regulations apply to any particular operation. The project team typically selects the use cluster with the greatest opportunities for environmental improvement for the detailed analysis of a CTSA. Once the use cluster is selected, team members identify substitutes within the use cluster, select a subset of these substitutes for evaluation in a CTSA, and establish the project baseline. The project baseline is typically the industry standard practice, to which other substitutes can be effectively compared. The next step is to set the boundaries of the evaluation by identifying the life cycle stages and types of environmental impacts (e.g., human health and environmental risk to workers, energy impacts, etc.) of greatest concern.

Each of these steps sets the stage for the detailed substitutes assessments that are performed in a CTSA. Following completion of a CTSA, the project partners develop a variety of outreach tools to communicate the results of the CTSA. These may include fact sheets, bulletins, pollution prevention case studies, software, videos, and training materials. The final phase of a project is to disseminate CTSA results to businesses and other stakeholders, who may not have the resources to develop the information on their own. By providing a clear picture of the trade-offs among environmental, economic, and performance concerns, CTSA projects encourage continuous environmental improvement.

The first task for the project team is to conduct research and analysis to identify use clusters within an industry and the use clusters that would provide a productive project focus. Two outcomes of these initial scoping exercises, the Industry and Use Cluster Profile and the Regulatory Profile, provide the foundation for selecting a use cluster and beginning a CTSA. The Industry and Use Cluster Profile gives market data for the industry, describes technological trends, and presents a summary of each of the use clusters within the industry. This information helps the project team to select a use cluster for evaluation in the CTSA. It also provides information to other sections of the CTSA, such as the exposure assessment. Table 11 lists some of the information typically included in an Industry and Use Cluster Profile and gives examples of how this information may be used in a CTSA.

Table 11. *Uses of Information from a Use Cluster and Industry Profile.*

Type of Information	Potential Uses in a CTSA
Geographic distribution of industry by size (number of employees, sales) and function.	Determine the aggregate number of workers exposed, information needed in the exposure assessment.
Key industry processes, individual steps within processes, and chemicals typically used in each step.	Identify traditional chemicals and processes within the focal use cluster; provide the foundation for the source release assessment, exposure scenarios, and exposure pathways.
The set of readily identifiable substitutes for each step, which make up the use clusters.	Preliminary pool of substitutes for evaluation in the CTSA.
Technology trends.	Identify potential substitutes; help select subset of substitutes for evaluation.

The Regulatory Profile identifies the principal federal environmental regulations that may affect the industry under study and the factors that determine which regulations apply to any particular operation. Such factors might include the size of the operation; the location of a facility (i.e., in an ozone non-attainment area); the types of chemical products it uses; and the types, quantity, and toxicity of the emissions and waste streams it generates. For the purposes of a CTSA, the Regulatory Profile helps focus the selection of alternatives by:

- Providing project participants with consistent information on the regulatory requirements affecting an industry.
- Determining if implementing a substitute would reduce the overall regulatory burden of a company.
- Determining if implementing a substitute would shift the environmental impact across environmental media, such as from air to water, or from water to land.
- Identifying impending chemical or technology bans, phase-outs or other regulatory actions that could affect the market availability and use of affected substitutes.

The Regulatory Profile also serves as a data source for the regulatory status section of the CTSA which evaluates in more detail the regulatory status of each of the potential substitutes selected for quantitative assessment in a CTSA.

Each use cluster constitutes an area where the relative human health and environmental risk, performance, cost, and resource conservation of alternatives can be compared. For example, Figure 3 in Chapter 4 illustrates the basic functional steps in printed wiring board (PWB) fabrication. To date, Regulatory Profile documents have not explicitly analyzed the regulatory effects of implementing a substitute, but the regulatory status data can be used by project partners to determine what the effects might be. Since a principal objective of the overall process is to identify and evaluate substitutes that have the greatest potential for reducing overall environmental impacts, attention is focused on finding alternatives that prevent pollution instead of simply shifting pollutants from one environmental medium to another. Factors to consider when selecting a use cluster for evaluation include the following:

The degree of risk associated with current practice in the use cluster: Use clusters that involve greater exposure to highly toxic chemicals may pose greater human health and environmental risk and offer greater potential for improvement. EPA uses a relative risk ranking methodology to screen the relative health and environmental effects of different use clusters. The Use Clusters Scoring System ranks use clusters into broad concern categories (high, medium, or low) based on use volumes, total environmental releases of chemicals, health and environmental hazards, exposure potential and other factors.

The degree of interest that industry and other stakeholders have in the use cluster: Project teams typically represent different stakeholder communities with differing values. Understanding the interests of each of the partners is important to building consensus. The level of interest in the use cluster of each of the partners will also be an important factor motivating their participation. For example, the cooperation of suppliers in providing information on or samples of their products has proven to be essential to the success of past projects.

The availability of potentially cleaner substitutes: The purpose of a CTSA is to evaluate the trade-offs among substitutes of human health and environmental risk, performance, cost, and other environmental effects. Viable substitutes within a use cluster that are in use or ready to be demonstrated are necessary for a CTSA to have the best potential for real environmental gains in the near-term. Processes or technologies that perform a similar function in other industries may also be viable substitutes. The project team may elect to include new technologies that are still in the research and development stage, even though tangible environmental improvements from the use of these technologies may be less immediate.

The degree to which a use cluster is tied to other process steps outside of the use cluster: In some cases, implementing a substitute product, process, or technology might require changes in process steps outside of the use cluster. If so, the project team may need to evaluate these other changes as well to ensure that selection of a substitute does not adversely affect performance or cost outside of the use cluster or shift the environmental impacts from one part of the process to another. Project teams need to consider the time and resources they have available for the evaluation as well as the potential improvement opportunities of these more complex use clusters.

The status of other ongoing projects related to a use cluster: If other projects are already evaluating a use cluster the project team should determine if a CTSA will add valuable information to information already being developed. In some cases, it may be possible to coordinate the work of a project team with other efforts that are not considering the full range of issues evaluated in a CTSA.

The Use Cluster and Industry Profile, with its preliminary list of chemicals, processes and technologies employed in each use cluster, provides the initial pool of substitutes for evaluation in a CTSA. The identification of substitutes is not limited to this preliminary stage of a CTSA, however. Additional substitutes are identified as a CTSA progresses and more information is gained about the characteristics of the use cluster and of the industry.

The project team begins to identify additional substitutes after the focal use cluster is selected. All stakeholder groups are potential sources of information about additional substitutes. Manufacturers and suppliers of chemical products and technologies play an important role in substitute identification, since they frequently have an up-to-date understanding of current industry trends, and emerging products or technologies. Also, the participation of suppliers in the CTSA process is essential to developing generic chemical product formulations which may be used in the risk characterization if necessary to protect proprietary formulation information. At the same time, trade associations may be tracking new developments; their laboratories and research facilities may be currently developing alternatives. Universities and other research organizations also may be involved in applied or basic research on new alternatives. Public interest groups concerned about human health risk or other environmental impacts may have independently searched for options to prevent pollution. International organizations may have information on alternatives used abroad. Project teams use all of these resources to develop a substitutes tree.

The Substitutes Tree

A substitutes tree is a graphical depiction of the substitute or alternative chemical products, technologies, or processes that form the use cluster and their relationship to each other within the functional category defined by the use cluster. In a project, the terms "substitute" and "alternative" are used interchangeably to mean any traditional or novel chemical product, technology, or process that can be used to perform a particular fractions. The substitutes tree developed for Dry Cleaning Project is illustrated in Figure 4. The diagram helps to illustrate the thought processes that are employed in identifying substitutes. The Dry Cleaning Project evolved from several years of work by EPA with the dry cleaning industry to examine ways to reduce exposure to perchloroethylene (PCE). PCE, a suspected carcinogen, is the chemical solvent most frequently used to dry clean clothes (EPA, 1995a). The dry cleaning process was originally developed to clean water-sensitive fabrics. If the function of dry cleaning is defined as solvent-based cleaning, a number of chemical substitutes can be readily identified that are currently used in dry cleaning facilities. When identifying alternatives in a use cluster, however, the project team must be careful to not define the function too narrowly or too broadly. The following discussion illustrates the limitations that would have been imposed on the dry cleaning project if the function had been defined as solvent based cleaning.

Recall that a goal of a CTSA is to evaluate both traditional and novel chemicals, processes, or technologies that can substitute for one another to perform a particular function. The substitutes tree shown in Figure 4 is too narrow in its scope since it only illustrates traditional chemicals. Figure 5 shows the substitutes tree expanded to include newly available professional dry cleaning technologies, and dry cleaning chemicals and technologies that are currently under development. This also proved to be too narrowly defined. Each of these substitutes or alternatives are dry cleaning processes, which is how the use cluster has been defined in Figure 4.

In the Dry Cleaning Project, however, the project gained momentum when an alternative process called multi-process wet cleaning came to the attention of the project partners. This process primarily uses controlled application of heat, steam, and soap to clean garments, including garments made from water-sensitive fabrics. If the function of the use cluster is redefined as professional garment cleaning (excluding water-washable garments that are usually home-laundered), which is the ultimate function that dry cleaners provide and the service that consumers seek, a whole new array of potential alternatives can be identified.

The Industry and Use Cluster Profile typically lists the categories of chemicals (e.g., adhesive, cleaning solvent, surfactant, etc.) and the major chemicals in each use cluster. Early in the CTSA, project team members begin collecting data on the chemical and physical properties of these chemicals. A process description of the use cluster is prepared to help define the chemical properties of the chemical products which enable them to perform the desired function (e.g., the chemical properties of an organic solvent make it suitable for dissolving oily residues on clothes) and to identify any functional groups in the use cluster.

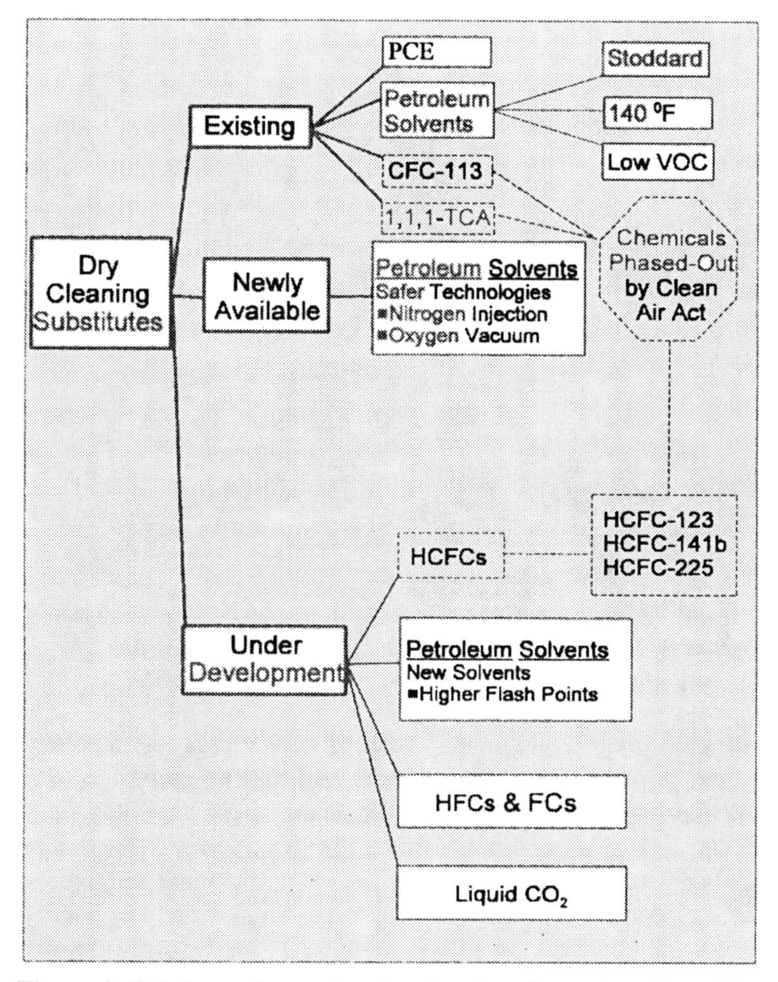

Figure 4. *Existing and emerging dry cleaning alternatives (from EPA Cleaner Technologies Assessment - EPA744-R-95-002, Dec. 1996).*

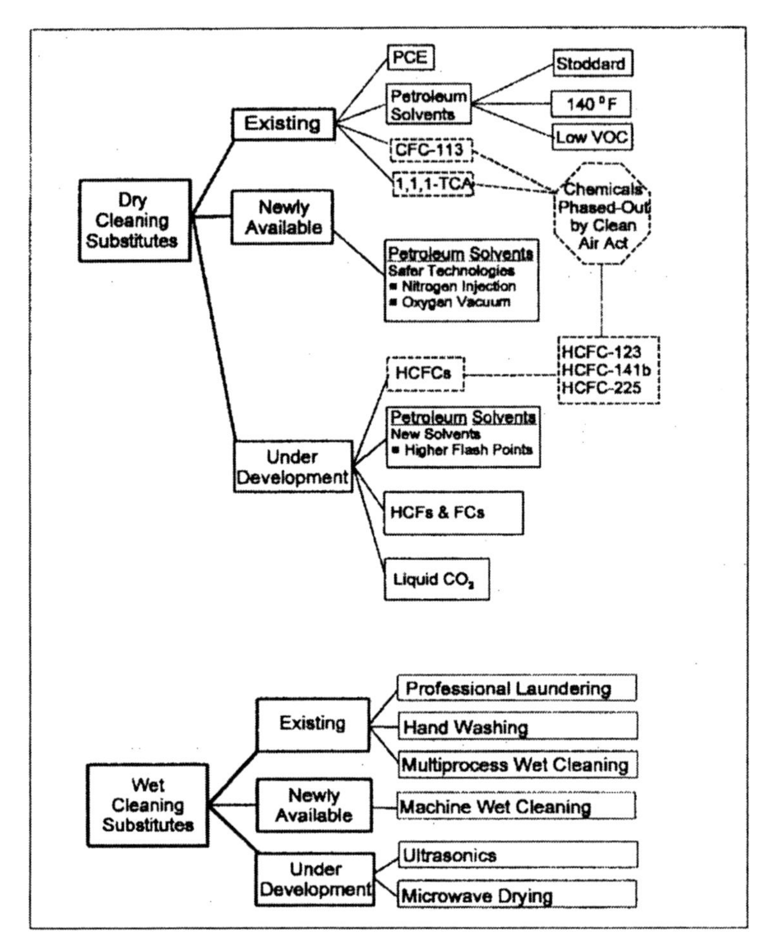

Figure 5. *Garment cleaning alternatives (from EPA Cleaner Technologies Assessment - EPA744-R-95-002, Dec. 1996).*

A functional group is a discrete, functional step of a multi-step process. The chemical components can substitute for one another to perform a particular function of a chemical mixture. For example, in the garment cleaning use cluster, the traditional dry cleaning process uses solvents to remove oils, stains, and odors. Although small amounts of water, detergent, and other additives may be used, chemical products in the process essentially employ one functional group: chemical cleaning solvents. Note that all of the chemical properties and data regarding the chemical properties which enable the chemicals to perform the desired function are analyzed together to identify alternative chemicals that have similar properties or that perform similar functions in other industries.

The CTSA will focus on identifying substitutes for the processes, the chemicals, and the technologies through a careful review of literature, including the patent literature and PMNs (Pre-Manufacturing Notifications).

Once several substitutes for technologies/processes/chemicals have been identified, the project team must decide which of these to evaluate. Traditional substitutes, those currently in widespread use, are usually selected for evaluation because they provide a baseline against which the risk, performance, and cost of all substitutes can be compared. In addition, dissimilar chemical formulations or methods within the range of traditional substitutes may pose vastly different risks. Nonetheless, if a substantial number of traditional substitutes are currently in use, the project team may have to place practical limits on the number evaluated. This is especially true for substitute chemical products. The project team should also consider one or more new alternatives, depending on the project resources. Factors to consider when selecting new or novel alternatives include the following:

- The ability of an alternative to meet regulatory requirements in the application under review.
- The potential for reducing human health and environmental risk or net environmental impacts.
- The cost required to evaluate the alternative relative to others.
- The viability of the alternative in terms of its known relative cost or performance.
- The degree to which the suppliers or developers of the alternative are willing to participate in the project. Participation may include providing information or samples to the project.
- The applicability of the alternative to the industry as a whole.
- The degree to which the alternative is ready to enter the market (e.g., the research and development stage of the alternative).
- Whether or not implementing an alternative would require changes in process steps outside of the use cluster that would also have to be evaluated in the CTSA.

Participation by the developer(s) or supplier(s) of an alternative can be crucial to the project's success. For example, developers or suppliers of chemical products will need to provide information on their specific product formulations to conduct the risk characterization and samples of their products and material safety data sheets (MSDSs) for the performance assessments. Developers or suppliers of technologies will need to provide operating instructions in order to train staff of demonstration facilities in the correct use of the technology. Furthermore, if the technology has not been introduced to the market, the developer may need to provide one or more complete sets of equipment for the performance assessment.

The chemical formulations of commercial products containing several distinct chemicals are frequently considered proprietary. When undertaking a risk characterization or performance evaluation of such chemical products, the confidential nature of these formulations can complicate a CTSA analysis. Manufacturers of these products typically prefer not to reveal their chemical formulations because a competitor can potentially use the disclosed formulation to sell the product, often at a lower price, since the competitor did not invest the research and development resources in originally formulating and testing the product.

To make the CTSA usable and flexible, a standard format for representing each chemical product with a generic product formulation is used. Each product can be assigned a code name and each supplier asked to give the confidential product formulation to EPA. While EPA uses the confidential formulations to conduct a detailed risk characterization of each chemical product that appears in the CTSA, the published CTSA represents a chemical product only by a code name and the generic formulation developed by EPA and the individual supplier. The generic formulations allow the users of the CTSA to compare different product systems while protecting the proprietary nature of the product formulation. While the generic formulations are important in obtaining supplier participation, they also make the CTSA a useful tool for evaluating other brand name products that may contain similar chemical constituents as those already evaluated. Given the formulation of a chemical product from a detailed MSDS, the human health risk, performance, and cost information can be compared with a product already evaluated in the CTSA. However, as a MSDS only lists chemical constituents which are hazardous to human health, environmental risks may not be able to be determined from the information presented solely on the MSDS.

A CTSA is a comparative evaluation requiring a baseline to compare the risk, performance, cost, and other environmental effects of alternatives (substitutes). Project teams select one or more alternatives that are currently in widespread use or familiar to most of the industry to serve as an industry standard(s) or project baseline(s). With a familiar baseline as the basis for comparison, the comparative data on risk, performance, cost, and conservation developed through the project will be understandable to the majority of industry.

The number of alternatives selected depends on a number of factors, including the following:

- *Is there a clear, industry-wide baseline?* For many industries, it may be difficult to establish a single product, process, or technology as the baseline.
- *Is the type of product, process, or technology used dependent on the size of a business?* The baseline may differ for small and large businesses. For example, automated technologies that are cost-effective for large companies may not be economically feasible for small businesses. The decision to include different project baselines for both small and large industry sectors will depend in part on the resources available to the project team and the primary environmental issues the project team plans to address.
- *Are different products, processes, or technologies required to meet end-user performance requirements?* Performance requirements and the alternatives typically employed to meet them may vary depending on the end-use of the product or service an industry sector provides.
- *Is the industry standard static or constantly changing?* Industry standard practice can change rapidly, especially in industries that are continuously evolving to meet increasing technological or other demands. If the industry standard changes rapidly, the project team needs to build flexibility into the project baseline to ensure that current and pertinent data are collected.
- *Are suppliers of the project, process or technology participating in the project and willing to provide data?* To provide an adequate basis for comparison, data on the baseline must be at least as complete as the data on the alternatives. Again, suppliers are a crucial link to obtaining adequate information.

The goal of designing for the environment is to design products and processes that minimize environmental impacts throughout their life cycles. Due to the complexity of the product life cycle, however, businesses often focus their environmental improvement efforts on the areas where the greatest environmental improvement opportunities lie and where they can most influence change. The CTSA methodology provides a flexible format that enables teams to use this concept to set the boundaries of the evaluation before embarking on a CTSA. Setting the boundaries of the evaluation involves the following considerations:

- *What are the life cycle stages where the most significant environmental effects are believed to occur?* Environmental effects occur in each stage of the life cycle of a product or process, from extraction and processing of raw materials through manufacturing, use, and disposal. For practical purposes, past projects have focused on the use and disposal stages of the life cycle, where the greatest environmental impacts were believed to occur and the most data were available. Other project teams may choose to focus on other life cycle stages.

- *What are the primary environmental issues associated with the use cluster?* Some partners may be concerned about the chemical risk from using toxic chemicals in their operations. Partners working on other industry sectors may identify other issues, such as energy or nonrenewable resource consumption, as the primary environmental issues associated with a use cluster.
- *To what degree can project partners influence change?* Projects are designed to promote continuous environmental improvement. Due to time and resource constraints, project partners typically elect to focus their efforts on the areas where they can most influence change.

Each of these considerations is related. For example, the product life cycle must be reviewed to identify the primary issues associated with a use cluster. Without participation by suppliers or representatives from up-stream processes, the project team may find their ability limited to gather data as well as influence change in the up-stream process. The life cycle concept and each of these considerations are discussed in more detail below.

The Life Cycle Concept

Businesses, whether manufacturers of consumer products, commercial products, or commercial service industries, have traditionally defined the life cycle of the product, goods, or service they provide as beginning with product conception and moving through design, manufacturing, use, and disposal. Performance, quality, and cost requirements for the manufacturing, use, and disposal phases of the product life cycle are established during product conception. The product designer is charged with ensuring that these requirements are met.

In the 1990s, the term "product life cycle" has taken on new meaning. Environmental decision makers in all stakeholder sectors have recognized that, to ensure the overall environmental improvement of a product or process, all stages of the life cycle where significant environmental impacts can occur should be considered. This can include the extraction and processing of the raw materials used to make the product, product manufacturing, transportation, use, recycling, and disposal. The concept of designing products and processes for the environment combines these two definitions of the product life cycle. The environmental effects of all significant stages of the product life cycle can be evaluated to incorporate environmental considerations into the design and redesign of products and processes.

"Extended product responsibility" is an emerging principle of pollution prevention that advocates this life cycle approach to identifying opportunities to prevent pollution and addresses the question, "How much can project partners influence change?" Under this principle, there is assumed responsibility for the environmental impacts of a product throughout the product's life cycle, also called the "product chain," including up-stream impacts inherent in the selection of materials for the product, impacts from the manufacturer's production process, and down-stream impacts from the use and disposal of the product. Thus, a shared "chain of responsibility" is borne by designers, manufacturers, distributors, users, and disposers of products. The greater the ability of the actor (i.e., designer, manufacturer, etc.) to influence the life cycle impacts of the product system, the greater the degree of responsibility for addressing those impacts should be. Because effective measures to reduce the life cycle environmental impacts of a product system usually involve changes in more than one link in the product chain, extended product responsibility creates a need and an opportunity for partnerships throughout the product chain (President's Council on Sustainable Development, 1996). To set the boundaries of the evaluation from a life cycle perspective, the project team might ask, "In which stage of the life cycle are the greatest environmental impacts believed to occur?" In some cases, this will be apparent, in others, it will not. For example, when considering the life cycle of the automobile, practitioners of life cycle assessment agree that significant environmental impacts occur during the use of the automobile, due to the substantial amount of energy consumed and the emissions of air pollutants. In the case of pesticides, the manufacturing of chemical ingredients and use by consumers may be equally important, since pesticide products are intentionally released to the environment during use. On a practical note, the time and resources available to conduct a CTSA may determine the degree to which up-stream or down-stream processes can be included in the evaluation. The following considerations may be helpful when identifying the life cycle stages on which to focus:

- *Are the natural resources used in the use cluster in abundant supply?* Resources that are being rapidly depleted are a serious concern. An industry dependent on scarce resources may wish to focus on the extraction and processing of raw materials to evaluate the environmental impacts, especially the social benefits and costs, of alternatives.

- *Do the natural resources occur only in low concentrations in their natural state?* The extracting and processing of raw materials that occur naturally in low concentrations may be of great environmental impact. For example, some metals that are found only in low concentrations in their ores may require more mining and processing of raw materials, more water and chemical use for extracting the metals, generate more mill tailings, and consume excessive energy.
- *Is use of the product likely to cause risk to consumers exposed to toxic chemicals?* Some products may have the greatest environmental impact during use by consumers. For example, the risk to workers manufacturing solvent-based paints could be small compared to the risk to persons using the paints who do not use personal protective equipment.
- *What are the environmental impacts of disposal of the product?* Some products are intentionally released to the environment by the consumer after use. For example, the aquatic toxicity of household cleaning products that are rinsed down the drain by the consumer could be of significant concern.

By focusing on the life cycle of the product, processes, or technologies in the use cluster, the project team will most likely identify many of the primary environmental issues associated with the use cluster, but in a holistic fashion.

By involving representatives from up and down the product chain as well as public-interest groups, labor organizations, and other stakeholder communities, partnerships provide an excellent forum for identifying the primary environmental issues associated with a use cluster. Diverse stakeholder groups bring different resources and unique perspectives to the table to ensure that important environmental issues are not overlooked. Examples of the issues the project team may elect to focus on include the following:

- Reducing risk to workers, surrounding populations (human and ecological), or consumers through use of substitutes, improved workplace practices that prevent pollution, or even pollution control technologies.
- Reducing energy impacts or conserving natural resources.
- Reducing workplace safety hazards.

Regardless of whether the focus is on alternative systems, technologies, or pollution control methods, the goal is to reduce risk, resource consumption, process safety hazards and/or other environmental effects, and provide tangible environmental improvements. The following are examples of questions a project team might ask to determine where the greatest improvement opportunities lie:

- *Where is a typical business located?* Facilities located in urban areas may have different impacts than those in rural areas. For example, dry cleaning facilities are typically located in or near residential areas. Therefore, the dry cleaning team elected to evaluate the risk to persons living near these shops.

- *Are many facilities located in areas with local or regional regulatory requirements?* Local or regional regulatory requirements may cause many businesses to seek alternative products or processes. For example, businesses that emit volatile organic compounds in non-ozone attainment areas may seek substitute chemical products that do not contribute to photochemical smog.

While these types of questions may identify the primary environmental issues associated with a use cluster, they will not necessarily identify the most significant problems for individual businesses. For example, a business located in a rural area where photochemical smog is not an overriding issue may be more concerned about the water releases to their septic system. Again, the flexible format of a CTSA is the key to providing sufficient information to enable individuals to make the best choices for their given situation. As noted, the aim of a CTSA is to develop as complete and systematic a picture as possible of the trade-offs among risk, competitiveness (i.e., performance, cost, etc.), and conservation associated with the substitutes in a use cluster. To accomplish this, a CTSA employs a modular approach to data collection and analysis utilizing "information modules. " An information module is a standard analysis or set of data designed to build on or feed into other information modules to form an overall assessment of the substitutes. A CTSA records and presents facts collected in the information modules, but does not make value judgements or advocate particular choices.

To evaluate the trade-off issues, project partners prepare data summaries related to risk (releases of pollutants to the environment, potential exposure levels, risk of chemical exposure to human health and the environment), competitiveness (performance, cost, market availability, regulatory status), and conservation (energy impacts and effects of resource conservation). All of this information is combined to evaluate the social benefits and costs of implementing an alternative. Finally, the risk, competitiveness, and conservation data summaries are organized together with the results of the social benefits/costs assessment in a decision information summary that records and presents facts, but does not make value judgements or advocate particular choices. The information module approach of the CTSA methodology is modeled after the risk management process that EPA conducts under the authority of the Toxic Substances Control Act (TSCA), with some important distinctions. Under TSCA, EPA has regulatory authority to perform the following activities regarding existing chemicals: (1) gather toxicity, production, use, disposal, and fate information; (2) assess human and environmental exposure; (3) determine if a chemical poses unreasonable risks; and (4) take appropriate actions to control these risks, based on a social benefits and costs analysis. TSCA is the only U.S. statute under which multi-media risk assessments are performed as part of the regulatory rulemaking process. To identify potential risk early in the screening process, EPA uses a two-phase risk management process. Phase I is a screening level risk assessment and fact-finding mechanism, intended to ensure that EPA only focuses on chemicals with the potential to present unreasonable risk to human health and the environment. If this initial investigation finds that unreasonable risk may exist, chemicals are evaluated further in Phase 2. Phase 2 is a more detailed and comprehensive risk assessment process that includes a thorough evaluation of the hazards and exposures to specific chemicals, identification of strategies to reduce or eliminate risk, and an evaluation of pollution prevention opportunities. To the extent possible, EPA bases the Phase 2 assessments on existing information, although new data may have to be generated. Each member of an EPA assessment team is responsible for completing one or more standardized analyses (information modules) on the chemicals, including Chemical Properties, Market Information, Chemistry of Use & Process Description, Source Release Assessment, Human Health and Environmental Hazards Summaries, Exposure Assessment, and Risk Characterization modules. These information modules build on or feed into each other to form an assessment of the chemical. EPA's standardized assessment process is designed to promote efficiency and consistency among results. The CTSA process is modeled after EPA's risk management process, with these important distinctions:

- *The CTSA process is designed to assist a voluntary decision-making process and, as such, is not as rigorous or detailed an evaluation as the regulatory rulemaking process.* In order to respond to a project team's needs in a timely manner and reduce resource needs, the CTSA process is designed to collect only the information necessary to adequately assist an individual making a voluntary business decision. As such, the data collection and analysis performed in a CTSA are quite detailed, but it is not necessary or intended to be as rigorous as the regulatory rulemaking process. For example, past CTSAs have qualitatively evaluated the social benefits and costs of implementing an alternative, but have not monetized overall social benefits and costs, which may be required for regulatory rulemaking.

- *A CTSA adds additional information modules to collect data on issues related to competitiveness, conservation, and pollution prevention.* A CTSA contains the risk related information modules in Phase 2 of EPA's risk management process, plus additional modules to address competitiveness issues (e.g., performance, cost, etc.) and conservation issues (energy impacts and resource conservation). A CTSA also compiles extensive information on pollution prevention opportunities, including improved workplace practices that prevent pollution, that may be more comprehensive than those compiled in the risk management process.

By building on EPA's risk management process, the CTSA process has a range of standardized data collection and analytical methods already available that can be tailored to the needs of a specific project.

REFERENCES

1. Bounicore, Anthony J., and Wayne T. Davis, eds. 1992. *Air Pollution Engineering Manual.* New York: Van Nostrand Reinhold.
2. European Fertilizer Manufacturers' Association. 1995a. "Production of NPK Fertilizers by the Nitrophosphate Route." Booklet 7 of 8. Brussels. and 1995b. "Production of NPK Fertilizers by the Mixed Acid Route." Booklet 8 of 8. Brussels.
3. Sauchelli, Vincent. 1960. *Chemistry and Technology of Fertilizers.* New York: Reinhold Publishing.
4. Sittig, Marshall. 1979. *Fertilizer Industry; Processes, Pollution Control and Energy Conservation.* Park Ridge, N.J.: Noyes Data Corporation.
5. UNIDO (United Nations Industrial Development Organization). 1978. *Process Technologies for Nitrogen Fertilizers.* New York and 1978. *Process Technologies for Phosphate Fertilizers.* New York.
6. European Union. 1996. "Best Available Technology Notes on Various Pesticides Manufacturing Processes." Brussels.
7. Sittig, Marshall. *Pesticide Manufacturing and Toxic Materials Control Encyclopedia.* Park Ridge, N.J.: Noyes Data Corporation.
8. UNIDO (United Nations Industrial Development Organization). 1992. *International Safety Guidelines for Pesticides Formulation in Developing Countries.* Vienna.
9. USEPA (U.S. Environmental Protection Agency). 1988. *Pesticide Waste Control Technology.* Park Ridge, N.J.: Noyes Data Corporation.
10. WHO (World Health Organization). 1996. *International Programme on Chemical Safety (IPCS): The WHO Recommended Classification of Pesticides*

by Hazard and Guidelines to Classification 1996-1997. Geneva.

11. World Bank. 1993. "Agricultural Pest Management." Guidelines and Best Practice, GB 4.03. *World Bank Operational Manual.* April, Washington.

12. 1996. "Pollution Prevention and Abatement: Pesticides Manufacturing." Draft Technical Background Document. Environment Department, Washington, D.C.

13. ACS (American Chemical Society). 1983. *Advances in Pesticide Formulation Technology.* ACS Symposium Series 254. Washington, D.C.

14. Seaman, D. 1990. "Trends in the Formulation of Pesticides: An Overview." *Pesticide Science* 29:437-49.

15. Bounicore, Anthony J., and Wayne T. Davis, eds. 1992. *Air Pollution Engineering Manual.* New York: Van Nostrand Reinhold.

16. European Community. 1993. "Technical Note on the Best Available Technologies to Reduce Emissions into Air from Coke Plants." Paper presented to BAT Exchange of Information Committee, Brussels, 1993. "Study on the Technical and Economic Aspects of Measures to Reduce the Pollution from the Industrial Emissions of Cokeries." Paper presented to BAT Exchange of Information Committee, Brussels.

17. USEPA (United States Environmental Protection Agency). 1982. *Development Document for Effluent Limitations Guidelines and Standards for the Iron and Steel Manufacturing Point Source Subcategory.* EPA440/ 1-82/024. Washington, D.C.

18. United States. 1992. *Federal Register,* vol. 57, no. 160, August 18. Washington, D.C.: Government Printing Office.

19. World Bank. 1995. "Industrial Pollution Prevention and Abatement: Coke Manufacturing." Draft Technical Background Document. Washington, D.C.

20. WHO (World Health Organization). 1989. *Management and Control of the Environment.* WHO/PEP/89.1. Geneva.

21. Kirk, Raymond E., and Donald F. Othmer. *1980. KirkOthmer Encyclopedia of Chemical Technology.* 3d ed. New York: John Wiley and Sons.

22. Austen, George T., R. N. Shreve, and Joseph A. Brink. *1984. Shreve's Chemical Process Industries.* New York: McGraw-Hill.

23. *Cleaner Technologies Substitute Assessments: A Methodology and Resource Guide*, USEPA, EPA Document Number EPA 744-R-95-002, December 1996.

Chapter 6
Pollution Prevention Practices in the Petroleum Refining Industry

INTRODUCTION

Petroleum refining is one of the leading manufacturing industries in the United States in terms of its share of the total value of shipments of the U.S. economy. In relation to its economic importance, however, the industry is comprised of relatively few companies and facilities. The number of refineries operating in the U.S. can vary significantly depending on the information source. For example, in 1992, the Census Bureau counted 232 facilities and the Department of Energy reported 199 facilities.

In addition, EPA's Toxic Release Inventory (TRI) for 1993 identified 159 refineries. The differences lie in each organization's definition of a refinery. The Census Bureau's definition is based on the type of product that a facility produces and includes a number of very small operations producing a specific petroleum product, such as lubricating oils from other refined petroleum products. These small facilities often employ fewer than 10 people and account for only one to two of the petroleum refining industry's total value of shipments. In comparison to the typically much more complex, larger and more numerous crude oil processing refineries, these facilities with their smaller and relatively simple operations do not warrant the same level of attention from an economic and environmental compliance standpoint, nor are the pollution prevention opportunities likely to be substantial, except on a collective basis. Refineries recognized by the Department of Energy tend to be only the larger facilities which process crude oil into refined petroleum products.

This chapter provides an overview of petroleum refining operations, and then focuses attention on the environmental problems and pollution prevention opportunities. It is important to note that P2 in this industry faces a tradeoff between environmental problems and energy conservation. It is also an industry that provides the raw materials to the chemical process and allied industries. As such, both pollution prevention and waste minimization take on special meanings in that this industry sector is a supplier. Downstream operations such as chemicals and polymers manufacturing are impacted indirectly (and in some cases directly) by environmental practices and the environmental liabilities that refinery products carry along with them.

INDUSTRY DESCRIPTION AND PRACTICES

This section describes the major industrial processes within the petroleum refining industry, including the materials and equipment used, and the processes employed. The section is necessary for an understanding of the industry, and for grasping the interrelationship between the industrial processes and pollutant outputs and pollution prevention opportunities. This section specifically contains a description of commonly used production processes, associated raw materials, the by-products produced or released, and the materials either recycled or transferred off-site. This discussion, coupled with schematic drawings of the identified processes, provide a concise description of where wastes may be produced in the process. This section also describes the potential fate (via air, water, and soil pathways) of these waste products.

Petroleum refining is the physical, thermal and chemical separation of crude oil into its major distillation fractions which are then further processed through a series of separation and conversion steps into finished petroleum products. The primary products of the industry fall into three major categories: fuels (motor gasoline, diesel and distillate fuel oil, liquefied petroleum gas, jet fuel, residual fuel oil, kerosene, and coke); finished nonfuel products (solvents, lubricating oils, greases, petroleum wax, petroleum jelly, asphalt, and coke); and chemical industry feedstocks (naphtha, ethane, propane, butane, ethylene, propylene, butylenes, butadiene, benzene, toluene, and xylene). These petroleum products comprise about 40 percent of the total energy consumed in the U.S. and are used as primary input to a vast number of products, including: fertilizers, pesticides, paints, waxes, thinners, solvents, cleaning fluids, detergents, refrigerants, anti-freeze, resins, sealants, insulations, latex, rubber compounds, hard plastics, plastic sheeting, plastic foam and synthetic fibers. About 90 percent of the petroleum products used in the U.S. are fuels with motor gasoline accounting for about 43 percent of the total. Figure 1 provides a conceptual breakdown of the products. The Standard Industrial Classification (SIC) code established by the Bureau of Census to track the flow of goods and services within the economy is 29 for the Petroleum Refining and Related Industries. The petroleum refining industry is classified as SIC 2911, which includes the production of petroleum products through distillation and fractionation of crude oil, re-distillation of unfinished petroleum derivatives, cracking, or other processes. The related industries under SIC 29 are: 2951, Asphalt Paving Mixtures and Blocks; 2952, Asphalt Felts and Coatings; 2992, Lubricating Oils and Greases; and 2999, Petroleum and Coal Products, Not Elsewhere Classified. Certain products that are produced by the petroleum refining industry are also produced by other industries, including: 2865, Cyclic Organic Crudes and Intermediates, and Organic Dyes and Pigments; 2869, Industrial Organic Chemicals; 2819, Industrial Inorganic Chemicals, Not Elsewhere Classified; 2821, Plastic Materials, Synthetic Resins, Nonvulcanizable Elastomers; 2873, Nitrogenous Fertilizers; 4613, Refined Petroleum Pipelines; and 5171, Petroleum Bulk Stations and Terminals. Most crude oil distillation capacity is owned by large, integrated companies with multiple high capacity refining facilities. Small refineries with capacities below 50,000 barrels per day, however, make up about half of all facilities, but only 14 percent of the total crude distillation capacity.

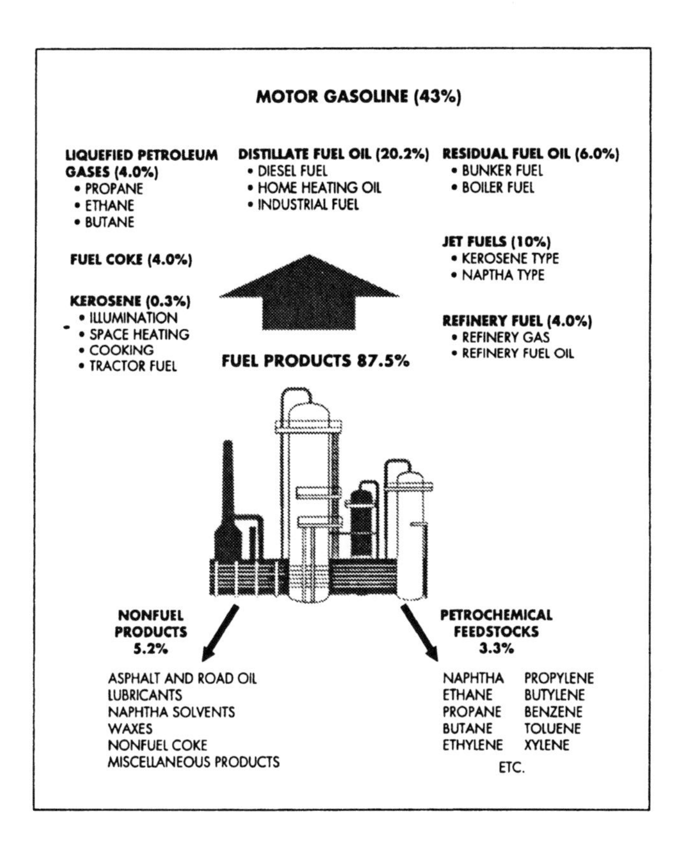

Figure 1. *Refinery products and yields.*

The United States is a net importer of crude oil and petroleum products. In 1994, imports accounted for more than 50 percent of the crude oil used in the U.S. and about 10 percent of finished petroleum products. The imported share of crude oil is expected to increase as U.S. demand for petroleum products increases and the domestic production of crude oil declines. Imported finished petroleum products serve specific market niches arising from logistical considerations, regional shortages, and long-term trade relations between suppliers and refiners. Exports of refined petroleum products, which primarily consist of petroleum coke, residual fuel oil, and distillate fuel oil, account for about four percent of the U.S. refinery output. Exports of crude oil produced in the U.S. account for about one percent of the total U.S. crude oil produced and imported.

The petroleum refining industry in the U.S. has felt considerable economic pressures in the past two decades arising from a number of factors including: increased costs of labor; compliance with new safety and environmental regulations; and the elimination of government subsidies through the Crude Oil Entitlements Program which had encouraged smaller refineries to add capacity throughout the 1970s. A rationalization period began after crude oil pricing and entitlements were decontrolled in early 1981. The market determined that there was surplus capacity and the margins dropped to encourage the closure of the least efficient capacity. Reflecting these pressures, numerous facilities have closed in recent years. Between 1982 and 1994, the number of U.S. refineries as determined by the Department of Energy dropped from 301 to 176. Most of these closures have involved small facilities refining less than 50,000 barrels of crude oil per day. Some larger facilities, however, have also closed in response to economic pressures. Industry representatives cited complying with the increasing environmental regulations, particularly, the requirements of the Clean Air Act Amendments of 1990, as the most important factor affecting petroleum refining in the 1990s. Despite the closing of refineries in recent years, total refinery output of finished products has remained relatively steady with slight increases in the past two to three years. Increases in refinery outputs are attributable to higher utilization rates of refinery capacity, and to incremental additions to the refining capacity at existing facilities as opposed to construction of new refineries.

Demand for refined petroleum products is expected to increase on the average by about 1.5 percent per year, which is slower than the expected growth of the economy. This slower rate of increase of demand will be due to increasing prices of petroleum products as a result of conservation, the development of substitutes for petroleum products, and rising costs of compliance with environmental and safety requirements.

Recent and future environmental and safety regulatory changes are expected to force the petroleum refining industry to make substantial investments in upgrading certain refinery processes to reduce emissions and alter product compositions. For example, industry estimates of the capital costs to comply with the 1990 Clean Air Act Amendments, which mandates specific product compositions are about $35 to $40 billion. There is concern that in some cases it may be more economical for some refineries to close down partially or entirely rather than upgrade facilities to meet these standards. In fact, the U.S. Departments of Energy and Commerce expect refinery shutdowns to continue through the early part of the new decade; however, total crude oil distillation capacity is expected to remain relatively stable as a result of increased capacity and utilization rates at existing facilities. Increases in demand for finished petroleum products will be filled by increased imports. Pressure to meet the Clean Air Amendments is a major driving force for pollution prevention programs in this industry sector.

Processes and Operations

Crude oil is a mixture of many different hydrocarbons and small amounts of impurities. The composition of crude oil can vary significantly depending on its source. Petroleum refineries are a complex system of multiple operations and the operations used at a given refinery depend upon the properties of the crude oil to be refined and the desired products. For these reasons, no two refineries are alike. Portions of the outputs from some processes are re-fed back into the same process, fed to new processes, fed back to a previous process, or blended with other outputs to form finished products. The major unit operations typically involved at petroleum refineries are described briefly below. In addition to those listed below, there are also many special purpose processes that cannot be described here and which may play an important role in a facility's efforts to comply with pollutant discharge and product specification requirements.

Refining crude oil into useful petroleum products can be separated into two phases and a number of supporting operations. The first phase is desalting of crude oil and the subsequent distillation into its various components or "fractions." The second phase is made up of three different types of "downstream" processes: combining, breaking, and reshaping.

Downstream processes convert some of the distillation fractions into petroleum products (residual fuel oil, gasoline, kerosene, etc.) through any combination of different cracking, coking, reforming, and alkylation processes. Supporting operations may include wastewater treatment, sulfur recovery, additive production, heat exchanger cleaning, blowdown systems, blending of products, and storage of products. Refinery pollutant outputs are discussed in more detail later.

Crude Oil Distillation and Desalting

One of the most important operations in a refinery is the initial distillation of the crude oil into its various boiling point fractions. Distillation involves the heating, vaporization, fractionation, condensation, and cooling of feedstocks. This subsection discusses the atmospheric and vacuum distillation processes which when used in sequence result in lower costs and higher efficiencies. This subsection also discusses the important first step of desalting the crude oil prior to distillation.

Desalting - Before separation into fractions, crude oil usually must first be treated to remove corrosive salts. The desalting process also removes some of the metals and suspended solids which cause catalyst deactivation. Desalting involves the mixing of heated crude oil with water (about three to 10 percent of the crude oil volume) so that the salts are dissolved in the water. The water must then be separated from the crude oil in a separating vessel by adding demulsifier chemicals to assist in breaking the emulsion and/or, more commonly, by applying a high potential electric field across the settling vessel to coalesce the polar salt water droplets. The desalting process creates an oily desalter sludge and a high temperature salt water waste stream which is typically added to other process wastewaters for treatment in the refinery wastewater treatment facilities. The water used in crude desalting is often untreated or partially treated water from other refining process water sources.

Atmospheric Distillation - The desalted crude oil is then heated in a heat exchanger and furnace to about 750°F and fed to a vertical, distillation column at atmospheric pressure where most of the feed is vaporized and separated into its various fractions by condensing on 30 to 50 fractionation trays, each corresponding to a different condensation temperature. The lighter fractions condense and are collected towards the top of the column. Heavier fractions, which may not vaporize in the column, are further separated later by vacuum distillation.

Within each atmospheric distillation tower, a number of side streams (at least four) of low-boiling point components are removed from the tower from different trays. These low-boiling point mixtures are in equilibrium with heavier components which must be removed. The side streams are each sent to a different small stripping tower containing four to 10 trays with steam injected under the bottom tray. The steam strips the light-end components from the heavier components and both the steam and light-ends are fed back to the atmospheric distillation tower above the corresponding side stream draw tray. Fractions obtained from atmospheric distillation include naphtha, gasoline, kerosene, light fuel oil, diesel oils, gas oil, lube distillate, and heavy bottoms. Most of these can be sold as finished products, or blended with products from downstream processes. Another product produced in atmospheric distillation, as well as many other refinery processes, is the light, noncondensible refinery fuel gas (mainly methane and ethane). Typically this gas also contains hydrogen sulfide and ammonia gases. The mixture of these gases is known as "sour gas" or "acid gas." The sour gas is sent to the refinery sour gas treatment system which separates the fuel gas so that it can be used as fuel in the refinery heating furnaces. Air emissions during atmospheric distillation arise from the combustion of fuels in the furnaces to heat the crude oil, process vents and fugitive emissions. Oily sour water (condensed steam containing hydrogen sulfate and ammonia) and oil is also generated in the fractionators (refer to Figure 2).

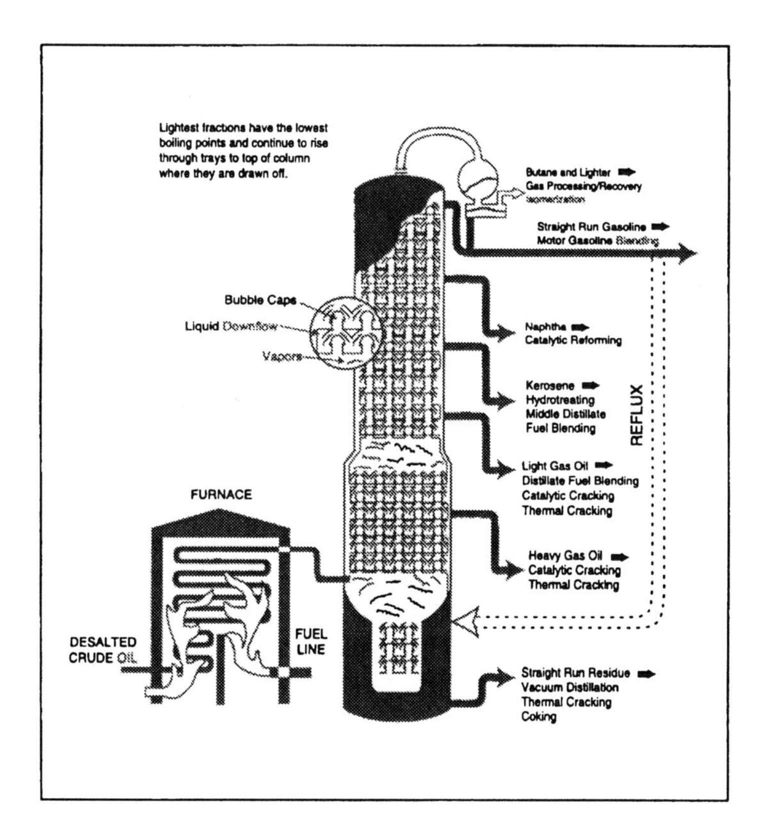

Figure 2. *Crude oil distillation.*

Vacuum Distillation - Heavier fractions from the atmospheric distillation unit that cannot be distilled without cracking under its pressure and temperature conditions are vacuum distilled. Vacuum distillation is simply the distillation of petroleum fractions at a very low pressure (0.2 to 0.7 psia) to increase volatilization and separation. In most systems, the vacuum inside the fractionator is maintained with steam ejectors and vacuum pumps, barometric condensers or surface condensers.

The injection of superheated steam at the base of the vacuum fractionator column further reduces the partial pressure of the hydrocarbons in the tower, facilitating vaporization and separation. The heavier fractions from the vacuum distillation column are processed downstream into more valuable products through either cracking or coking operations.

A potential source of emissions from distillation of crude oil are the combustion of fuels in the furnace and some light gases leaving the top of the condensers on the vacuum distillation column. A certain amount of noncondensable light hydrocarbons and hydrogen sulfide pass through the condenser to a hot well, and then are discharged to the refinery sour fuel system or are vented to a process heater, flare or another control device to destroy hydrogen sulfide. The quantity of these emissions depends on the size of the unit, the type of feedstock, and the cooling water temperature. If barometric condensers are used in vacuum distillation, significant amounts of oily wastewater can be generated. Vacuum pumps and surface condensers have largely replaced barometric condensers in many refineries to eliminate this oily wastewater stream. Oily sour water is also generated in the fractionators.

Downstream Processing

Certain fractions from the distillation of crude oil are further refined in thermal cracking (visbreaking), coking, catalytic cracking, catalytic hydrocracking, hydrotreating, alkylation, isomerization, polymerization, catalytic reforming, solvent extraction, Merox, dewaxing, propane deasphalting and other operations. These downstream processes change the molecular structure of hydrocarbon molecules either by breaking them into smaller molecules, joining them to form larger molecules, or reshaping them into higher quality molecules. For many of the operations, a number of different techniques are used in the industry.

Thermal Cracking/Visbreaking

Thermal cracking, or visbreaking, uses heat and pressure to break large hydrocarbon molecules into smaller, lighter molecules. The process has been largely replaced by catalytic cracking and some refineries no longer employ thermal cracking. Both processes reduce the production of less valuable products such as heavy fuel oil and cutter stock and increase the feed stock to the catalytic cracker and gasoline yields. In thermal cracking, heavy gas oils and residue from the vacuum distillation process are typically the feed stocks. The feed stock is heated in a furnace or other thermal unit to up to 1,000°F and then fed to a reaction chamber which is kept at a pressure of about 140 psig. Following the reactor step, the process stream is mixed with a cooler recycle stream, which stops the cracking reactions. The product is then fed to a flasher chamber, where pressure is reduced and lighter products vaporize and are drawn off. The lighter products are fed to a fractionating tower where the various fractions are separated. The "bottoms" consist of heavy residue, part of which is recycled to cool the process stream leaving the reaction chamber; the remaining bottoms are usually blended into residual fuel (refer to Figure 3).

Air emissions from thermal cracking include emissions from the combustion of fuels in the process heater, vents, and fugitive emissions." A sour water stream is generated in the fractionator.

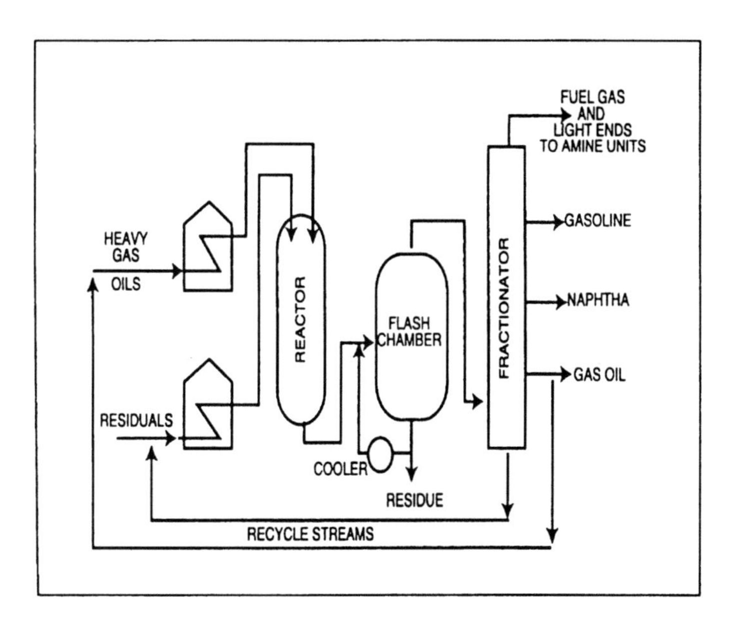

Figure 3. *Thermal cracker flow diagram.*

Coking - A coker flow diagram is shown in Figure 4. Coking is a cracking process used primarily to reduce refinery production of low-value residual fuel oils to transportation fuels, such as gasoline and diesel. As part of the upgrading process, coking also produces petroleum coke, which is essentially solid carbon with varying amounts of impurities, and is used as a fuel for power plants if the sulfur content is low enough. Coke also has nonfuel applications as a raw material for many carbon and graphite products including anodes for the production of aluminum, and furnace electrodes for the production of elemental phosphorus, titanium dioxide, calcium carbide and silicon carbide. A number of different processes are used to produce coke; both *delayed coking* and *fluid coking* are the most widely used processes. Fluid coking produces a higher grade of coke. In delayed coking operations, the same basic process as thermal cracking is used except feed streams are allowed to react longer without being cooled. The delayed coking feed stream of residual oils from various upstream processes is first introduced to a fractionating tower where residual lighter materials are drawn off and the heavy ends are condensed. The heavy ends are removed and heated in a furnace to about 900 to 1,000 °F and then fed to an insulated vessel called a coke drum where the coke is formed. When the coke drum is filled with product, the feed is switched to an empty parallel drum. Hot vapors from the coke drums, containing cracked lighter hydrocarbon products, hydrogen sulfide, and ammonia, are fed back to the fractionator where they can be treated in the sour gas treatment system or drawn off as intermediate products.

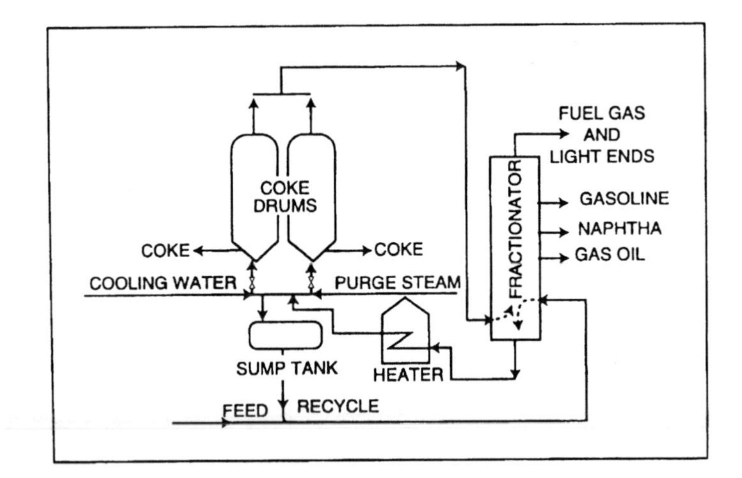

Figure 4. *Coker flow diagram.*

Steam is then injected into the full coke drum to remove hydrocarbon vapors, water is injected to cool the coke, and the coke is removed. Typically, high pressure water jets are used to cut the coke from the drum.

Air emissions from coking operations include the process heater flue gas emissions, fugitive emissions and emissions that may arise from the removal of the coke from the coke drum. The injected steam is condensed and the remaining vapors are typically flared. Wastewater is generated from the coke removal and cooling operations and from the steam injection. In addition, the removal of coke from the drum can release particulate emissions and any remaining hydrocarbons to the atmosphere.

Catalytic Cracking - Catalytic cracking uses heat, pressure and a catalyst to break larger hydrocarbon molecules into smaller, lighter molecules. Catalytic cracking has largely replaced thermal cracking because it is able to produce more gasoline with a higher octane and less heavy fuel oils and light gases. Feed stocks are light and heavy oils from the crude oil distillation unit which are processed primarily into gasoline as well as some fuel oil and light gases. Most catalysts used in catalytic cracking consist of mixtures of crystalline synthetic silica-alumina, termed "zeolites," and amorphous synthetic silica alumina. The catalytic cracking processes, as well as most other refinery catalytic processes, produce coke which collects on the catalyst surface and diminishes its catalytic properties. The catalyst, therefore, needs to be regenerated continuously or periodically essentially by burning the coke off the catalyst at high temperatures. The method and frequency in which catalysts are regenerated are a major factor in the design of catalytic cracking units. A number of different catalytic cracking designs are currently in use, including fixed-bed reactors, moving-bed reactors, fluidized-bed reactors, and once-through units. The fluidized- and moving-bed reactors are by far the most prevalent.

Fluidized-bed catalytic cracking units (FCCUs) are the most common catalytic cracking units. In the fluidized-bed process, oil and oil vapor preheated to 500 to 800 °F is contacted with hot catalyst at about 1,300 °F either in the reactor itself or in the feed line (called the 'riser') to the reactor. The catalyst is in a fine, granular form which, when mixed with the vapor, has many of the properties of a fluid. The fluidized catalyst and the reacted hydrocarbon vapor separate mechanically in the reactor and any oil remaining on the catalyst is removed by steam stripping. The cracked oil vapors are then fed to a fractionation tower where the various desired fractions are separated and collected. The catalyst flows into a separate vessel(s) for either single- or two-stage regeneration by burning off the coke deposits with air.

In the moving-bed process, oil is heated to up to 1,300 °F and is passed under pressure through the reactor where it comes into contact with a catalyst flow in the form of beads or pellets. The cracked products then flow to a fractionating tower where the various compounds are separated and collected. The catalyst is regenerated in a continuous process where deposits of coke on the catalyst are burned off. Some units also use steam to strip remaining hydrocarbons and oxygen from the catalyst before being fed back to the oil stream. In recent years moving-bed reactors have largely been replaced by fluidized-bed reactors.

Catalytic cracking is one of the most significant sources of air pollutants at refineries. Figure 5 shows a flow scheme for a catalytic cracking unit. Air emissions from catalytic cracking operations include: the process heater flue gas emissions, fugitive emissions, and emissions generated during regeneration of the catalyst. Relatively high concentrations of carbon monoxide can be produced during regeneration of the catalyst which is typically converted to carbon dioxide either in the regenerator or further downstream in a carbon monoxide waste heat boiler. In addition, a significant amount of fine catalyst dust is produced in FCCUs as a result of the constant movement of the catalyst grains against each other. Much of this dust, consisting primarily of alumina and relatively small amounts of nickel, is carried with the carbon monoxide strewn to the carbon monoxide burner.

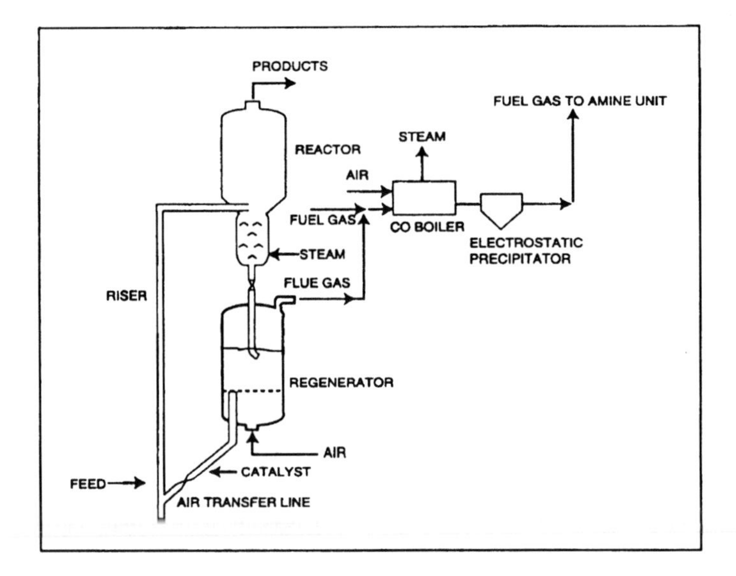

Figure 5. Catalytic cracking flow diagram.

The catalyst dust is then separated from the resulting carbon dioxide stream via cyclones and/or electrostatic precipitators and is sent off-site for disposal or treatment. Generated wastewater is typically sour water from the fractionator containing some oil and phenols. Wastewater containing metal impurities from the feed oil can also be generated from the steam used to purge and regenerate catalysts.

Catalytic Hydrocracking - Catalytic hydrocracking normally utilizes a fixed-bed catalytic cracking reactor with cracking occurring under substantial pressure (1,200 to 2,000 psig) in the presence of hydrogen. Feedstocks to hydrocracking units are often those fractions that are the most difficult to crack and cannot be cracked effectively in catalytic cracking units. The feedstocks include: middle distillates, cycle oils, residual fuel oils and reduced crudes.

The hydrogen suppresses the formation of heavy residual material and increases the yield of gasoline by reacting with the cracked products. However, this process also breaks the heavy, sulfur and nitrogen bearing hydrocarbons and releases these impurities to where they could potentially foul the catalyst. For this reason, the feedstock is often first hydrotreated to remove impurities before being sent to the catalytic hydrocracker. Sometimes hydrotreating is accomplished by using the first reactor of the hydrocracking process to remove impurities. Water also has a detrimental effect on some hydrocracking catalysts and must be removed before being fed to the reactor. The water is removed by passing the feed stream through a silica gel or molecular sieve dryer. Depending on the products desired and the size of the unit, catalytic hydrocracking is conducted in either single stage or multi-stage reactor processes. Most catalysts consist of a crystalline mixture of silica-alumina with small amounts of rare earth metals. Hydrocracking feedstocks are usually first hydrotreated to remove the hydrogen sulfide and ammonia that will poison the catalyst. Sour gas and sour water streams are produced at the fractionator, however, if the hydrocracking feedstocks are first hydrotreated to remove impurities, both streams will contain relatively low levels of hydrogen sulfide and ammonia. Hydrocracking catalysts are typically regenerated off-site after two to four years of operation. Therefore, little or no emissions are generated from the regeneration processes. Air emissions arise from the process heater, vents, and fugitive emissions.

Hydrotreating/Hydroprocessing - Hydrotreating and hydroprocessing are similar processes used to remove impurities such as sulfur, nitrogen, oxygen, halides and trace metal impurities that may deactivate process catalysts. Hydrotreating also upgrades the quality of fractions by converting olefins and diolefins to paraffins for the purpose of reducing gum formation in fuels. Hydroprocessing, which typically uses residuals from the crude distillation units, also cracks these heavier molecules to lighter more saleable products. Both hydrotreating and hydroprocessing units are usually placed upstream of those processes in which sulfur and nitrogen could have adverse effects on the catalyst, such as catalytic reforming and hydrocracking units. The processes utilize catalysts in the presence of substantial amounts of hydrogen under high pressure and temperature to react the feedstocks and impurities with hydrogen. The reactors are nearly all fixed-bed with catalyst replacement or regeneration done after months or years of operation often at an off-site facility. In addition to the treated products, the process produces a stream of light fuel gases, hydrogen sulfide, and ammonia. The treated product and hydrogen-rich gas are cooled after they leave the reactor before being separated. The hydrogen is recycled to the reactor. The off-gas stream may be very rich in hydrogen sulfide and light fuel gas. The fuel gas and hydrogen sulfide are typically sent to the sour gas treatment unit and sulfur recovery unit. Catalysts are typically cobalt or molybdenum oxides on alumina, but can also contain nickel and tungsten. Air emissions from hydrotreating may arise from process heater flue gas, vents, and fugitive emissions. Figure 6 provides a simplified flow diagram.

Alkylation - Alkylation is used to produce a high octane gasoline blending stock from the isobutane formed primarily during catalytic cracking and coking operations, but also from catalytic reforming, crude distillation and natural gas processing. Alkylation joins an olefin and an isoparaffin compound using either a sulfuric acid or hydrofluoric acid catalyst. The products are alkylates including propane and butane liquids. When the concentration of acid becomes less than 88 percent, some of the acid must be removed and replaced with stronger acid. In the hydrofluoric acid process, the slip stream of acid is redistilled.

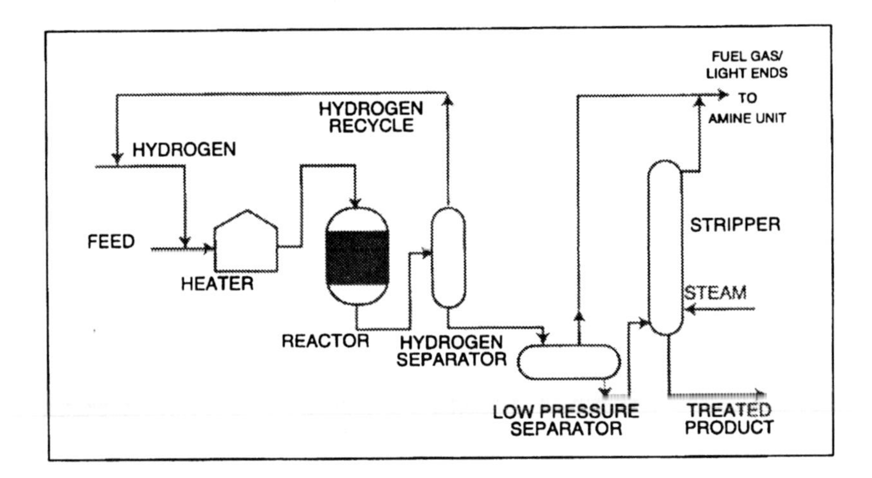

Figure 6. *Flow diagram for hydrotreating.*

Dissolved polymerization products are removed from the acid as a thick dark oil. The concentrated hydrofluoric acid is recycled and the net consumption is about 0.3 pounds per barrel of alkylates produced. Hydrofluoric acid alkylation units require special engineering design, operator training and safety equipment precautions to protect operators from accidental contact with hydrofluoric acid which is an extremely hazardous substance. In the sulfuric acid process, the sulfuric acid removed must be regenerated in a sulfuric acid plant which is generally not a part of the alkylation unit and may be located off-site. Spent sulfuric acid generation is substantial; typically in the range of 13 to 30 pounds per barrel of alkylate. Air emissions from the alkylation process may arise from process vents and fugitive emissions.

Isomerization - Isomerization is used to alter the arrangement of a molecule without adding or removing anything from the original molecule. Typically, paraffins (butane or pentane from the crude distillation unit) are converted to isoparaffins having a much higher octane. Isomerization reactions take place at temperatures in the range of 200 to 400 °F in the presence of a catalyst that usually consists of platinum on a base material. Two types of catalysts are currently in use. One requires the continuous addition of small amounts of organic chlorides which are converted to hydrogen chloride in the reactor. In such a reactor, the feed must be free of oxygen sources including water to avoid deactivation and corrosion problems. The other type of catalyst uses a molecular sieve base and does not require a dry and oxygen free feed. Both types of isomerization catalysts require an atmosphere of hydrogen to minimize coke deposits; however, the consumption of hydrogen is negligible. Catalysts typically need to be replaced about every two to three years or longer. Platinum is then recovered from the used catalyst off site. Light ends are stripped from the product stream leaving the reactor and are then sent to the sour gas treatment unit. Some isomerization units utilize caustic treating of the light fuel gas stream to neutralize any entrained hydrochloric acid. This will result in a calcium chloride (or other salts) waste stream. Air emissions may arise from the process heater, vents and fugitive emissions. Wastewater streams include caustic wash and sour water.

Polymerization - Polymerization is occasionally used to convert propene and butene to high octane gasoline blending components. The process is similar to alkylation in its feed and products, but is often used as a less expensive alternative to alkylation. The reactions typically take place under high pressure in the presence of a phosphoric acid catalyst. The feed must be free of sulfur, which poisons the catalyst; basic materials, which neutralize the catalyst; and oxygen, which affects the reactions. The propene and butene feed is washed first with caustic to remove mercaptans (molecules containing sulfur), then with an amine solution to remove hydrogen sulfide, then with water to remove caustics and amines, and finally dried by passing through a silica gel or molecular sieve dryer. Air emissions of sulfur dioxide may arise during the caustic washing operation. Spent catalyst, which typically is not regenerated, is occasionally disposed as a solid waste. Wastewater streams will contain caustic wash and sour water with amines and mercaptans.

Catalytic Reforming - Catalytic reforming uses catalytic reactions to process primarily low octane heavy straight run (from the crude distillation unit) gasolines and naphthas into high octane aromatics (including benzene). There are four major types of reactions which occur during reforming processes: (1) dehydrogenation of naphthenes to aromatics; (2) dehydrocyclization of paraffins to aromatics; (3) isomerization; and (4) hydrocracking. The dehydrogenation reactions are very endothermic, requiring that the hydrocarbon stream be heated between each catalyst bed. All but the hydrocracking reaction release hydrogen which can be used in the hydrotreating or hydrocracking processes. Fixed-bed or moving bed processes are utilized in a series of three to six reactors. Feedstocks to catalytic reforming processes are usually hydrotreated first to remove sulfur, nitrogen and metallic contaminants. In continuous reforming processes, catalysts can be regenerated one reactor at a time, once or twice per day, without disrupting the operation of the unit. In semi regenerative units, regeneration of all reactors can be carried out simultaneously after three to 24 months of operation by first shutting down the process. Because the recent reformulated gasoline rules have limited the allowable amount of benzene in gasoline, catalytic reforming is being used less as an octane enhancer than in past years.

Air emissions from catalytic reforming arise from the process heater gas and fugitive emissions. The catalysts used in catalytic reforming processes are usually very expensive and extra precautions are taken to ensure that catalyst is not lost. When the catalyst has lost its activity and can no longer be regenerated, the catalyst is usually sent off-site for recovery of the metals. Subsequent air emissions from catalyst regeneration is, therefore, relatively low. Relatively small volumes of wastewater containing sulfides, ammonia, and mercaptans may be generated from the stripping tower used to remove light ends from the reactor effluent.

Solvent Extraction - Solvent extraction uses solvents to dissolve and remove aromatics from lube oil feed stocks, improving viscosity, oxidation resistance, color and gum formation. A number of different solvents are used with the two most common being furfural and phenol. Typically, feed lube stocks are contacted with the solvent in a packed tower or rotating disc contactor. Each solvent has a different solvent-to-oil ratio and recycle ratio within the tower. Solvents are recovered from the oil stream through distillation and steam stripping in a fractionator. The stream extracted from the solvent will likely contain high concentrations of hydrogen sulfide, aromatics, naphthenes and other hydrocarbons, and is often fed to the hydrocracking unit. The water stream leaving the fractionator will likely contain some oil and solvents.

Chemical Treating - In petroleum refining, chemical treating is used to remove or change the undesirable properties associated with sulfur, nitrogen, or oxygen compound contaminates in petroleum products. Chemical treating is accomplished by either extraction or oxidation (also known as sweetening), depending upon the product. Extraction is used to remove sulfur from the very light petroleum fractions, such as propane/propylene (PP) and butane/butylene (BB). Sweetening, though, is more effective on gasoline and middle distillate products. A typical extraction process is "Merox" extraction. Merox extraction is used to remove mercaptans (organic sulfur compounds) from PP and BB streams. PP streams may undergo amine treating before the Merox extraction to remove excess H_2S which tends to fractionate with PP and interferes with the Merox process. A caustic prewash of the PP and BB removes any remaining trace H_2S prior to Merox extraction. The PP and BB streams are passed up through the trays of an extraction tower. Caustic solution flowing down the extraction tower absorbs mercaptan from the PP and BB streams. The rich caustic is then regenerated by oxidizing the mercaptans to disulfide in the presence of aqueous Merox catalyst and the lean caustic recirculated to the extraction tower. The disulfide is insoluble in the caustic and can be separated.

Oxidation or "sweetening" is used on gasoline and distillate fractions. A common oxidation process is also a Merox process that uses a solid catalyst bed. Air and a minimum amount of alkaline caustic ("mini-alky" operation) is injected into the hydrocarbon stream. As the hydrocarbon passes through the Merox catalyst bed, sulfur mercaptans are oxidized to disulfide. In the sweetening Merox process, the caustic is not regenerated. The disulfide can remain with the gasoline product, since it does not possess the objectionable odor properties of mercaptans; hence, the product has been "sweetened."

In the extraction process, a waste oily disulfide stream leaves the separator. Air emissions arise from fugitive hydrocarbons and the process vents on the separator which may contain disulfides.

Dewaxing - Dewaxing of lubricating oil base stocks is necessary to ensure that the oil will have the proper viscosity at lower ambient temperatures. Two types of dewaxing processes are used: selective hydrocracking and solvent dewaxing. In selective hydrocracking, one or two zeolite catalysts are used to selectively crack the wax paraffins. Solvent dewaxing is more prevalent. In solvent dewaxing, the oil feed is diluted with solvent to lower the viscosity, chilled until the wax is crystallized, and then filtered to remove the wax. Solvents used for the process include propane and mixtures of methyl ethyl ketone (MEK) with methyl isobutyl ketone (MIBK) or MEK with toluene. Solvent is recovered from the oil and wax through heating, two-stage flashing, followed by steam stripping. The solvent recovery stage results in solvent contaminated water which typically is sent to the wastewater treatment plant. The wax is either used as feed to the catalytic cracker or is deoiled and sold as industrial wax. Air emissions may arise from fugitive emissions of the solvents.

Propane Deasphalting - Propane deasphalting produces lubricating oil base stocks by extracting asphaltenes and resins from the residuals of the vacuum distillation unit. Propane is usually used to remove asphaltenes due to its unique solvent properties. At lower temperatures (100 to 140 °F), paraffins are very soluble in propane and at higher temperatures (about 200 °F) all hydrocarbons are almost insoluble in propane. The propane deasphalting process is similar to solvent extraction in that a packed or baffled extraction tower or rotating disc contactor is used to mix the oil feed stocks with the solvent. In the tower method, four to eight volumes of propane are fed to the bottom of the tower for every volume of feed flowing down from the top of the tower. The oil, which is more soluble in the propane dissolves and flows to the top. The asphaltene and resins flow to the bottom of the tower where they are removed in a propane mix. Propane is recovered from the two streams through two-stage flash systems followed by steam stripping in which propane is condensed and removed by cooling at high pressure in the first stage and at low pressure in the second stage. The asphalt recovered can be blended with other asphalts or heavy fuels, or can be used as feed to the coker. The propane recovery stage results in propane contaminated water which typically is sent to the wastewater treatment plant. Air emissions may arise from fugitive propane emissions and process vents.

Supporting Operations

Many important refinery operations are not directly involved in the production of hydrocarbon fuels but serve in a supporting role. Some of the major supporting processes are described below.

Wastewater Treatment

Relatively large volumes of water are used by the petroleum refining industry. Four types of wastewater are produced: surface water runoff, cooling water, process water, and sanitary wastewater. Surface water runoff is intermittent and will contain constituents from spills to the surface, leaks in equipment and any materials that may have collected in drains. Runoff surface water also includes water coming from crude and product storage tank roof drains.

A large portion of water used in petroleum refining is used for cooling. Cooling water typically does not come into direct contact with process oil streams and therefore contains less contaminants than process wastewater. Most cooling water is recycled over and over with a bleed or blowdown stream to the wastewater treatment unit to control the concentration of contaminants and the solids content in the water. Cooling towers within the recycle loop cool the water using ambient air. Some cooling water, termed "once through," is passed through a process unit once and is then discharged directly without treatment in the wastewater treatment plant. The water used for cooling often contains chemical additives such as chromates, phosphates, and antifouling biocides to prevent scaling of pipes and biological growth. It should be noted, however, that many refineries in the United States no longer use chromates in cooling water as anti-fouling agents, however this is not the case in other parts of the world. Although cooling water usually does not come into direct contact with oil process streams, it also may contain some oil contamination due to leaks in the process equipment.

Water used in processing operations also accounts for a significant portion of the total wastewater. Process wastewater arises from desalting crude oil, steam stripping operations, pump gland cooling, product fractionator reflux drum drains and boiler blowdown. Because process water often comes into direct contact with oil, it is usually highly contaminated. Petroleum refineries typically utilize primary and secondary wastewater treatment technologies. Primary wastewater treatment consists of the separation of oil, water and solids in two stages. During the first stage, an API separator, a corrugated plate interceptor, or other separator design is used. Wastewater moves very slowly through the separator allowing free oil to float to the surface and be skimmed off, and solids to settle to the bottom and be scraped off to a sludge collecting hopper. The second stage utilizes physical or chemical methods to separate emulsified oils from the wastewater. Physical methods may include the use of a series of settling ponds with a long retention time, or the use of dissolved air flotation (DAF). In DAF, air is bubbled through the wastewater, and both oil and suspended solids are skimmed off the top. Chemicals, such as ferric hydroxide or aluminum hydroxide, can be used to coagulate impurities into a froth or sludge which can be more easily skimmed off the top. Some wastes associated with the primary treatment of wastewater at petroleum refineries may be considered hazardous and include: API separator sludge, primary treatment sludge, sludges from other gravitational separation techniques, float from DAF units, and wastes from settling ponds.

After primary treatment, the wastewater can be discharged to a publicly owned treatment works or undergo secondary treatment before being discharged directly to surface waters under a National Pollution Discharge Elimination System (NPDES) permit. In secondary treatment, dissolved oil and other organic pollutants may be consumed biologically by microorganisms. Biological treatment may require the addition of oxygen through a number of different techniques, including activated sludge units, trickling filters, and rotating biological contactors. Secondary treatment generates bio-mass waste which is typically treated anaerobically, and then dewatered.

Some refineries employ an additional stage of wastewater treatment called polishing to meet discharge limits. The polishing step can involve the use of activated carbon, anthracite coal, or sand to filter out any remaining impurities, such as biomass, silt, trace metals and other inorganic chemicals, as well as any remaining organic chemicals.

Certain refinery wastewater streams are treated separately, prior to the wastewater treatment plant, to remove contaminants that would not easily be treated after mixing with other wastewater. One such waste stream is the sour water drained from distillation reflux drums. Sour water contains dissolved hydrogen sulfide and other organic sulfur compounds and ammonia which are stripped in a tower with gas or steam before being discharged to the wastewater treatment plant.

Wastewater treatment plants are also a significant source of refinery air emissions and solid wastes. Air releases arise from fugitive emissions from the numerous tanks, ponds and sewer system drains. Solid wastes are generated in the form of sludges from a number of the treatment units.

Gas Treatment and Sulfur Recovery

Sulfur is removed from a number of refinery process off-gas streams (sour gas) in order to meet the SO_x emissions limits of the CAA and to recover saleable elemental sulfur. Process off-gas streams, or sour gas, from the coker, catalytic cracking unit, hydrotreating units and hydroprocessing units can contain high concentrations of hydrogen sulfide mixed with light refinery fuel gases. Before elemental sulfur can be recovered, the fuel gases (primarily methane and ethane) need to be separated from the hydrogen sulfide. This is typically accomplished by dissolving the hydrogen sulfide in a chemical solvent. Solvents most commonly used are amines, such as diethanolamine (DEA). Dry adsorbents such as molecular sieves, activated carbon, iron sponge and zinc oxide are also used. In the amine solvent processes, DEA solution or another amine solvent is pumped to an absorption tower where the gases are contacted and hydrogen sulfide is dissolved in the solution. The fuel gases are removed for use as fuel in process furnaces in other refinery operations. The amine-hydrogen sulfide solution is then heated and steam stripped to remove the hydrogen sulfide gas.

Current methods for removing sulfur from the hydrogen sulfide gas streams are typically a combination of two processes: the Claus Process followed by the Beaven Process, Scot Process, or the Wellman-Land Process. The Claus process consists of partial combustion of the hydrogen sulfide-rich gas stream (with one-third the stoichiometric quantity of air) and then reacting the resulting sulfur dioxide and unburned hydrogen sulfide in the presence of a bauxite catalyst to produce elemental sulfur. Refer to the process flow diagram in Figure 7.

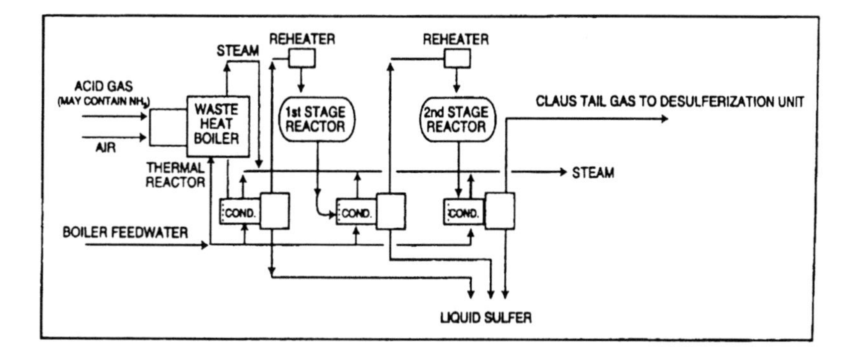

Figure 7. *Claus sulfur recovery flow diagram.*

Since the Claus process by itself removes only about 90 percent of the hydrogen sulfide in the gas stream, the Beaven, SCOT, or Wellman-Lord processes are often used to further recover sulfur. In the Beaven process, the hydrogen sulfide in the relatively low concentration gas stream from the Claus process can be almost completely removed by absorption in a quinone solution.

The dissolved hydrogen sulfide is oxidized to form a mixture of elemental sulfur and hydro-quinone. The solution is injected with air or oxygen to oxidize the hydro-quinone back to quinone. The solution is then filtered or centrifuged to remove the sulfur and the quinone is then reused.

The Beaven process is also effective in removing small amounts of sulfur dioxide, carbonyl sulfide, and carbon disulfide that are not affected by the Claus process. These compounds are first converted to hydrogen sulfide at elevated temperatures in a cobalt molybdate catalyst prior to being fed to the Beaven unit.

Air emissions from sulfur recovery units will consist of hydrogen sulfide, SO_x, and NO_x in the process tail gas as well as fugitive emissions and releases from vents.

The SCOT process is also widely used for removing sulfur from the Claus tail gas. The sulphur compounds in the Claus tail gas are converted to hydrogen sulfide by heating and passing it through a cobalt-molybdenum catalyst with the addition of a reducing gas. The gas is then cooled and contacted with a solution of di-isopropanolamine (DIPA) which removes all but trace amounts of hydrogen sulfide. The sulfide-rich DIPA is sent to a stripper where hydrogen sulfide gas is removed and sent to the Claus plant. The DIPA is returned to the absorption column.

Additive Production

A number of chemicals (mostly alcohols and ethers) are added to motor fuels to either improve performance or meet federal and state environmental requirements. Since the 1970s, alcohols (methanol and ethanol) and ethers have been added to gasoline to increase octane levels and reduce carbon monoxide generation in place of the lead additives which were being phased out as required by the 1970 Clean Air Act. In 1990, the more stringent Clean Air Act Amendments established minimum and maximum amounts of chemically combined oxygen in motor fuels as well as an upper limit on vapor pressure. As a result, alcohol additives have been increasingly supplemented or replaced with a number of different ethers which are better able to meet both the new oxygen requirements and the vapor pressure limits.

The most common ethers being used as additives are methyl tertiary butyl ether (MTBE), and tertiary amyl methyl ether (TAME). Many of the larger refineries manufacture their own supplies of MTBE and TAME by reacting isobutylene and/or isoamylene with methanol. Smaller refineries usually buy their supplies from chemical manufacturers or the larger refineries.

Isobutylene is obtained from a number of refinery sources including: the light naphtha from the FCCU and coking units, the by-product from steam cracking of naphtha or light hydrocarbons during the production of ethylene and propylene, catalytic dehydrogenation of isobutane, and conversion of tertiary butyl alcohol recovered as a by-product in the manufacture of propylene oxides. Several different processes are currently in use to produce MTBE and TAME from isobutylene and methanol. Most processes use a two stage acidic ion exchange resin catalyst. The reaction is exothermic and cooling to the proper reaction temperature is critical in obtaining the optimal conversion efficiency. The process usually produces an MTBE or TAME stream and a relatively small stream of unreacted hydrocarbons and methanol. The methanol is extracted in a water wash and the resulting methanol-water mixture is distilled to recover the methanol for recycling.

Heat Exchanger Cleaning

Heat exchangers are used abundantly throughout petroleum refineries to heat or cool petroleum process streams. The heat exchangers consist of bundles of pipes, tubes, plate coils, or steam coils enclosing heating or cooling water, steam, or oil to transfer heat indirectly to or from the oil process stream. The bundles are cleaned periodically to remove accumulations of scales, sludge and any oily residues.

Because chromium has almost been eliminated as a cooling water additive, wastes generated from the cleaning of heat exchanger bundles no longer account for a significant portion of the hazardous wastes generated at refining facilities. The sludge generated may contain lead or chromium, although some refineries which do not produce leaded gasoline and which use non-chrome corrosion inhibitors typically do not generate sludge that contains these constituents. Oily wastewater is also generated during heat exchanger cleaning.

Blowdown System

Most refinery process units and equipment are manifolded into a collection unit, called the blowdown system. Blowdown systems provide for the safe handling and disposal of liquid and gases that are either automatically vented from the process units through pressure relief valves, or that are manually drawn from units. Recirculated process streams and cooling water streams are often manually purged to prevent the continued build up of contaminants in the stream. Part or all of the contents of equipment can also be purged to the blowdown system prior to shutdown before normal or emergency shutdowns.

Blowdown systems utilize a series of flash drums and condensers to separate the blowdown into its vapor and liquid components. The liquid is typically composed of mixtures of water and hydrocarbons containing sulfides, ammonia, and other contaminants, which are sent to the wastewater treatment plant.

The gaseous component typically contains hydrocarbons, hydrogen sulfide, ammonia, mercaptans, solvents, and other constituents, and is either discharged directly to the atmosphere or is combusted in a flare. The major air emissions from blowdown systems are hydrocarbons in the case of direct discharge to the atmosphere and sulfur oxides when flared.

Blending

Blending is the final operation in petroleum refining. It consists of mixing the products in various proportions to meet specifications such as vapor pressure, specific gravity, sulfur content, viscosity, octane number, initial boiling point, and pour point. Blending can be carried out inline or in batch blending tanks. Air emissions from blending are fugitive VOCs from blending tanks, valves, pumps and mixing operations.

Storage Tanks

Storage tanks are used throughout the refining process to store crude oil and intermediate process feeds for cooling and further processing. Finished petroleum products are also kept in storage tanks before transport off site. Storage tank bottoms are mixtures of iron rust from corrosion, sand, water, and emulsified oil and wax, which accumulate at the bottom of tanks. Liquid tank bottoms (primarily water and oil emulsions) are periodically drawn off to prevent their continued build up.

Tank bottom liquids and sludge are also removed during periodic cleaning of tanks for inspection. Tank bottoms may contain amounts of tetraethyl or tetramethyl lead (although this is increasingly rare due to the phaseout of leaded products), other metals, and phenols. Solids generated from leaded gasoline storage tank bottoms are listed as a RCRA hazardous waste. Even if equipped with floating tops, storage tanks account for considerable VOC emissions at petroleum refineries. A study of petroleum refinery emissions found that the majority of tank losses occurred through tank seals on gasoline storage tanks.

Cooling Towers

Cooling towers cool heated water by circulating the water through a tower with a predetermined flow of ambient air pushed with large fans. A certain amount of water exits the system through evaporation, mist droplets and as bleed or blowdown to the wastewater treatment system. Therefore, make-up water in the range of about five percent of the circulation rate is required.

MATERIAL BALANCE INFORMATION

Raw material input to petroleum refineries is primarily crude oil; however, petroleum refineries use and generate an enormous number of chemicals, many of which leave the facilities as discharges of air emissions, wastewater, or solid waste. Pollutants generated typically include VOCs, carbon monoxide (CO), sulfur oxides (SO_x), nitrogen oxides (NO_x), particulates, ammonia (NH_3) hydrogen sulfide (H_2S) metals, spent acids, and numerous toxic organic compounds.

When discussing material outputs of the petroleum refining industry, it is important to note the relationship between the outputs of the industry itself and the outputs resulting from the use of refinery products. Petroleum refineries play an important role in the U.S. economy, supplying approximately 40 percent of the total energy used in the U.S. and virtually all of the energy consumed in the transportation sector.

The pollutant outputs from the refining facilities, however, are modest in comparison to the pollutant outputs realized from the consumption of petroleum products by the transportation sector, electric utilities, chemical manufacturers and other industrial and commercial users.

Air emissions from refineries include fugitive emissions of the volatile constituents in crude oil and its fractions, emissions from the burning of fuels in process heaters, and emissions from the various refinery processes themselves. Fugitive emissions occur throughout refineries and arise from the thousands of potential fugitive emission sources such as valves, pumps, tanks, pressure relief valves, flanges, etc.

While individual leaks are typically small, the sum of all fugitive leaks at a refinery can be one of its largest emission sources. Fugitive emissions can be reduced through a number of techniques, including improved leak resistant equipment, reducing the number of tanks and other potential sources and, perhaps the most effective method, an ongoing Leak Detection and Repair (LDAR) program.

The numerous process heaters used in refineries to heat process streams or to generate steam (boilers) for heating or steam stripping, can be potential sources of SO_x, NO_x, CO, particulate matter and hydrocarbons emissions. When operating properly and when burning cleaner fuels such as refinery fuel gas, fuel oil or natural gas, these emissions are relatively low. If, however, combustion is not complete, or heaters are fired with refinery fuel pitch or residuals, emissions can be significant.

The majority of gas streams exiting each refinery process contain varying amounts of refinery fuel gas, hydrogen sulfide and ammonia. These streams are collected and sent to the gas treatment and sulfur recovery units to recover the refinery fuel gas and sulfur. Emissions from the sulfur recovery unit typically contains some H_2S, SO_x, and NO_x.

Other emissions sources from refinery processes arise from periodic regeneration of catalysts. These processes generate streams that may contain relatively high levels of carbon monoxide, particulates and VOCs. Before being discharged to the atmosphere, such off-gas streams may be treated first through a carbon monoxide boiler to burn carbon monoxide and any VOCs, and then through an electrostatic precipitator or cyclone separator to remove particulate matter.

Wastewaters consist of cooling water, process water, sanitary sewage water, and storm water. Wastewaters are treated in onsite wastewater treatment facilities and then discharged to POTWs or discharged to surfaces waters under NPDES permits. In addition, some facilities use underground injection of some wastewater streams.

Many refineries unintentionally release, or have unintentionally released in the past, liquid hydrocarbons to ground water and surface waters. At some refineries contaminated ground water has migrate off-site and resulted in continuous "seeps" to surface waters. While the actual volume of hydrocarbons released in such a manner are relatively small, there is the potential to contaminate large volumes of ground water and surface water possibly posing a substantial risk to human health and the environment.

There are a variety of other wastes that are generated from many of the refining processes, petroleum handling operations, as well as wastewater treatment. Both hazardous and nonhazardous wastes are generated, treated and disposed.

Residual refinery wastes are typically in the form of sludges, spent process catalysts, filter clay, and incinerator ash. Treatment of these wastes includes incineration, land treating off-site, land filling onsite, land filling off-site, chemical fixation, neutralization, and other treatment methods.

A significant portion of the nonpetroleum product outputs of refineries is transported off-site and sold as by-products. These outputs include sulfur, acetic acid, phosphoric acid, and recovered metals.

Metals from catalysts and from the crude oil that have deposited on the catalyst during the production often are recovered by third party recovery facilities.

Table 1 provides a summary of the typical material outputs from major petroleum refining operations. Where possible, typical quantities and concentrations of pollutants are reported. These should be considered very approximate figures since no two refinery operations are identical. However, they do provide a general idea of the quantities, flows, and levels of different types of priority pollutants handled by refinery operations.

Table 1. *Typical Material Outputs from Selected Petroleum Refining Processes.*

AIR EMISSIONS	PROCESS WASTE WATER	RESIDUAL WASTES GENERATED
	Process: Crude oil desalting	
Heater stack gas (CO, SO_x, NO_x, hydrocarbons and particulates), fugitive emissions (hydrocarbons).	Flow=2.1 Gal/Bbl Oil, H_2S, NH, phenol, high levels of SS, dissolved solids, high BOD, high temperature.	Crude oil/desalter sludge (iron rust, clay, sand, water, emulsified oil and wax, metals).
	Process: Atmospheric distillation and Vacuum Distillation	
Heater stack gas (CO, SO_x, NO_x, hydrocarbons and PM), vents and fugitive emissions (hydrocarbons) Steam ejector emissions (hydrocarbons), heater stack gas (CO, SO_x, NO_x, hydrocarbons and PM), vents and fugitive emissions (hydrocarbons).	Flow=26.0 Gal/Bbl Oil, H_2S NH, suspended solids, chlorides, mercaptans, phenol, elevated pH.	Typically, little or no residual waste generated.
	Process: Thermal Cracking/Visbreaking	
Heater stack gas (CO, SO_x, NO_x, HCs and PM), vents and fugitive emissions (HCs).	Flow=2.0 Gal/Bbl Oil, H_2S NH, phenol, suspended solids, high pH, BOD, COD.	Typically, little or no residual waste generated.
AIR EMISSIONS	**PROCESS WASTE WATER**	**RESIDUAL WASTES GENERATED**
	Process: Coking	
Heater stack gas (CO, SO_x, NO_x, hydrocarbons and PM), vents and fugitive emissions (HCs) and decoking emissions (HCs and PM).	Flow=1.0 Gal/Bbl High pH, H_2S, NH_3, SS, COD.	Coke dust (carbon particles and HCs).
	Process: Catalytic Cracking	
Heater stack gas (CO, SO_x, NO_x, HCs and PM), fugitive emissions (HCs) and catalyst regeneration (CO, NO_x, SO_x, and PM).	Flow=15.0 Gal/Bbl High levels of oil, SS, phenols, cyanides, H_2S, NH3, high pH, BOD, COD.	Spent catalysts (metals from crude oil and hydrocarbons), spent catalyst fines from ESPs (aluminum silicate and metals).

Table 1. continued

AIR EMISSIONS	PROCESS WASTE WATER	RESIDUAL WASTES GENERATED
	Process: Catalytic Hydrocracking	
Heater stack gas (CO, SO_x, NO_x, hydrocarbons and PM), fugitive emissions (hydrocarbons) and catalyst regeneration (CO, NO_x, SO_x, and catalyst dust).	Flow=2.0 Gal/Bbl High COD, SS, H_2S, relatively low levels of BOD.	Spent catalysts fines (metals from crude oil, and hydrocarbons).
	Process: Hydrotreating/Hydroprocessing	
Heater stack gas (CO, SO_x, NO_x, hydrocarbons and PM), vents and fugitive emissions (HCs) and catalyst regeneration (CO, NO_x, SO_x).	Flow=1.0 Gal/Bbl H_2S, NH, High pH, phenols, SS, BOD, COD.	Spent catalyst fines (aluminum silicate and metals).
	Process: Alkylation	
Heater stack gas (CO, SO_x, NO_x, HCs and PM), vents and fugitive emissions (HCs).	Low pH, SS, dissolved solids, COD, H_2S, spent sulfuric acid.	Neutralized alkylation sludge (sulfuric acid or calcium fluoride, HCs).
	Process: Isomerization	
Heater stack gas (CO, SO_x, NO_x, HCs and PM), HCl (potentially in light ends), vents and fugitive emissions (HCs).	Low pH, chloride salts, caustic wash, relatively low H_2S, and NH_3.	Calcium chloride sludge from neutralized HCl gas.
	Process: Polymerization	
H_2S, from caustic washing.	H_2S, NH_3, caustic wash, mercaptans and ammonia, high pH.	Spent catalyst containing phosphoric acid.
	Process: Catalytic Reforming	
Heater stack gas (CO, SO_x, NO_x, HCs and PM), fugitive emissions (hydrocarbons) and catalyst regeneration (CO, NO_x, SO_x).	Flow=6.0 Gal/Bbl High levels oil, SS, COD. Relatively low H_2S.	Spent catalyst fines from electrostatic precipitators (alumina silicate and metals).
	Process: Solvent Extraction	
Fugitive solvents.	Oil and solvents.	Little or no residual wastes generated.
	Process: Dewaxing	
Fugitive solvents, heaters.	Oil and solvents.	Little or no residual wastes generated.

Table 1. continued

AIR EMISSIONS	PROCESS WASTE WATER	RESIDUAL WASTES GENERATED
	Process: Propane Deasphalting	
Heater stack gas (CO, SO_x, NO_x, HCs and PM), fugitive propane.	Oil and propane.	Little or no residual wastes generated.
	Process: Merox treating	
Vents and fugitive emissions (HCs and disulfides).	Little or no wastewater generated.	Spent Merox caustic solution, waste oil-disulfide mixture.
	Process: Wastewater treatment	
Fugitive emissions (H_2S, NH_3, and HCs).	Not Applicable.	API separator sludge (phenols, metals and oil), chemical precipitation sludge (chemical coagulants, oil), DAF floats, biological sludges (metals, oil, SS), spent lime.
	Process: Gas Treatment and Sulfur Recovery	
SO_x, NO_x, and H_2S from vent and tail gas emissions.	H_2S, NH,, amines, Stretford solution.	Spent catalyst.
	Process: Blending	
Fugitive emissions (HCs).	Little or no wastewater generated.	Little or no residual waste generated.
	Process: Heat exchanger cleaning	
Periodic fugitive emissions (HCs.)	Oily wastewater generated.	Heat exchanger sludge (oil, metals, and SS).
	Process: Storage Tanks	
Fugitive emissions (hydrocarbons).	Water drained from tanks contaminated with tank product.	Tank bottom sludge (iron rust, clay, sand, water, emulsified oil and wax, metals).
	Process: Blowdown and Flaring Operations	
Combustion products (CO, SO_x, NO_x, and HCs) from flares, fugitive emissions.	Little or no wastewater generated.	Little or no residual waste generated.

Sources: Assessment of Atmospheric Emissions from Petroleum Refining, Radian Corp., 1980; Petroleum Refining Hazardous Waste Generation, U.S. EPA, Office of Solid Waste, 1994.

POLLUTION PREVENTION PRACTICES AND OPPORTUNITIES

The Pollution Prevention Act of 1990 (PPA) requires facilities to report information about the management of TRI chemicals in waste and efforts made to eliminate or reduce those quantifies. These data have been collected annually in the TRI reporting Form R beginning with the 1991 reporting year. The following discussions are based on a review of data between the years 1992-1995 and is meant to provide a basic understanding of the quantities of waste handled by the industry, the methods typically used to manage this waste, and recent trends, practices and opportunities for further pollution prevention.

From the yearly data it is apparent that the portion of the wastes reported as recycled on-site has increased and the portions treated or managed through energy recovery on-site have decreased between 1992 and 1995. Table 2 shows that the petroleum refining industry managed about 1.6 billion pounds of production-related waste (total quantity of TRI chemicals in the waste from routine production operations) in 1993 (column B).

Column C reveals that of this production-related waste, 30 percent was either transferred off-site or released to the environment. Column C is calculated by dividing the total TRI transfers and releases by the total quantity of production-related waste. In other words, about 70 percent of the industry's TRI wastes were managed on-site through recycling, energy recovery, or treatment as shown in columns E, F, and G, respectively. The majority of waste that is released or transferred off-site can be divided into portions that are recycled off-site, recovered for energy off-site, or treated off-site as shown in columns H, I, and J, respectively. The remaining portion of the production related wastes (4 percent), shown in column D, is either released to the environment through direct discharges to air, land, water, and underground injection, or it is disposed of off-site. In general, toxic chemical releases have been declining. Reported releases dropped by 42.7 percent between 1988 and 1993, with declining trends continued throughout the 1990s. Although on-site releases have decreased, the total amount of reported toxic waste has not declined because the amount of toxic chemicals transferred off-site has increased. Transfers have increased from 3.7 billion pounds in 1991 to 4.7 billion pounds in 1993. Better management practices have led to increases in off-site transfers of toxic chemicals for recycling. The amount of TRI chemicals generated by the petroleum refining industry provides a gross profile of the types and relative amounts of toxic chemical outputs from refining processes. Additional information, which can be related back to possible compliance requirements, is available from the distribution of chemical releases across specific media within the environment.

Table 2. *Source Reduction and Recycling Activity for Petroleum Industry.*

A	B	C	D	ON-SITE			OFF-SITE		
				E	F	G	H	I	J
Year	Quantity of Production-Related Waste (10^6lbs.)[a]	% Released and Transferred[b]	% Released and Disposed Off-site[c]	% Recycled	% Energy Recovery	% Treated	% Recycled	% Energy Recovery	% Treated
1992	1,476	24%	3%	10%	37%	22%	27%	<1%	<1 %
1993	1,600	30%	4%	14%	36%	20%	26%	<1%	<1 %
1994	1,867	---	4%	19%	37%	15%	25%	<1%	<1 %
		---	4%	21%	32%	17%	27%	<1%	

a - Within this industry sector, non-production related waste < 1 percent of production related wastes for 1993.
b - Total TRI transfers and releases as reported in Sections 5 and 6 of Form R as a percentage of production related wastes.
c - Percentage of production related waste released to the environment and transferred off-site for disposal.

The TRI data requires filers to list releases to air, water, and land separately. The distribution across media can also be compared to the profile of other industry sectors.

The petroleum refining industry releases 75 percent of its total TRI poundage to the air, 24 percent to the water (including 20 percent to underground injection and 4 percent to surface waters), and 1 percent to the land. This release profile differs from other TRI industries which average approximately 59 percent to air, 30 percent to water, and 10 percent to land. Examining the petroleum refining industry's TRI reported toxic chemical releases highlights the likely origins of the large air releases for the industry.

According to TRI data, the petroleum refining industry releases (discharges to the air, water, or land without treatment) and transfers (shipped off-site) a total of 482 million pounds of pollutants per year, made up of 103 different chemicals. This represents about 11 percent of the total pounds of TRI chemicals released and transferred by all manufacturers in a year. In comparison, the chemical industry generates on the average 2.5 billion pounds per year, accounting for 33 percent of all releases and transfers.

Overall, the petroleum refining industry's releases declined between 1988 and 1993. Between 1991 and 1993 the decrease in releases was 6.7 percent compared to the average for all industries of 18 percent. In the same period, however, transfers were reported to increase 65 percent which is higher than the average increase in transfers of 25 percent for all manufacturing industries.

A large portion of the increases were in the form of transfers to recycling. Spent sulfuric acid generated in the alkylation process makes up about half of all transfers of TRI listed chemicals off-site. At the facility level, the industry reported a level of pollution prevention activities of 42 percent of all refineries which is slightly higher than the overall average of about 35 percent of TRI reporting facilities.

Comparisons of the reported pounds released or transferred per facility demonstrate that the petroleum refining industry is far above average in its pollutant releases and transfers per facility when compared to other TRI industries. Of the twenty manufacturing SIC codes listed in the TRI database, the mean amount of pollutant release per facility (including petroleum refining) was approximately 120,000 pounds.

The TRI releases of the average petroleum refining facility (SIC 2911) were 404,000 pounds, making the industry 3.4 times higher in per facility releases than for other industries. For transfers, the mean of petroleum refining facilities was about 13 times as much that of all TRI manufacturing facilities (202,000 pounds transferred off-site per facility compared to 2,626,000 per refinery). These high releases and transfers per facility reflect the large volumes of material processed at a relatively small number of facilities.

Of the top ten most frequently reported toxic chemicals on the TRI list, the prevalence of volatile chemicals explains the air intensive toxic chemical loading of the refining industry. Nine of the ten most commonly reported toxic chemicals are highly volatile. Seven of the ten are aromatic hydrocarbons (benzene, toluene, ethylbenzene, xylene, cyclohexane, 1,2,4-trimethylbenzene, and ethylbenze).

Aromatic hydrocarbons are highly volatile compounds and make up a portion of both crude oil and many finished petroleum products. Ammonia, the ninth most commonly reported toxic chemical, is also released and transferred from petroleum refineries in large quantities. Ammonia may be found in high concentrations in process water streams from steam distillation processes and in refinery sour gas. The primary means of release to the environment is through underground injection of wastewater and emissions to air.

Gasoline blending additives (i.e., methanol, ethanol, and MTBE) and chemical feedstocks (propylene, ethylene, and naphthalene) are also commonly reported to TRI. Additives and chemical feedstocks are, for the most part, released as air emissions due to their high volatility. A significant portion of the remaining chemicals of the reported TRI toxic chemicals are metals compounds, which are typically transferred off-site for recovery or as a component of hazardous wastes.

Although it is not the most frequently reported toxic chemical released or transferred, sulfuric acid is, by far, generated in the largest quantities. Spent sulfuric acid is primarily generated during the alkylation process. The acid is typically transferred off-site for regeneration.

Table 3 provides a summary of the toxicity and fate information for the principal hazardous chemicals released by petroleum refinery operations. The table provides descriptions of the most common routes by which these pollutants enter the environment as a result of common refinery practices and operations.

Table 3. *Toxicity and Environmental Fate Information.*

Ammonia CAS #7664-41-7
Sources. Ammonia is formed from the nitrogen bearing components of crude oil and can be found throughout petroleum refineries in both the gaseous and aqueous forms. Gaseous ammonia often leaves distillation, cracking and treating processes mixed with the sour gas or acid gas along with refinery fuel gases and hydrogen sulfide. Aqueous ammonia is present in the sourwater generated in the vacuum distillation unit and steam strippers or fractionators. Some release sources include, fugitive emissions, sour gas stripper, sulfur unit, and wastewater discharges. **Toxicity.** Anhydrous ammonia is irritating to the skin, eyes, nose, throat, and upper respiratory system. Ecologically, ammonia is a source of nitrogen (an essential element for aquatic plant growth), and may therefore contribute to eutrophication of standing or slow-moving surface water, particularly in nitrogen-limited waters such as the Chesapeake Bay. In addition, aqueous ammonia is moderately toxic to aquatic organisms. **Carcinogenicity.** There is currently no evidence to suggest that this chemical is carcinogenic. **Environmental Fate.** Ammonia combines with sulfate ions in the atmosphere and is washed out by rainfall, resulting in rapid return of ammonia to the soil and surface waters. Ammonia is a central compound in the environmental cycling of nitrogen. Ammonia in lakes, rivers, and streams is converted to nitrate. **Physical Properties.** Ammonia is a corrosive and severely irritating gas with a pungent odor.

Table 3. continued

Toluene CAS #108-88-3
Sources. Toluene is a component of crude oil and is therefore present in many refining operations. Toluene is also produced during catalytic reforming and is sold as one of the large volume aromatics used as feedstocks in chemical manufacturing. Its volatile nature makes fugitive emissions its largest release source. Point air sources may arise during the process of separating toluene from other aromatics and from solvent dewaxing operations where toluene is often used as the solvent. **Toxicity.** Inhalation or ingestion of toluene can cause headaches, confusion, weakness, and memory loss. Toluene may also affect the way the kidneys and liver function. Reactions of toluene in the atmosphere contribute to ozone formation. Ozone can affect the respiratory system, especially in sensitive individuals such as asthma or allergy sufferers. Unborn animals were harmed when high levels of toluene were inhaled by their mothers, although the same effects were not seen when the mothers were fed large quantities of toluene. Note that these results may reflect similar difficulties in humans. **Carcinogenicity.** There is currently no evidence to suggest carcinogenicity. **Environmental Fate.** A portion of releases of toluene to land and water will evaporate. Toluene may also be degraded by microorganisms. Once volatilized, toluene in the lower atmosphere will react with other atmospheric components contributing to the formation of ground-level ozone and other air pollutants. **Physical Properties.** Toluene is a volatile organic chemical.

Xylenes (Mixed Isomers) CAS #1330-20-7
Sources. Xylene isomers are a component of crude oil and are therefore present in many refining operations. Xylenes are also produced during catalytic reforming and are sold as one of the large volume aromatics used as feedstocks in chemical manufacturing. Xylene's volatile nature make fugitive emissions the largest release source. Point air sources may arise during the process of separating xylene from other aromatics. **Toxicity.** Xylenes are rapidly absorbed into the body after inhalation, ingestion, or skin contact. Short-term exposure of humans to high levels of xylene can cause irritation of the skin, eyes, nose, and throat, difficulty in breathing, impaired lung function, impaired memory, and possible changes in the liver and kidneys. Both short- and long-term exposure to high concentrations can cause effects such as headaches, dizziness, confusion, and lack of muscle coordination. Reactions of xylene in the atmosphere contribute to the formation of ozone in the lower atmosphere. Ozone can affect the respiratory system, especially in sensitive individuals such as asthma or allergy sufferers. **Carcinogenicity.** There is currently no evidence to suggest that this chemical is carcinogenic. **Environmental Fate.** A portion of releases to land and water will quickly evaporate, although some degradation by microorganisms will occur. Xylene are moderately mobile in soils and may leach into groundwater, where they may persist for many years. Xylenes are VOCs. As such, xylene will react with other atmospheric components, contributing to the formation of ground-level ozone and other air pollutants.

Table 3. continued

Methyl Ethyl Ketone CAS #78-93-3
Sources. Methyl ethyl ketone (MEK) is used in some refineries as a solvent in lube oil dewaxing. Its extremely volatile characteristic makes fugitive emissions its primary source of releases to the environment. **Toxicity.** Breathing moderate amounts of methyl ethyl ketone (MEK) for short periods of time can cause adverse effects on the nervous system ranging from headaches, dizziness, nausea, and numbness in the fingers and toes to unconsciousness. Its vapors are irritating to the skin, eyes, nose, and throat and can damage the eyes. Repeated exposure to moderate to high amounts may cause liver and kidney effects. **Carcinogenicity.** No agreement exists over the carcinogenicity of MEK. One source believes MEK is a possible carcinogen in humans based on limited animal evidence. Other sources believe that there is insufficient evidence to make any statements about possible carcinogenicity. **Environmental Fate.** Most of the MEK released to the environment will end up in the atmosphere. MEK can contribute to the formation of air pollutants in the lower atmosphere. It can be degraded by microorganisms living in water and soil. **Physical Properties.** Methyl ethyl ketone is a flammable liquid.
Propylene CAS # 115-07-1
Sources. Propylene (propene) is one of the light ends formed during catalytic and thermal cracking and coking operations. It is usually collected and used as a feedstock to the alkylation unit. Propylene is volatile and soluble in water making releases to both air and water significant. **Toxicity.** At low concentrations, inhalation of propylene causes mild intoxication, a tingling sensation, and an inability to concentrate. At higher concentrations, unconsciousness, vomiting, severe vertigo, reduced blood pressure, and disordered heart rhythms may occur. Skin or eye contact with propylene causes freezing burns. Reaction of propylene (see environmental fate) in the atmosphere contributes to the formation of ozone in the lower atmosphere. Ozone can affect the respiratory system, especially in sensitive individuals such as asthma or allergy sufferers. Ecologically, similar to ethylene, propylene has a stimulating effect on plant growth at low concentrations, but inhibits plant growth at high levels. **Carcinogenicity.** There is currently no evidence to suggest that this chemical is carcinogenic. **Environmental Fate.** Propylene is degraded principally by hydroxyl ions in the atmosphere. Propylene released to soil and water is removed primarily through volatilization. Hydrolysis, bioconcentration, and soil adsorption are not expected to be significant fate processes of propylene in soil or aquatic ecosystems. Propylene is readily biodegraded by microorganisms in surface water. **Physical Properties.** Propylene is a volatile organic chemical.
Benzene CAS # 71-43-2
Sources. Benzene is a component of crude oil. It is also produced during catalytic reforming and is sold as one of the large volume aromatics used as feedstocks in chemical manufacturing. Benzene's volatile nature makes fugitive emissions the largest release source.

Table 3. continued

Toxicity. Short-term inhalation of benzene affects the central nervous system and respiratory system. Chronic exposure causes bone marrow toxicity in animals and humans, causing suppression of the immune system and development of leukemia. Ingestion of benzene is rare. Reactions of benzene in the atmosphere contributes to the formation of ozone in the lower atmosphere (troposphere).
Carcinogenicity. Benzene is a known human carcinogen.
Environmental Fate. A portion of benzene releases to soil and surface waters evaporate rapidly. Benzene is highly mobile in the soil and may leach to groundwater. Once in groundwater, it is likely biodegraded by microorganisms only in the presence of oxygen. Benzene is not expected to significantly adsorb to sediments, bioconcentrate in aquatic organisms or break down in water. Atmospheric benzene is broken down through reacting with chemical ions in the air; this process is greatly accelerated in the presence of other air pollutants such as nitrogen oxides or sulfur dioxide. Benzene is fairly soluble in water and is removed from the atmosphere in rain. As a volatile chemical, benzene in the lower atmosphere will react with other atmospheric components, contributing to the formation of ground-level ozone and other air pollutants, which can contribute to respiratory illnesses in both the general and highly susceptible populations, such as asthmatics and allergy-sufferers.

In addition to chemicals covered under TRI, many other chemicals are released. For example, the EPA Office of Air Quality Planning and Standards has compiled air pollutant emission factors for determining the total air emissions of priority pollutants (e.g., VOCs, SO_x, NO_x CO, particulates, etc.) from many refinery sources. The EPA Office of Aerometric Information Retrieval System (AIRS) contains a wide range of information related to stationary sources of air pollution, including the emissions of a number of air pollutants which may be of concern within a particular industry. With the exception of volatile organic compounds (VOCs), there is little overlap with the TRI chemicals reported above.

Control technologies employed for the handling of air emissions normally includes the capture and recycling or combustion of emissions from vents, product transfer points, storage tanks, and other handling equipment. Boilers, heaters, other combustion devices, cokers, and catalytic units may require particulate matter controls. Use of a carbon monoxide boiler is normally a standard practice in the fluidized catalytic cracking units. Catalytic cracking units should be provided with particular removal devices. Steam injection in flaring stacks can reduce particulate matter emissions.

Refinery wastewaters often require a combination of treatment methods to remove oil and contaminants before discharge. Separation of different streams, such as stormwater, cooling water, process water, sanitary, sewage, etc., is essential for minimizing treatment requirements. A typical system may include sour water stripper, gravity separation of oil and water, dissolved air flotation, biological treatment, and clarification. A final polishing step using filtration, activated carbon, or chemical treatment may also be required. Achievable pollutant loads per ton of crude processed include BOD, 6 g; COD, 50 g; suspended solids, 10 g; and oil and grease, 2 g.

Sludge treatment is usually performed using land application (bioremediation) or solvent extraction followed by combustion of the residue or by use for asphalt, where feasible. In some cases, the residue may require stabilization prior to disposal to reduce the leachability of toxic metals. Oil is recovered from slops using separation techniques such as gravity separators and centrifuges.

Implementation of pollution prevention measures can yield both economic and environmental benefits. However, a balance on energy usage and environmental impacts may have to be struck. New refineries should be designed to maximize energy conservation and reduce hydrocarbon losses. A good target for simple refineries (i.e., refineries with distillation, catalytic reforming, hydrotreating, and offsite facilities) is that the total quantity of oil consumed as fuel and lost in production operations should not exceed 3.5% of the throughput. For refineries with secondary conversion units (i.e., hydrocrackers or lubricating oil units), the target should be 5-6% (and, in some cases, up to 10%) of the throughput. Fugitive VOC emissions from the process units can be reduced to 0.05% of the throughput, with total VOC emissions of less than 1 kg per ton of crude (or 0.1% of throughput). Methods of estimating these figures include emissions monitoring, mass balance, and inventories of emissions sources. Design assumptions should be recorded to allow for subsequent computation and reduction of losses. Vapor recovery systems to control losses of VOCs from storage tanks and loading areas should achieve 90-100% recovery.

Plant operators should aim at using fuel with less than 0.5% sulfur (or an emissions level corresponding to 0.5% sulfur in fuel). High-sulfur fuels should be directed to units equipped with SO_x controls. Fuel blending is another option. A sulfur recovery system that achieves at least 97% (but preferably over 99%) sulfur recovery should be used when the hydrogen sulfide concentration in tail gases exceeds 230 mg/Nm^3. The total release of sulfur dioxide should be below 0.5 kg per ton for a hydroskimming refinery and below 1 kg per ton for a conversion refinery. A wastewater generation rate of 0.4 m^3/t of crude processed is achievable with good design and operation, and new refineries should achieve this target as a minimum.

The generation rate of solid wastes and sludges should be less than 0.5% of the crude processed, with a target of 0.3%.

As already noted, petroleum refineries are complex plants, and the combination and sequence of processes is usually very specific to the characteristics of the raw materials (crude oil) and the products. Specific pollution prevention or source reduction measures can often be determined only by the technical staff of the specific refinery operation. However, there are a number of general areas where improvements are often possible, and site-specific waste reduction measures in these areas should be designed into the plant and targeted by management of operating plants. Areas where efforts should be concentrated are summarized in Table 4.

Although numerous cases have been documented where petroleum refineries have simultaneously reduced pollution outputs and operating costs through pollution prevention techniques, there are often barriers to their implementation.

The primary barrier to most pollution prevention projects is cost. Many pollution prevention options simply do not pay for themselves, or the economics often appear marginal. Corporate investments typically must earn an adequate return on invested capital for the shareholders and some pollution prevention options at some facilities may not meet the requirements set by company policies. Additionally, the equipment used in the petroleum refining industry are very capital intensive and have very long lifetimes. This reduces the incentive to make process modifications to (expensive) installed equipment that is still useful.

It should be emphasized however, that pollution prevention techniques are, nevertheless, often more cost-effective than pollution reduction through end-of-pipe treatment technologies. A case study based on the Amoco/EPA joint study claimed that the same pollution reduction currently realized through end-of-pipe regulatory requirements at the Amoco facility could be achieved at 15 percent the current costs using pollution prevention techniques.

To better understand some of the broad areas of pollution prevention practices listed in Table 4, the following provides a summary of these widespread pollution prevention techniques found to be effective at petroleum refineries.

Process and Equipment Modifications Practices

Place secondary seals on storage tanks - One of the largest sources of fugitive emissions from refineries is storage tanks containing gasoline and other volatile products. These losses can be significantly reduced by installing secondary seals on storage tanks. An Amoco/EPA joint study estimated that VOC losses from storage tanks could be reduced 75 to 93 percent. Equipping an average tank with a secondary seal system was estimated to cost about $20,000.

Table 4. *Recommended Pollution Prevention Practices.*

AREA OF OPPORTUNITY	RECOMMENDED POLLUTION PREVENTION PRACTICE
Reduction of Air Emissions	1. Minimize losses from storage tanks and product transfer areas by methods such as vapor recovery systems and double seals. 2. Minimize SO_x emissions either through desulfurization of fuels, to the extent feasible, or by directing the use of high-sulfur fuels to units equipped with SO_x emissions controls. 3. Recover sulfur from tail gases in high-efficiency sulfur recovery units. 4. Recover nonsilica-based (i.e., metallic) catalysts and reduce particulate emissions. 5. Use low-NO_x burners to reduce nitrogen oxide emissions. 6. Avoid and limit fugitive emissions by proper process design and maintenance. 7. Keep fuel usage to a minimum.
Elimination or Reduction of Pollutants	8. Consider reformate and other octane boosters instead of tetraethyl lead and other organic lead compounds for octane boosting. 9. Use nonchrome-based inhibitors in cooling water, where inhibitors are needed. 10. Use long-life catalysts and regenerate to extend the catalysts' life cycle.
Recycling and Reuse	11. Recycle cooling water and, where cost-effective, treated wastewater. 12. Maximize recovery of oil from oily waste waters and sludges. Minimize losses of oil to the effluent system. 13. Recover and reuse phenols, caustics, and solvents from their spent solutions. 14. Return oily sludges to coking units or crude distillation units.
Operating Procedures	15. Segregate oily waste waters from stormwater systems. 16. Reduce oil losses during tank drainage carried out to remove water before product dispatch. 17. Optimize frequency of tank and equipment cleaning to avoid accumulating residue at the bottom of the tanks. 18. Prevent solids and oily wastes from entering the drainage system. 19. Institute dry sweeping instead of washdown to reduce wastewater volumes. 20. Establish and maintain an emergency preparedness and response plan and carry out frequent training. 21. Practice corrosion monitoring, prevention, and control in underground piping and tank bottoms. 22. Establish leak detection and repair programs.

Establish leak detection and repair program - Fugitive emissions are one of the largest sources of refinery hydrocarbon emissions. A leak detection and repair (LDAR) program consists of using a portable VOC detecting instrument to detect leaks during regularly scheduled inspections of valves, flanges, and pump seals. Leaks are then repaired immediately or are scheduled for repair as quickly as possible. A LDAR program could reduce fugitive emissions 40 to 64 percent, depending on the frequency of inspections.

Regenerate or eliminate filtration clay - Clay from refinery filters must periodically be replaced. Spent clay often contains significant amounts of entrained hydrocarbons and, therefore, must be designated as hazardous waste. Back washing spent clay with water or steam can reduce the hydrocarbon content to levels so that it can be reused or handled as a nonhazardous waste. Another method used to regenerate clay is to wash the clay with naphtha, dry it by steam heating and then feed it to a burning kiln for regeneration. In some cases clay filtration can be replaced entirely with hydrotreating.

Reduce the generation of tank bottoms - Tank bottoms from crude oil storage tanks constitute a large percentage of refinery solid waste and pose a particularly difficult disposal problem due to the presence of heavy metals. Tank bottoms are comprised of heavy hydrocarbons, solids, water, rust and scale. Minimization of tank bottoms is carried out most cost effectively through careful separation of the oil and water remaining in the tank bottom. Filters and centrifuges can also be used to recover the oil for recycling.

Minimize solids leaving the desalter - Solids entering the crude distillation unit are likely to eventually attract more oil and produce additional emulsions and sludges. The amount of solids removed from the desalting unit should, therefore, be maximized. A number of techniques can be used such as: using low shear mixing devices to mix desalter wash water and crude oil; using lower pressure water in the desalter to avoid turbulence; and replacing the water jets used in some refineries with mud rakes which add less turbulence when removing settled solids.

Minimize cooling tower blowdown - The dissolved solids concentration in the recirculating cooling water is controlled by purging or blowing down a portion of the cooling water stream to the wastewater treatment system. Solids in the blowdown eventually create additional sludge in the wastewater treatment plant. However, the amount of cooling tower blowdown can be lowered by minimizing the dissolved solids content of the cooling water. A significant portion of the total dissolved solids in the cooling water can originate in the cooling water makeup stream in the form of naturally occurring calcium carbonates. Such solids can be controlled either by selecting a source of cooling tower makeup water with less dissolved solids or by removing the dissolved solids from the makeup water stream. Common treatment methods include: cold lime softening, reverse osmosis, or electrodialysis.

Install vapor recovery for barge loading - Although barge loading is not a factor for all refineries, it is an important emissions source for many facilities. One of the largest sources of VOC emissions identified during the Amoco/EPA study was fugitive emissions from loading of tanker barges. It was estimated that these emissions could be reduced 98 percent by installing a marine vapor loss control system. Such systems could consist of vapor recovery or VOC destruction in a flare.

Minimize FCCU decant oil sludge - Decant oil sludge from the fluidized bed catalytic cracking unit (FCCU) can contain significant concentrations of catalyst fines. These fines often prevent the use of decant oil as a feedstock or require treatment which generates an oily catalyst sludge. Catalysts in the decant oil can be minimized by using a decant oil catalyst removal system. One system incorporates high voltage electric fields to polarize and capture catalyst particles in the oil. The amount of catalyst fines reaching the decant oil can be minimized by installing high efficiency cyclones in the reactor to shift catalyst fines losses from the decant oil to the regenerator where they can be collected in the electrostatic precipitator.

Control of heat exchanger cleaning solids - In many refineries, using high pressure water to clean heat exchanger bundles generates and releases water and entrained solids to the refinery wastewater treatment system. Exchanger solids may then attract oil as they move through the sewer system and may also produce finer solids and stabilized emulsions that are more difficult to remove. Solids can be removed at the heat exchanger cleaning pad by installing concrete overflow weirs around the surface drains or by covering drains with a screen. Other ways to reduce solids generation are by using anti-foulants on the heat exchanger bundles to prevent scaling and by cleaning with reusable cleaning chemicals that also allow for the easy removal of oil.

Control of surfactants in wastewater - Surfactants entering the refinery wastewater streams will increase the amount of emulsions and sludges generated. Surfactants can enter the system from a number of sources including: washing unit pads with detergents; treating gasolines with an end point over 400°F thereby producing spent caustics; cleaning tank truck tank interiors; and using soaps and cleaners for miscellaneous tasks. In addition, the overuse, and mixing of the organic polymers used to separate oil, water and solids in the wastewater treatment plant can actually stabilize emulsions. The use of surfactants should be minimized by educating operators, routing surfactant sources to a point downstream of the DAF unit and by using dry cleaning, high pressure water or steam to clean oil surfaces of oil and dirt.

Thermal treatment of applicable sludges - The toxicity and volume of some deoiled and dewatered sludges can be further reduced through thermal treatment. Thermal sludge treatment units use heat to vaporize the water and volatile components in the feed and leave behind a dry solid residue. The vapors are condensed for separation into the hydrocarbon and water components. Non-condensible vapors are either flared or sent to the refinery amine unit for treatment and use as refinery fuel gas.

Eliminate use of open ponds - Open ponds used to cool, settle out solids and store process water can be a significant source of VOC emissions. Wastewater from coke cooling and coke VOC removal is occasionally cooled in open ponds where VOCs easily escape to the atmosphere. In many cases, open ponds can be replaced with closed storage tanks.

Remove unnecessary storage tanks from service - Since storage tanks are one of the largest sources of VOC emissions, a reduction in the number of these tanks can have a significant impact. The need for certain tanks can often be eliminated through improved production planning and more continuous operations. By minimizing the number of storage tanks, tank bottom solids and decanted wastewater may also be reduced.

Replace old boilers - Older refinery boilers can be a significant source of SO_x, NO_x and particulate emissions. It is possible to replace a large number of old boilers with a single new cogeneration plant with emissions controls.

Modify the FCCU to allow the use of catalyst fines - Some FCCUs can be modified to recycle some of the catalyst fines generated.

Reduce the use of 55-gallon drums - Replacing 55-gallon drums with bulk storage can minimize the chances of leaks and spills.

Install rupture discs and plugs - Rupture discs on pressure relieve valves and plugs in open ended valves can reduce fugitive emissions.

Install high pressure power washer - Chlorinated solvent vapor degreasers can be replaced with high pressure power washers which do not generate spent solvent hazardous wastes.

Refurbish or eliminate underground piping - Underground piping can be a source of undetected releases to the soil and groundwater. Inspecting, repairing or replacing underground piping with surface piping can reduce or eliminate these potential sources.

Waste Segregation and Separation Practices

Segregate process waste streams - A significant portion of refinery waste arises from oily sludges found in combined process/storm sewers. Segregation of the relatively clean rainwater runoff from the process streams can reduce the quantity of oily sludges generated. Furthermore, there is a much higher potential for recovery of oil from smaller, more concentrated process streams.

Control solids entering sewers - Solids released to the wastewater sewer system can account for a large portion of a refinery's oily sludges. Solids entering the sewer system (primarily soil particles) become coated with oil and are deposited as oily sludges in the API oil/water separator. Because a typical sludge has a solids content of 5 to 30 percent by weight, preventing one pound of solids from entering the sewer system can eliminate 3 to 20 pounds of oily sludge. An Amoco/EPA study estimated that at the Yorktown facility 1,000 tons of solids per year enter the refinery sewer system. Methods used to control solids include: using a street sweeper on paved areas, paving unpaved areas, planting ground cover on unpaved areas, re-lining sewers, cleaning solids from ditches and catch basins, and reducing heat exchanger bundle cleaning solids by using antifoulants in cooling water.

Improve recovery of oils from oily sludges - Because oily sludges make up a large portion of refinery solid wastes, any improvement in the recovery of oil from the sludges can significantly reduce the volume of waste. There are a number of technologies currently in use to mechanically separate oil, water and solids, including: belt filter presses, recessed chamber pressure filters, rotary vacuum filters, scroll centrifuges, disc centrifuges, shakers, thermal driers and centrifuge-drier combinations.

Identify benzene sources and install upstream water treatment - Benzene in wastewater can often be treated more easily and effectively at the point it is generated rather than at the wastewater treatment plant after it is mixed with other wastewater.

Recycling Practices

Recycle and regenerate spent caustics - Caustics used to absorb and remove hydrogen sulfide and phenol contaminants from intermediate and final product streams can often be recycled. Spent caustics may be saleable to chemical recovery companies if concentrations of phenol or hydrogen sulfide are high enough. Process changes in the refinery may be needed to raise the concentration of phenols in the caustic to make recovery of the contaminants economical. Caustics containing phenols can also be recycled on-site by reducing the pH of the caustic until the phenols become insoluble thereby allowing physical separation. The caustic can then be treated along with the refinery waste waters.

Use oily sludges as feedstock - Many oily sludges can be sent to a coking unit or the crude distillation unit where it becomes part of the refinery products.

Sludge sent to the coker can be injected into the coke drum with the quench water, injected directly into the delayed coker, or injected into the coker blowdown contactor used in separating the quenching products. Use of sludge as a feedstock has increased significantly in recent years and is currently carried out by most refineries. The quantity of sludge that can be sent to the coker is restricted by coke quality specifications which may limit the amount of sludge solids in the coke. Coking operations can be upgraded, however, to increase the amount of sludge that they can handle.

Control and reuse FCCU and coke fines - Significant quantities of catalyst fines are often present around the FCCU catalyst hoppers and reactor and regeneration vessels. Coke fines are often present around the coker unit and coke storage areas. The fines can be collected and recycled before being washed to the sewers or migrating off-site via the wind. Collection techniques include dry sweeping the catalyst and coke fines and sending the solids to be recycled or disposed of as nonhazardous waste. Coke fines can also be recycled for fuel use. Another collection technique involves the use of vacuum ducts in dusty areas (and vacuum hoses for manual collection) which run to a small baghouse.

Recycle lab samples - Lab samples can be recycled to the oil recovery system.

Training and Supervision

Train personnel to reduce solids in sewers - A facility training program which emphasizes the importance of keeping solids out of the sewer systems will help reduce that portion of wastewater treatment plant sludge arising from the everyday activities of refinery personnel.

Train personnel to prevent soil contamination - Contaminated soil can be reduced by educating personnel on how to avoid leaks and spills.

Material Substitution

Use nonhazardous degreasers - Spent conventional degreaser solvents can be reduced or eliminated through substitution with less toxic and/or biodegradable products.

Eliminate chromates as an anti-corrosive - Chromate containing wastes can be reduced or eliminated in cooling tower and heat exchanger sludges by replacing chromates with less toxic alternatives such as phosphates.

Use high quality catalysts - By using catalysts of a higher quality, process efficiencies can be increased while the required frequency of catalyst replacement can be reduced.

Replace ceramic catalyst support with activated alumina supports Activated alumina supports can be recycled with spent alumina catalyst.

CLOSING REMARKS

The best way to reduce pollution is to prevent it in the first place. Some companies have creatively implemented pollution prevention techniques that improve efficiency and increase profits while at the same time minimizing environmental impacts. This can be done in many ways such as reducing material inputs, re-engineering processes to reuse by-products, improving management practices, and employing substitution of toxic chemicals. Some smaller facilities are able to actually get below regulatory thresholds just by reducing pollutant releases through aggressive pollution prevention policies.

Pollution prevention in the petroleum refining industry is expected to become increasingly important as federal, state and municipal regulations become more stringent and as waste disposal costs rise. According to the American Petroleum Institute, the industry currently spends a significant amount of money every year on environmental quality and protection." This provides the industry with a strong incentive to find ways to reduce the generation of waste and to lessen the burden of environmental compliance investments. For the petroleum refining industry, pollution prevention will primarily be realized through improved operating procedures, increased recycling, and process modifications.

Chapter 7
Pollution Prevention Practices in the Metallurgical Industries

INTRODUCTION

This chapter describes the major industrial processes within the iron and steel, and allied metallurgical industries (both ferrous and nonferrous), including the materials and equipment used, and the processes employed. An understanding of the basic industry practices and processes is essential to developing approaches to managing pollutant outputs and pollution prevention opportunities. The chapter specifically contains a description of commonly used production processes, associated raw materials, the byproducts produced or released, and the materials either recycled or transferred off-site. In addition, this chapter provides a concise description of where wastes may be produced in the process. The chapter also describes the potential fate (via air, water, and soil pathways) of these waste products.

It is important to note that in light of the high cost of most new equipment and the relatively long lead times necessary to bring new equipment on line in the metallurgical industry, changes in production methods and products, particularly in the steel industry, are typically made gradually. Installation of major pieces of new equipment may cost many millions of dollars and require additional retrofitting of other equipment. Even new process technologies that fundamentally improve productivity, such as the continuous casting process for steelmaking (described below), are adopted only over long periods of time. Given the recent financial performance of the ferrous and nonferrous industries, the ability to raise the capital needed to purchase such equipment tends to be limited.

Despite this, if these industries are to be sustained, they must invest in cost-saving and quality enhancing technologies. In the long term, these industries will likely continue to move towards more simplified and continuous manufacturing technologies that reduce the capital costs for new construction and allow smaller mills and manufacturing operations to operate efficiently. The companies that excel will be those that have the resources and foresight to invest in such technologies, including pollution prevention practices.

IRON AND STEEL MANUFACTURING

Industry Description and Practices

Steel is an alloy of iron usually containing less than one percent carbon. The process of steel production requires several sequential steps. The two types of steelmaking technology in use today are the basic oxygen furnace (BOF) and the electric arc furnace (EAF). Although these two technologies use different input materials, the output for both furnace types is molten steel which is subsequently formed into steel mill products. The BOF input materials are molten iron, scrap, and oxygen. In the EAF, electricity and scrap are the input materials used. BOFs are typically used for high tonnage production of carbon steels, while EAFs are used to produce carbon steels and low tonnage alloy and specialty steels. The processes leading up to steelmaking in a BOF are very different than the steps preceding steelmaking in an EAF; the steps after each of these processes producing molten steel are the same.

Steel manufacturing may be defined as the chemical reduction of iron ore, using an integrated steel manufacturing process or a direct reduction process. In the conventional integrated steel manufacturing process, the iron from the blast furnace is converted to steel in a BOF. As noted, it can also be made in an electric arc furnace (EAF) from scrap steel and, in some cases, from direct reduced iron. An emerging technology, direct steel manufacturing, produces steel directly from iron ore. In the BOF process, coke making and iron making precede steelmaking; these steps are not necessary with an EAF. Pig iron is manufactured from sintered, pelletized, or lump iron ores using coke and limestone in a blast furnace. It is then fed to a BOF in molten form along with scrap metal, fluxes, alloys, and high-purity oxygen to manufacture steel. In some integrated steel mills, sintering (heating without melting) is used to agglomerate fines and so recycle iron-rich material such as mill scale (see photographs in Figure 1 for an example).

When making steel using a BOF, coke-making and iron-making precede steelmaking; these steps are not needed for steelmaking with an EAF. Coke, which is the fuel and carbon source, is produced by heating coal in the absence of oxygen at high temperatures in coke ovens. Hence, merchant coke plants are needed to support industry based on this technology. Pig iron is then produced by heating the coke, iron ore, and limestone in a blast furnace. In the BOF, molten iron from the blast furnace is combined with flux and scrap steel where high-purity oxygen is injected. This process, with coke-making, iron-making, steelmaking, and subsequent forming and finishing operations is referred to as fully integrated production.

Figure 1. *Rotating mixers used in sintering stage at the Azovstal Steel plant in Ukraine.*

Alternatively, in an EAF, the input material is primarily scrap steel, which is melted and refined by passing an electric current from the electrodes through the scrap. The molten steel from either process is formed into ingots or slabs that are rolled into finished products. Rolling operations may require reheating, rolling, cleaning, and coating the steel. Descriptions of both steelmaking processes follows:

Basic Oxygen Furnace Technology

The process of making steel in a BOF is preceded by coke-making and ironmaking operations. In coke-making, coke is produced from coal. In ironmaking, molten iron is produced from iron ore and coke. Each of these processes and the subsequent steelmaking process in the BOF are briefly described below.

Coke-making - Coal processing typically involves producing coke, coke gas and by-product chemicals from compounds released from the coal during the coke-making process. Coke is carbon-rich and is used as a carbon source and fuel to heat and melt iron ore in ironmaking. The coke-making process starts with bituminous pulverized coal charge which is fed into the coke oven through ports in the top of the oven. After charging, the oven ports are sealed and the coal is heated at high temperatures (1600 to 2300° F) in the absence of oxygen. Coke manufacturing is done in a batch mode where each cycle lasts for 14 to 36 hours.

A coke oven battery comprises a series of 10 to 100 individual ovens, side-by-side, with a heating flue between each oven pair. Volatile compounds are driven from the coal, collected from each oven, and processed for recovery of combustible gases and other coal by-products. The solid carbon remaining in the oven is the coke. The necessary heat for distillation is supplied by external combustion of fuels (e.g., recovered coke oven gas, blast furnace gas) through flues located between ovens.

At the end of the heating cycle, the coke is pushed from the oven into a rail quench car (see Figure 2 for example). The quench car takes it to the quench tower, where the hot coke is cooled with a water spray. The coke is then screened and sent to the blast furnace or to storage. In the by-products recovery process, volatile components of the coke oven gas stream are recovered including the coke oven gas itself (which is used as a fuel for the coke oven), naphthalene, ammonium compounds, crude light oils, sulfur compounds, and coke breeze (coke fines).

During the coke quenching, handling, and screening operation, coke breeze is produced. Typically, the coke breeze is reused in other manufacturing processes on-site (e.g., sintering) or sold off-site as a by-product.

Coke-making is perhaps the major environmental concern in this industry. Both air emissions and quench water are the key problems. As a result, many steelmakers have turned in recent years to pulverized coal injection, which substitutes coal for coke in the blast furnace. The use of pulverized coal injection can replace roughly 25-40 percent of the coke in the blast furnace thereby reducing the amount of coke needed and the associated emissions. It is also common practice to inject other fuels, such as natural gas, oil, and tar/pitch to replace a portion of the coke.

Quench water from coke-making is also an area of significant environmental concern. In Europe, many plants have implemented technology to shift from water quenching to dry quenching which eliminates suspected carcinogenic PM and VOCs. However, major construction changes are required for such a solution and considering the high capital costs of coke batteries, combined with the depressed state of the steel industry and increased regulations for coke-making, it is unlikely that new facilities will be constructed. Instead, many countries with steelmaking industries are experiencing increases in the amount of coke imported.

Figure 2. *Loading coke onto rail quench cars.*

Iron ore, coke, and limestone are fed into the top of the blast furnace. Heated air is forced into the bottom of the furnace through a bustle pipe and tuyeres (orifices) located around the circumference of the furnace. The carbon monoxide from the burning of the coke reduces iron ore to iron.

The acid part of the ores reacts with the limestone to create a slag which is drawn periodically from the furnace. This slag contains unwanted impurities in the ore. Among the most common impurities is sulfur from the fuels. When the furnace is tapped, iron is removed through one set of runners and molten slag via another. The molten iron is tapped into refractory-lined cars for transport to the steelmaking furnaces. Residuals from the process are mainly sulfur dioxide or hydrogen sulfide, which are driven off from the hot slag. The slag is the largest by-product generated from the ironmaking process and is reused extensively in the construction industry.

Blast furnace flue gas is cleaned and used to generate steam to preheat the air coming into the furnace, or it may be used to supply heat to other plant processes. The cleaning of the gas may generate air pollution control dust in removing coarse particulates (which may be reused in the sintering plant or landfilled), and water treatment plant sludge in removing fine particulates by venturi scrubbers. Sintering (briefly described earlier) is the process that agglomerates fines (including iron ore fines, dusts, coke breeze, water treatment plant sludge, coke breeze, and flux) into a porous mass for charging to the blast furnace. By means of the sintering operations, a mill can recycle iron-rich material, such as mill scale and processed slag. Not all mills have sintering capabilities. The input materials are mixed together, placed on a slow- moving grate or rotating/tilting mixer and ignited. Windboxes under the grate draw air through the materials to deepen the combustion throughout the traveling length of the grate. The coke breeze provides the carbon source for sustaining the controlled combustion. In the process, the fine materials are fused into the sinter agglomerates, which can be reintroduced into the blast furnace along with ore. Air pollution control equipment removes the particulate matter generated during the thermal fusing process. For wet scrubbers, water treatment plant sludge are generally land disposed waste. If electrostatic precipitators or baghouses are used as the air pollution control equipment, the dry particulate matter that is captured are typically recycled as sinter feedstock, or are landfilled as solid waste.

Steelmaking - Molten iron from the blast furnace, flux, alloy materials, and scrap are placed in the BOF, melted and refined by injecting high-purity oxygen. A chemical reaction occurs, where the oxygen reacts with carbon and silicon generating the heat necessary to melt the scrap and oxidize the impurities. The operation is performed as a batch process with a cycle time of about 45 minutes. Slag is produced from impurities removed by the combination of the fluxes with the injected oxygen. Various alloys are added to produce different grades of steel. The molten steel is typically cast into slabs, beams or billets.

Waste products from the basic oxygen steelmaking process include slag, carbon monoxide, and oxides of iron emitted as dust. Also, when the hot iron is poured into ladles or the furnace, iron oxide fumes are released and some of the carbon in the iron is precipitated as graphite (called kish, see Figure 3 for an example). The BOF slag can be processed to recover the high metallic portions for use in sintering or blast furnaces, but its applications as a saleable construction materials are more limited than the blast furnace slag. Basic oxygen furnaces are equipped with air pollution control systems for containing, cooling, and cleaning the volumes of hot gases and sub-micron fumes that are released during the process. Water is used to quench or cool the gases and fumes to temperatures at which they can be effectively treated by the gas cleaning equipment. The resulting waste streams from the pollution control operations include dust and water treatment plant sludge.

Figure 3. *Shows kish build-up on a valve near the BOF.*

About 1,000 gallons of water per ton of steel is typically used in a wet scrubber in order to effectively remove air pollutants. The primary pollutants captured from the off-gases are TSS and metals such as zinc and lead.

Electric Arc Furnace Technology

In the steelmaking process that uses an electric arc furnace (EAF), the primary raw material is scrap metal. The scrap metal is melted and refined using electrical energy. During melting, oxidation of phosphorus, silicon, manganese, carbon and other materials occurs and a slag containing some of these oxidation products forms on top of the molten metal. Oxygen is used to decarburize the molten steel and to provide thermal energy. This is a batch process with a cycle time of about two to three hours. Since scrap metal is used instead of molten iron, there are no coke-making or ironmaking operations associated with steel production that uses an EAF.

This technology results in the production of metal dusts, slag, and gaseous products. Particulate matter and gases evolve together during the steelmaking process and are conveyed into a gas cleaning system. Emissions are cleaned using a wet or dry system. The particulate matter that is removed as emissions in the dry system is referred to as EAF dust, or EAF sludge if it is from a wet system. This waste is a listed hazardous waste under the RCRA. The composition of EAF dust can vary greatly depending on the scrap composition and furnace additives. The primary component is iron or iron oxides, and it may also contain flux (lime and/or fluorspar), zinc, chromium and nickel oxides (when stainless steel is being produced) and other metals associated with the scrap. The two primary hazardous constituents of EAF emission dust are lead and cadmium. Roughly 20 pounds of dust per ton of steel is expected, but as much as 40 pounds of dust per ton of steel may be generated, depending on production practices. Oils are burned off "charges" of oil-bearing scrap in the furnace. Small but not insignificant amounts of nitrogen oxides and ozone are generated during the melting process. The furnace is extensively cooled by water; however, this water is recycled through cooling towers.

Forming and Finishing Operations

Steel Forming - Whether the molten steel is produced using a BOF or an EAF, to convert it into a product, it must be solidified into a shape suitable and finished. The traditional forming method is called ingot teeming, and involves pouring the metal into ingot molds, allowing the steel to cool and solidify. The alternative method of forming steel is called continuous casting. This process bypasses several steps of the conventional ingot teeming process by casting steel directly into semifinished shapes. Molten steel is poured into a reservoir from which it is released into the molds of the casting machine. The metal is cooled as it descends through the molds, and before emerging, a hardened outer shell is formed. As the semifinished shapes proceed on the runout table, the center also solidifies, allowing the cast shape to be cut into lengths. Process contact water cools the continuously cast steel and is collected in settling basins along with oil, grease, and mill scale generated in the casting process. The scale settles out and is removed and recycled for sintering operations, if the mill has a Sinter Plant. Waste treatment plant sludge is also generated during the operation. The steel is further processed to produce slabs, strips, bars, or plates through various forming steps. The most common hot forming operation is hot rolling, where heated steel is passed between two rolls revolving in opposite directions. Hot rolling units may have as many as 13 stands, each producing an incremental reduction in thickness. The final shape and characteristics of a hot formed piece depend on the rolling temperature, the roll profile, and the cooling process after rolling. Wastes generated from hot rolling include waste treatment plant sludge and scale.

In subsequent cold forming, the cross-sectional area of unheated steel is progressively reduced in thickness as the steel passes through a series of rolling stands. Generally, wires, tubes, sheet and strip steel products are produced by cold rolling operations. Cold forming is used to obtain improved mechanical properties, better machinability, special size accuracy, and the production of thinner gages than hot rolling can accomplish economically. During cold rolling, the steel becomes hard and brittle. To make the steel more ductile, it is heated in an annealing furnace. Process contact water is used as a coolant for rolling mills to keep the surface of the steel clean between roller passes. Cold rolling operations also produce a waste treatment plant sludge, primarily due to the lubricants applied during rolling. Grindings from resurfacing of the worn rolls and disposal of used rolls can be a significant contributor to the wastestream.

Finishing Stages - One of the most important aspects of a finished product is the surface quality. To prevent corrosion, a protective coating is usually applied to the steel product. Prior to coating, the surface of the steel must be cleaned so the coating will adhere to the steel. Mill scale, rust, oxides, oil, grease, and soil are chemically removed from the surface of steel using solvent cleaners, pressurized water or air blasting, cleaning with abrasives, alkaline agents or acid pickling. In the pickling process, the steel surface is chemically cleaned of scale, rust, and other materials. Inorganic acids such as hydrochloric or sulfuric acid are most commonly used for pickling. Stainless steels are pickled with hydrochloric, nitric, and hydrofluoric acids. Spent pickle liquor is a hazardous waste if it contains considerable residual acidity and high concentrations of dissolved iron salts. Pickling prior to coating may use a mildly acidic bath which is not necessarily hazardous. Steel generally passes from the pickling bath through a series of rinses. Alkaline cleaners may also be used to remove mineral oils and animal fats and oils from the steel surface prior to cold rolling. Common alkaline cleaning agents include: caustic soda, soda ash, alkaline silicates, phosphates. Steel products are often given a coating to inhibit oxidation and extend the life of the product. Coated products can also be painted to further inhibit corrosion. Common coating processes include: galvanizing (zinc coating), tin coating, chromium coating, aluminizing, and terne coating (lead and tin). Metallic coating application processes include hot dipping, metal spraying, metal cladding (to produce bi-metal products), and electroplating. Galvanizing is a common coating process where a thin layer of zinc is deposited on the steel surface.

Material Balance Information

There are a large number of outputs that are produced as a result of the manufacturing of coke, iron, and steel, the forming of metals into basic shapes, and the cleaning and scaling of metal surfaces. Many of these outputs, categorized by process are listed in Table 1.

Table 1. *Inputs and Outputs from Steelmaking Processes.*

INPUTS	OUTPUTS
	COKE-MAKING
Coal, heat, quench water	Process residues from coke by-product recovery
	Coke oven gas by-products such as coal tar, light oil, ammonia liquor, and the remainder of the gas stream is used as fuel. Coal tar is typically refined to produce commercial and industrial products including pitch, creosote oil, refined tar, naphthalene, and bitumen.
	Charging emissions (fine particles of coke generated during oven pushing, conveyor transport, loading and unloading of coke that are captured by pollution control equipment. Approximately one pound per ton of coke produced are captured and generally land disposed).
	Ammonia, phenol, cyanide and hydrogen sulfide.
	Lime sludge, generated from the ammonia still.
	Decanter tank tar sludge.
	Benzene releases in coke by-product recovery operations.
	Naphthalene residues, generated in the final cooling tower.
	Tar residues.
	Sulfur compounds, emitted from the stacks of the coke ovens.
	Wastewater from cleaning and cooling (contains zinc, ammonia still lime, or decanter tank tar, tar distillation residues).
	Coke oven gas condensate from piping and distribution system.

Inputs	Outputs
IRONMAKING	
Iron ore (primarily in the form of taconite pellets), coke, sinter, coal, limestone, heated air.	Slag, which is either sold as a by-product, primarily for use in the construction industry, or landfilled.
	Residual sulfur dioxide or hydrogen sulfide.
	Particulates captured in the gas, including the air pollution control dust or waste treatment plant sludge.
	Iron is the predominant metal found in the process wastewater.
	Blast furnace gas (CO).
STEELMAKING	
In the steelmaking process that uses a basic oxygen furnace (BOF), inputs include molten iron, metal scrap, and high-purity oxygen. In the steelmaking process that uses an electric arc furnace (EAF), the primary inputs are scrap metal, electric energy and graphite electrodes. For both processes, fluxes and alloys are added, and may include: fluorspar, dolomite, and alloying agents such as aluminum, manganese, and others.	Basic Oxygen Furnace emission control dust and sludge, a metals bearing waste.
	Electric Arc Furnace emission control dust and sludge; generally, 20 pounds of dust per ton of steel is expected, but as much as 40 pounds of dust per ton of steel may be generated depending on the scrap that is used.
	Metal dusts (consisting of iron particulate, zinc, and other metals associated with the scrap and flux (lime and/or fluorspar) not associated with the EAF.
	Slag.
	Carbon monoxide.
	NO_x and ozone, which are generated during the melting process.
FORMING, CLEANING, AND DESCALING	
Carbon steel is pickled with hydrochloric acid; stainless steels are pickled with hydrochloric, nitric and hydrofluoric acids.	Wastewater sludge from rolling, cooling, descaling, and rinsing operations which may contain cadmium, chromium, and lead.
Various organic chemicals are used in the pickling process.	Oils and greases from hot and cold rolling.
Alkaline cleaners are used to remove mineral oils and animal fats and oils from the steel surface. Common alkaline cleaning agents include: caustic soda, soda ash, alkaline silicates, phosphates.	Spent pickle liquor.
	Spent pickle liquor rinse water sludge from cleaning operations.
	Wastewater from the rinse baths. Rinse water from coating processes may contain zinc, lead, cadmium, or chromium.
	Grindings from roll refinishing may be RCRA characteristic waste from chromium.
	Zinc dross.

Sintering operations can emit significant dust levels of about 20 kilograms per metric ton (kg/t) of steel. Pelletizing operations can emit dust levels of about 15 kg/t of steel. Air emissions from pig iron manufacturing in a blast furnace include PM, ranging from less than 10 kg/t of steel manufactured to 40 kg/t;

sulfur oxides SO_x mostly from sintering or pelletizing operations (1.5 kg/t of steel); nitrogen oxides NO_x mainly from sintering and heating (1.2 kg/t of steel); hydrocarbons; carbon monoxide; in some cases dioxins (mostly from sintering operations); and hydrogen fluoride.

Air emissions from steel manufacturing using the BOF may include PM (ranging from less than 15 kg/t to 30 kg/t of steel). For closed systems, emissions come from the desulfurization step between the blast furnace and the BOF; the particulate matter emissions are about 10 kg/t of steel.

In the conventional process without recirculation, wastewaters, including those from cooling operations, are generated at an average rate of 80 m^3/t of steel manufactured. Major pollutants present in untreated wastewaters generated from pig iron manufacture include total organic carbon (typically 100-200 mg/1); total suspended solids (7,000 mg/l, 137 kg/t); dissolved solids; cyanide (15 mg/1); fluoride (1,000 mg/1); chemical oxygen demand, or COD (500 mg/1); and zinc (35 mg/1).

Major pollutants in wastewaters generated from steel manufacturing using the BOF include total suspended solids (up to 4,000 mg/l, 1030 kg/t), lead (8 mg/1), chromium (5 mg/1), cadmium (0.4 mg/1), zinc (14 mg/1), fluoride (20 mg/1), and oil and grease. Mill scale may amount to 33 kg/t. The process generates effluents with high temperatures.

Process solid waste from the conventional process, including furnace slag and collected dust, is generated at an average rate ranging from 300 kg/t of steel manufactured to 500 kg/t, of which 30 kg may be considered hazardous depending on the concentration of heavy metals present. Approximately, 65% of BOF slag from steel manufacturing can be recycled in various industries such as building materials and, in some cases, mineral wool.

Fate of Selected Chemicals

Table 2 provides a synopsis of current scientific toxicity and fate information for the top chemicals (by weight) that steel facilities self-report as being released based upon the TRI (Toxic Release Inventory) in the United States. The descriptions provided in Table 2 are taken directly from 1993 *Toxics Release Inventory Public Data Release* (EPA, 1994), and the Hazardous Substances Data Bank (HSDB), assessed via TOXNET. TOXNET is a computer system run by the National Library of Medicine. It includes a number of toxicological databases managed by EPA, the National Cancer Institute, and the National Institute for Occupational Safety and Health. HSDB contains chemical-specific information on manufacturing and use, chemical and physical properties, safety and handling, toxicity and biomedical effects, pharmacology, environmental fate and exposure potential, exposure standards and regulations, monitoring and analysis methods, and additional references. The information contained in Table 2 is based upon exposure assumptions that have been conducted using standard scientific procedures. The effects listed must be taken in context of these exposure assumptions that are more fully explained within the full chemical profiles in HSDB.

Table 2. *Toxicity and Environmental Fate Information.*

Ammonia CAS #7664-41-7
Sources. In coke-making, ammonia is produced by the decomposition of the nitrogen-containing compounds which takes place during the secondary thermal reaction (at temperatures greater than 700°C (1296°F)). The ammonia formed during coking exists in both the water and gas that form part of the volatile products. Recovery can be accomplished by several different processes where the by-product ammonium sulfate is formed by the reaction between the ammonia and sulfuric acid. **Toxicity.** Anhydrous ammonia is irritating to the skin, eyes, nose throat, and upper respiratory system. Ecologically, ammonia is a source of nitrogen (an essential element for aquatic plant growth), and may contribute to eutrophication of standing or slow-moving surface water, particularly in nitrogen-limited waters. In addition, aqueous ammonia is moderately toxic to aquatic organisms. **Carcinogenicity.** There is currently no evidence to suggest that this chemical is carcinogenic. **Environmental Fate.** Ammonia combines with sulfate ions in the atmosphere and is washed out by rainfall, resulting in rapid return of ammonia to the soil and surface waters. Ammonia is a central compound in the environmental cycling of nitrogen. Ammonia in lakes, rivers, and streams is converted to nitrate. **Physical Properties.** Ammonia is a corrosive and severely irritating gas with a pungent odor.
Hydrochloric Acid CAS #7647-01-1
Sources. During hot rolling, a hard black iron oxide is formed on the surface of the steel. This "scale" is removed chemically in the pickling process which commonly uses hydrochloric acid. **Toxicity.** HCl is primarily a concern in its aerosol form. Acid aerosols have been implicated in causing and exacerbating a variety of respiratory ailments. Dermal exposure and ingestion of highly concentrated HCl can result in corrosivity. Ecologically, accidental releases of solution forms of HCl may adversely affect aquatic life by including a transient lowering of the pH (i.e., increasing the acidity) of surface waters. **Carcinogenicity.** There is currently no evidence to suggest carcinogenicity. **Environmental Fate.** Releases to surface waters and soils will be neutralized to an extent due to the buffering capacities of both systems. The extent of these reactions will depend on the characteristics of the specific environment. **Physical Properties.** Concentrated hydrochloric acid is highly corrosive.

Manganese and Manganese Compounds CAS #7439-96-5; 20-12-2
Sources. Manganese is found in the iron charge and is used as an addition agent added to alloy steel to obtain desired properties in the final product. In carbon steel, Mg is used to combine with sulfur to improve steel ductility. An alloy steel with Mg is used for applications involving relatively small sections which are subject to severe service conditions, or in larger sections where the weight saving derived from the higher strength of the alloy steels is needed.
Toxicity. There is currently no evidence that human exposure to Mg at levels commonly observed in ambient atmosphere results in adverse health effects. However, recent EPA review of the fuel additive MMT (methylcyclopentadienyl magnesium tricarbonyl) concluded that use of MMT in gasoline could lead to ambient exposures to Mg at a level sufficient to result in adverse neurological effects. Chronic Mg poisoning bears some similarity to chronic lead poisoning. Both occur via inhalation of dust or fumes, and primarily involves the central nervous system. Early symptoms include languor, speech disturbances, sleepiness, and cramping and weakness in legs. A stolid mask-like appearance of face, emotional disturbances such as absolute detachment broken by uncontrollable laughter, euphoria, and a spastic gait with a tendency to fall while walking are seen in more advanced cases. Chronic Mg poisoning is reversible if treated early and exposure stopped. Populations at greatest risk of Mg toxicity are the very young and those with iron deficiencies. Ecologically, although Mg is an essential nutrient for both plants and animals, in excessive concentrations it inhibits plant growth.
Carcinogenicity. There is currently no evidence to suggest that this chemical is carcinogenic.
Environmental Fate. Mg is an essential nutrient for plants and animals. It accumulates in the top layers of soil or surface water sediments and cycles between the soil and living organisms. It occurs mainly as a solid under environmental conditions, though may also be transported in the atmosphere as a vapor or dust.

1,1,1-Trichloroethane CAS #71-55-6
Sources. Used, for surface cleaning of steel prior to coating.
Toxicity. Repeated contact with skin may cause serious skin cracking and infection. Vapors cause a slight smarting of the eyes or respiratory system if present in high concentrations. Exposure to high concentrations of TCE causes reversible mild liver and kidney dysfunction, central nervous system depression, gait disturbances, stupor, coma, respiratory depression, and even death. Exposure to lower concentrations of TCE leads to light-headedness, throat irritation, headache, disequilibrium, impaired coordination, drowsiness, convulsions and mild changes in perception.
Carcinogenicity. There is currently no evidence to suggest that this chemical is carcinogenic.
Environmental Fate. Releases of TCE to surface water or land will almost entirely volatilize. Releases to air may be transported long distances and may partially return to earth in rain. In the lower atmosphere, TCE degrades very slowly by photooxidation and slowly diffuses to the upper atmosphere where photodegradation is rapid. Any TCE that does not evaporate from soils leaches to groundwater. Degradation in soils and water is a very slow process. TCE does not hydrolyze in water, nor does it significantly bioconcentrate in aquatic organisms.

Zinc and Zinc Compounds CAS #7440-66-6; 20-19-9
Sources. To protect steel from rusting, it is coated with a material that will protect it from moisture and air. In the galvanizing process, steel is coated with zinc. **Toxicity.** Zinc is a nutritional trace element; toxicity from ingestion is low. Severe exposure to zinc might give rise to gastritis with vomiting due to swallowing of zinc dusts. Short-term exposure to very high levels of zinc is linked to lethargy, dizziness, nausea, fever, diarrhea, and reversible pancreatic and neurological damage. Long-term zinc poisoning causes irritability, muscular stiffness and pain, loss of appetite, and nausea. Zinc chloride fumes cause injury to mucous membranes and to the skin. Ingestion of soluble zinc salts may cause nausea, vomiting, and purging. **Carcinogenicity.** There is currently no evidence to suggest that this chemical is carcinogenic. **Environmental Fate.** Significant zinc contamination of soil is only seen in the vicinity of industrial point sources. Zinc is a relatively stable soft metal, though it burns in air (pyrophoric). Zinc bioconcentrates in aquatic organisms.

The toxic chemical release data obtained from TRI provides detailed information on the majority of facilities in the iron and steel industry in the United States. It also allows for a comparison across years and industry sectors. Reported chemicals are limited however to the 316 reported chemicals. Most of the hydrocarbon emissions from iron and steel facilities are not captured by TRI. The EPA Office of Air Quality Planning and Standards has compiled air pollutant emission factors for determining the total air emissions of priority pollutants (e.g., total hydrocarbons, SO_x, NO_x, CO, particulates, etc.) from many iron and steel manufacturing sources. The Aerometric Information Retrieval System (AIRS) contains a wide range of information related to stationary sources of air pollution, including the emissions of a number of air pollutants which may be of concern within a particular industry. With the exception of VOCs, there is little overlap with the TRI chemicals reported above. By way of comparison to other industry sectors, the steel industry in the U.S. emits about 1.5 million short tons/year of carbon monoxide, which is more than twice as much as the next largest releasing industry, pulp and paper. The iron and steel industry also ranks as one of the top five releasers for NO_2, PM_{10}, and SO_2. Carbon monoxide releases occur during ironmaking (in the burning of coke, CO produced reduces iron oxide ore), and during steelmaking (in either the basic oxygen furnace or the electric arc furnace). Nitrogen dioxide is generated during steelmaking. Particulate matter may be emitted from the coke-making (particularly in quenching operations), ironmaking, basic oxygen furnace (as oxides of iron that are emitted as sub-micron dust), or from the electric arc furnace (as metal dust containing iron particulate, zinc, and other materials associated with the scrap). Sulfur dioxide can be released in ironmaking or sintering.

Pollution Prevention Practices and Opportunities

Most of the pollution prevention practices have concentrated on reducing coke-making emissions, Electric Arc Furnace (EAF) dust, and spent acids used in finishing operations. Due to the complexity, size, and age of the equipment used, projects that have the highest pollution prevention potential often require major capital investments, which make many pollution prevention projects difficult to justify. Despite this, the industry must seek ways to become more cost-competitive, which requires investing in more cost-effective, less polluting technologies. Table 3 provides a summary of P2 practices and opportunities. To supplement this list, the following discussions should be considered.

With regard to coke-making, this process is seen by industry experts as one of the steel industry's areas of greatest environmental concern, with coke oven air emissions and quenching waste water as the major problems. In response to expanding regulatory constraints in the U.S., including the Clean Air Act National Emission Standards for coke ovens, U.S. steelmakers are turning to new technologies to decrease the sources of pollution from, and their reliance on, coke. Pollution prevention in coke-making has focused on two areas: reducing coke oven emissions and developing coke-less ironmaking techniques. Although these processes have not yet been widely demonstrated on a commercial scale, they may provide important benefits, especially for the integrated segment of the industry, by potentially lowering air emissions and wastewater discharges. Several technologies are available or are under development to reduce the emissions from coke ovens. Typically, these technologies reduce the quantity of coke needed by changing the method by which coke is added to the blast furnace or by substituting a portion of the coke with other fuels.

The reduction in the amount of coke produced proportionally reduces the coking emissions. Some of the most prevalent or promising coke reduction technologies are listed in Table 3.

Coke-less technologies (refer to Table 3 for examples) substitute coal for coke in the blast furnace, hence eliminating the need for coke-making. Such technologies have enormous potential to reduce pollution generated during the steelmaking process. The drawbacks with these technologies are (1) the capital investment required for retrofits is very significant, and (2) some countries whose economies are dependent upon the steel industry need to undergo significant industry rationalization and restructuring in order to justify investments into these technologies. For example, Russia and Ukraine, which have significant steel production and export capabilities heavily depend on a labor intensive coking industry. The elimination of the coking industry in these countries would likely result in significant social implications.

Table 3. *Recommended Pollution Prevention Practices.*

AREA OF OPPORTUNITY	RECOMMENDED POLLUTION PREVENTION PRACTICE
Eliminating coke with coke-less technologies	1. *The Japanese Direct Iron Ore Smelting (DIOS) process.* This process produces molten iron directly with coal and sinter feed ore. A 500 ton per day pilot plant was started up in October, 1993 and the designed production rates were attained as a short term average. Data generated is being used to determine economic feasibility on a commercial scale. 2. *HIsmelt process.* A plant using the HIsmelt process for molten iron production, developed by HIsmelt Corporation of Australia, was started up in late 1993. The process, using ore fines and coal, has achieved a production rate of 8 tons per hour using ore directly in the smelter. Developers anticipate reaching the production goal of 14 tons per hour. The data generated is being used to determine economic feasibility on commercial scale. If commercial feasibility is realized, Midrex is expected to become the U.S. engineering licensee of the HIsmelt process. 3. *Corex process.* The Corex or Cipcor process has integral coal desulfurizing, is amenable to a variety of coal types, and generates electrical power in excess of that required by an iron and steel mill which can be sold to local power grids. A Corex plant is in operation in South Africa, and other plants are expected to be operational in South Korea and India.
Reducing coke oven emissions with other technologies	4. *Pulverized coal injection.* This technology substitutes pulverized coal for a portion of the coke in the blast furnace. Use of pulverized coal injection can replace about 25 to 40 percent of coke in the blast furnace, substantially reducing emissions associated with coke-making operations. This reduction ultimately depends on the fuel injection rate applied to the blast furnaces which will, in turn be dictated by the aging of existing coking facilities, fuel costs, oxygen availability, capital requirements for fuel injection, and available hot blast temperature. 5. *Nonrecovery coke battery.* As opposed to the by-product recovery coke plant, the nonrecovery coke battery is designed to allow combustion of the gasses from the coking process, thus consuming the by-products that are typically recovered. The process results in lower air emissions and substantial reductions in coking process wastewater discharges. 6. *The Davy Still Auto-process.* In this pre-combustion cleaning process for coke ovens, coke oven battery process water is utilized to strip ammonia and hydrogen sulfide from coke oven emissions. 7. *Alternative fuels.* Steel producers can inject other fuels, such as natural gas, oil, and tar/pitch, instead of coke into the blast furnace, but these fuels can only replace coke in limited amounts.

Reducing Wastewater	8. In Europe, some plants have implemented technology to shift from water quenching to dry quenching in order to reduce energy costs. However, major construction changes are required for such a solution.
Recycling of Coke By-products	9. Improvements in the in-process recycling of tar decanter sludge are common practice. Sludge can either be injected into the ovens to contribute to coke yield, or converted into a fuel that is suitable for the blast furnace.
Electric Arc Furnace Dust	10. EAF dust is a hazardous waste because of its high concentrations of lead and cadmium. With 550,000 tons of EAF dust generated annually in the U.S., there is great potential to reduce the volume of this hazardous waste. U.S. steel companies typically pay a disposal fee of $150 to $200 per ton of dust. With an average zinc concentration of 19 percent, much of the EAF dust is sent off-site for zinc recovery. Most of the EAF dust recovery options are only economically viable for dust with a zinc content of at least 15 to 20 percent. Facilities that manufacture specialty steels such as stainless steel with a lower zinc content, still have opportunities to recover chromium and nickel from the EAF dust. In-process recycling of EAF dust involves pelletizing and then reusing the pellets in the furnace, however, recycling of EAF dust on-site has not proven to be technically or economically competitive for all mills. Improvements in technologies have made off-site recovery a cost effective alternative to thermal treatment or secure landfill disposal.
Pig Iron Manufacturing	11. Improve blast furnace efficiency by using coal and other fuels (such as oil or gas) for heating instead of coke, thereby minimizing air emissions. 12. Recover the thermal energy in the gas from the blast furnace before using it as a fuel. 13. Increase fuel efficiency and reduce emissions by improving blast furnace charge distribution. 14. Recover energy from sinter coolers and exhaust gases. 15. Use dry SO_x removal systems such as caron absorption for sinter plants or lime spraying in flue gases. 16. Recycle iron-rich materials such as iron ore fines, pollution control dust, and scale in a sinter plant. 17. Use low- NO_x burners to reduce NO_x emissions from burning fuel in ancillary operations. 18. Improve productivity by screening the charge and using better taphole practices. 19. Reduce dust emissions at furnaces by covering iron runners when tapping the blast furnace and by using nitrogen blankets during tapping. 20. Use pneumatic transport, enclosed conveyor belts, or self-closing conveyor belts, as well as wind barriers and other dust suppression measures, to reduce the formation of fugitive dust.

Steel Manufacturing	21. Use dry dust collection and removal systems to avoid the generation of wastewater. Recycle collected dust. 22. Use BOF gas as fuel. 23. Use enclosures for BOF. 24. Use a continuous process for casting steel to reduce energy consumption.
Finishing Stages	25. *Pickling Acids* - In finishing, pickling acids are recognized as an area where pollution prevention efforts can have a significant impact in reducing the environmental impact of the steel mill. The pickling process removes scale and cleans the surface of raw steel by dipping it into a tank of hydrochloric or sulfuric acid. If not recovered, the spent acid may be transported to deep injection wells for disposal, but as those wells continue to close, alternative disposal costs are rising. 26. *Pickling Acids* - Large-scale steel manufacturers recover HCl in their finishing operations, however the techniques used are not suitable for small- to medium-sized steel plants. Currently, a recovery technique for smaller steel manufacturers and galvanizing plants is in pilot scale testing. The system removes iron chloride (a saleable product) from the HCl, reconcentrates the acid for reuse, and recondenses the water to be reused as a rinse water in the pickling process. Because the only by-product of the hydrochloric acid recovery process is a non-hazardous, marketable metal chloride, this technology generates no hazardous wastes. The manufacturer projects industry-wide HCl waste reduction of 42,000 tons/year by 2010. This technology is less expensive than transporting and disposing waste acid, plus it eliminates the associated long-term liability. The total savings for a small- to medium-sized galvanizer is projected to be $260,000 each year. 27. *Pickling Acids* - To reduce spent pickling liquor and simultaneously reduce fluoride in the plant effluent, one facility modified their existing treatment process to recover the fluoride ion from rinse water and spent pickling acid raw water waste streams. The fluoride is recovered as calcium fluoride (fluorspar), an input product for steelmaking. The melt shop in the same plant had been purchasing 930 tons of fluorspar annually for use as a furnace flux material in the EAF at a cost of $100 per ton. The recovered calcium fluoride is expected to be a better grade than the purchased fluorspar, which would reduce the amount of flux used by approximately 10 percent. Not only would the generation rate of sludge from spent pickling liquor treatment be reduced (resulting in a savings in off-site sludge disposal costs), but a savings in chemical purchases would be captured.
Process Modifications	28. Replacing single-pass wastewater systems with closed-loop systems to minimize chemical use in wastewater treatment and to reduce water use.

	29. Continuous casting, now used for about 90% of crude steel cast in the U.S., offers great improvements in process efficiency when compared to the traditional ingot teeming method. This increased efficiency also results in a considerable savings in energy and some reduction in the volume of mill wastewater.
Materials Substitution	30. Use scrap steel with low lead and cadmium content as a raw material, if possible. 31. Eliminate the generation of reactive desulfurization slag generated in foundry work by replacing calcium carbide with a less hazardous material.
Recycling Miscellaneous Materials	32. Recycle or reuse oils and greases. 33. Recover acids by removing dissolved iron salts from spent acids. 34. Use thermal decomposition for acid recovery from spent pickle liquor. 35. Use a bipolar membrane/electrodialytic process to separate acid from metal by-products in spent NO_3-HF pickle liquor. 36. Recover sulfuric acid using low temperature separation of acid and metal crystals. 37. Use blast furnace slag in construction materials. Slag containing free lime can be used in ironmaking.

Mini Steel Mills

Mini steel mills normally use the electric arc furnace (EAF) to produce steel from returned steel, scrap, and direct reduced iron. EAF is a batch process with a cycle time of about two to three hours. Since the process uses scrap metal instead of molten iron, coke-making and ironmaking operations are eliminated. EAFs can economically serve small, local markets. Further processing of steel can include continuous casting, hot rolling and forming, cold rolling, wire drawing, coating, and pickling. As already noted, the continuous casting process bypasses several steps of the conventional ingot teeming process by casting steel directly into semifinished shapes. The casting, rolling, and steel finishing processes are also used in iron and steel manufacturing. Hot steel is transformed in size and shape through a series of hot rolling and forming steps to manufacture semifinished and finished steel products. The hot rolling process consists of slabheating (as well as billet and bloom), rolling, and forming operations. Several types of hot forming mills (primary, section, flat, pipe and tube, wire, rebar, and profile) manufacture a variety of steel products. For the manufacture of a very thin strip or a strip with a high-quality finish, cold rolling must follow the hot rolling operations. Lubricants emulsified in water are usually used to achieve high surface quality and to prevent overheating of the product.

Wire drawing includes heat treatment of rods, cleaning, and sometimes coating. Water, oil, or lead baths are used for cooling and to impart desired features. To prepare the steel for cold rolling or drawing, acid pickling is performed to chemically remove oxides and scale from the surface of the steel through use of inorganic acid water solutions. Mixed acids (nitric and hydrofluoric) are used for stainless steel pickling; sulfuric or hydrochloric acid is used for other steels. Other methods for removing scale include salt pickling, electrolytic pickling, and blasting; blasting is environmentally desirable, where feasible. EAFs produce metal dusts, slag, and gaseous emissions. The primary hazardous components of EAF dust are zinc, lead, and cadmium; nickel and chromium are present when stainless steels are manufactured. The composition of EAF dust can vary greatly, depending on scrap composition and furnace additives. EAF dust usually has a zinc content of more than 15%, with a range of 5-35%. Other metals present in EAF dust include lead (2-7%), cadmium (generally 0.1-0.2% but can be up to 2.5% where stainless steel cases of nickel-cadmium batteries are melted), chromium (up to 15%), and nickel (up to 4%). Generally, an EAF produces 10 kilograms of dust per metric ton (kg/t) of steel, with a range of 5-30 kg/t, depending on factors such as furnace characteristics and scrap quality. Major pollutants present in the air emissions include particulates (1,000 milligrams per normal cubic meter, mg/Nm^3), nitrogen oxides from cutting, scarfing, and pickling operations, and acid fumes (3,000 mg/Nm^3) from pickling operations. Both nitrogen oxides and acid fumes vary with steel quality.

Mini mills generate up to 80 cubic meters of wastewater per metric ton (m^3/t) of steel product. Untreated wastewaters contain high levels of total suspended solids (up to 3,000 milligrams per liter, mg/l), copper (up to 170 mg/l), lead (10 mg/l), total chromium (3,500 mg/l), hexavalent chromium (200 mg/l), nickel (4,600 mg/l), and oil and grease (130 mg/l). Chrome and nickel concentrations result mainly from pickling operations. The characteristics of the wastewater depend on the type of steel, the forming and finishing operations, and the quality of scrap used as feed to the process. Solid wastes, excluding EAF dust and wastewater treatment sludges, are generated at a rate of 20 kg/t of steel product. Sludges and scale from acid pickling, especially in stainless steel manufacturing, contain heavy metals such as chromium (up to 700 mg/kg), lead (up to 700 mg/kg), and nickel (400 mg/kg). These levels may be even higher for some stainless steels. Table 4 provides a list of pollution prevention practices and recommendations for mini steel mills.

Standard treatment technologies for air emissions are as follows. Dust emission control technologies include cyclones, baghouses, and ESPs. Scrubbers are used to control acid mists.

Table 4. *P2 Practices and Recommendations for Mini Steel Mills.*

Locate EAFs in enclosed buildings.
Improve feed quality by using selected scrap to reduce the release of pollutants to the environment.
Use dry dust collection methods such as fabric filters.
Replace ingot teeming with continuous casting.
Use continuous casting for semifinished and finished products wherever feasible. In some cases, continuous charging may be feasible and effective for controlling dust emissions.
Use bottom tapping of EAFs to prevent dust emissions.
Control water consumption by proper design of spray nozzles and cooling water systems.
Segregate wastewaters containing lubricating oils from other wastewater streams and remove oil.
Recycle mill scale to the sinter plant in an integrated steel plant.
Use acid-free methods (mechanical methods such as blasting) for descaling, where feasible.
In the pickling process, use countercurrent flow of rinse water; use indirect methods for heating and pickling baths.
Use closed-loop systems for pickling; regenerate and recover acids from spent pickling liquor using resin bed, retorting, or other regeneration methods such as vacuum crystallization of sulfuric acid baths.
Use electrochemical methods in combination with pickling to lower acid consumption.
Reduce nitrogen oxide emissions by use of natural gas as fuel, use low-NO_x burners, and use hydrogen peroxide and urea in stainless steel pickling baths.
Recycle slags and other residuals from manufacturing operations for use in construction and other industries.
High water use is associated with cooling. Recycle wastewaters to reduce the discharge rate to less than 5 m^3/t of steel produced, including indirect cooling waters.
Recover zinc from EAF dust containing more than 15% total zinc; recycle EAF dust to the extent feasible.

Fugitive emissions from charging and tapping of EAFs should be controlled by locating the EAF in an enclosed building or using hoods and by evacuating the dust to dust arrestment equipment to achieve an emissions level of less than 0.25 kg/t.

For liquid wastes - spent pickle liquor containing hydrochloric acid is treated by spraying it into a roasting chamber and scrubbing the vapors. If hexavalent chrome is present in salt pickling or electrolytic pickling baths, it can be reduced with a sulfide reagent, iron salts, or other reducing agents. The remaining wastewaters are typically treated using oil-water separation flotation, precipitation, chemical flocculation, sedimentation/parallel plate separation/hydrocycloning, and filtration. Methods such as ultrafiltration may be used for oil emulsions. For continuous casting and cold rolling, oil should be less than 5 g/t and total suspended solids less than 10 g/t. For hot rolling, the corresponding values are 10 g/t and 50 g/t, respectively.

LEAD AND ZINC SMELTING

Industry Description and Practices

Lead and zinc can be produced pyrometallurgically or hydrometallurgically, depending on the type of ore used as a charge. In the pyrometallurgical process, ore concentrate containing lead, zinc, or both is fed, in some cases after sintering, into a primary smelter. Lead concentrations can be 50-70%, and the sulfur content of sulfidic ores is in the range of 15-20%. Zinc concentration is in the range of 40-60%, with sulfur content in sulfidic ores in the range of 26-34%. Ores with a mixture of lead and zinc concentrate usually have lower respective metal concentrations. During sintering, a blast of hot air or oxygen is used to oxidize the sulfur present in the feed to sulfur dioxide. Blast furnaces are used in conventional processes for reduction and refining of lead compounds to produce lead. Modem direct smelting processes include QSL, Kivcet, AUSMELT, and TBRC.

Primary Lead Processing

The conventional pyrometallurgical primary lead production process consists of four steps: sintering, smelting, drossing, and refining. A feedstock made up mainly of lead concentrate is fed into a sintering machine. Other raw materials may be added, including iron, silica, limestone flux, coke, soda, ash, pyrite, zinc, caustic, and particulates gathered from pollution control devices. The sintering feed, along with coke, is fed into a blast furnace for reducing, where the carbon also acts as a fuel and smelts the lead-containing materials. The molten lead flows to the bottom of the furnace, where four layers form: "speiss" (the lightest material, basically arsenic and antimony), "matte" (copper sulfide and other metal sulfides), blast furnace slag (primarily silicates), and lead bullion (98% by weight). All layers are then drained off. The speiss and matte are sold to copper smelters for recovery of copper and precious metals. The blast furnace slag, which contains zinc, iron, silica, and lime, is stored in piles and is partially recycled. Sulfur oxide emissions are generated in blast furnaces from small quantities of residual lead sulfide and lead sulfates in the sinter feed.

Rough lead bullion from the blast furnace usually requires preliminary treatment in kettles before undergoing refining operations. During drossing, the bullion is agitated in a drossing kettle and cooled to just above its freezing point, 370-425°C (700-800°F). A dross composed of lead oxide, along with copper, antimony, and other elements, floats to the top and solidifies above the molten lead. The dross is removed and is fed into a dross furnace for recovery of the nonlead mineral values.

The lead bullion is refined using pyrometallurgical methods to remove any remaining nonlead materials (e.g., gold, silver, bismuth, zinc, and metal oxides such as oxides of antimony, arsenic, tin, and copper). The lead is refined in a cast-iron kettle in five stages. First, antimony, tin, and arsenic are removed. Next, gold and silver are removed by adding zinc. The lead is then refined by vacuum removal of zinc. Refining continues with the addition of calcium and magnesium, which combine with bismuth to form an insoluble compound that is skimmed from the kettle. In the final step, caustic soda, nitrates, or both may be added to remove any remaining traces of metal impurities. The refined lead will have a purity of 99.90 to 99.99%. It may be mixed with other metals to form alloys, or it may be directly cast into shapes.

Secondary Lead Processing

The secondary production of lead begins with the recovery of old scrap from worn-out, damaged, or obsolete products and with new scrap. The chief source of old scrap is lead-acid batteries; other sources include cable coverings, pipe, sheet, and other lead-bearing metals. Solder, a tin-based alloy, may be recovered from the processing of circuit boards for use as lead charge.

Prior to smelting, batteries are usually broken up and sorted into their constituent products. Fractions of cleaned plastic (such as polypropylene) case are recycled into battery cases or other products. The dilute sulfuric acid is either neutralized for disposal or recycled to the local acid market. One of the three main smelting processes is then used to reduce the lead fractions and produce lead bullion.

Most domestic battery scrap is processed in blast furnaces, rotary furnaces, or reverberatory furnaces. A reverberatory furnace is more suitable for processing fine particles and may be operated in conjunction with a blast furnace.

Blast furnaces produce hard lead from charges containing siliceous slag from previous runs (about 4.5% of the charge), scrap iron (about 4.5%), limestone (about 3%), and coke (about 5.5%). The remaining 82.5% of the charge is made up of oxides, pot furnace refining drosses, and reverberatory slag. The proportions of rerun slags, limestone, and coke vary but can run as high as 8% for slags, 10% for limestone, and 8% for coke. The processing capacity of the blast furnace ranges from 20 to 80 metric tons per day (tpd).

Newer secondary recovery plants use lead paste desulfurization to reduce sulfur dioxide emissions and generation of waste sludge during smelting. Battery paste containing lead sulfate and lead oxide is desulfurized with soda ash, yielding market-grade sodium sulfate as a by-product. The desulfurized paste is processed in a reverberatory furnace, and the lead carbonate product may then be treated in a short rotary furnace. The battery grids and posts are processed separately in a rotary smelter.

Zinc Manufacturing

In the most common hydrometallurgical process for zinc manufacturing, the ore is leached with sulfuric acid to extract the lead/zinc. These processes can operate at atmospheric pressure or as pressure leach circuits. Lead/zinc is recovered from solution by electrowinning, a process similar to electrolytic refining. The process most commonly used for low-grade deposits is heap leaching. Imperial smelting is also used for zinc ores.

Pollution Prevention Practices and Opportunities

The principal air pollutants emitted from the processes are particulate matter and sulfur dioxide. Fugitive emissions occur at furnace openings and from launders, casting molds, and ladles carrying molten materials, which release sulfur dioxide and volatile substances into the working environment. Additional fugitive particulate emissions occur from materials handling and transport of ores and concentrates. Some vapors are produced in hydrometallurgy and in various refining processes.

The principal constituents of the particulate matter are lead/zinc and iron oxides, but oxides of metals such as arsenic, antimony, cadmium, copper, and mercury are also present, along with metallic sulfates. Dust from raw materials handling contains metals, mainly in sulfidic form, although chlorides, fluorides, and metals in other chemical forms may be present. Off-gases contain fine dust particles and volatile impurities such as arsenic, fluorine, and mercury.

Air emissions for processes with few controls may be of the order of 30 kilograms lead or zinc per metric ton (kg/t) of lead or zinc produced. The presence of metals in vapor form is dependent on temperature. Leaching processes will generate acid vapors, while refining processes result in products of incomplete combustion (PICs). Emissions of arsine, chlorine, and hydrogen chloride vapors and acid mists are associated with electrorefining.

Wastewaters are generated by wet air scrubbers and cooling water. Scrubber effluents may contain lead/zinc, arsenic, and other metals. In the electrolytic refining process, by-products such as gold and silver are collected as slimes and are subsequently recovered. Sources of wastewater include spent electrolytic baths, slimes recovery, spent acid from hydrometallurgy processes, cooling water, air scrubbers, washdowns, and stormwater. Pollutants include dissolved and suspended solids, metals, and oil and grease.

The larger proportion of the solid waste is discarded slag from the smelter. Discarded slag may contain 0.5-0.7% lead/zinc and is frequently used as fill or for sandblasting. Slags with higher lead/zinc content, say 15% zinc, can be sent for metals recovery. Leaching processes produce residues, while effluent treatment results in sludges that require appropriate disposal. The smelting process typically produces less than 3 tons of solid waste per ton of lead/zinc produced.

The most effective pollution prevention option is to choose a process that entails lower energy usage and lower emissions. Modern flash-smelting processes save energy, compared with the conventional sintering and blast furnace process. Process gas streams containing over 5% sulfur dioxide are usually used to manufacture sulfuric acid. The smelting furnace will generate gas streams with SO_2 concentrations ranging from 0.5% to 10%, depending on the method used. It is important, therefore, to select a process that uses oxygen-enriched air or pure oxygen. The aim is to save energy and raise the SO_2 content of the process gas stream by reducing the total volume of the stream, thus permitting efficient fixation of sulfur dioxide. Processes should be operated to maximize the concentration of the sulfur dioxide. An added benefit is the reduction (or elimination) of nitrogen oxides NO_x. Table 5 provides a list of pollution prevention practices and opportunities. Regarding standard treatment technologies, ESPs and baghouses are used for product recovery and for the control of particulate emissions. Dust that is captured but not recycled will need to be disposed of in a secure landfill or in another acceptable manner.

Table 5. *Summary of Pollution Prevention Practices.*

Use doghouse enclosures where appropriate; use hoods to collect fugitive emissions.
Mix strong acidic gases with weak ones to facilitate production of sulfuric acid from sulfur oxides, thereby avoiding the release of weak acidic gases.
Maximize the recovery of sulfur by operating the furnaces to increase the SO_x content of the flue gas and by providing efficient sulfur conversion. Use a double-contact, double-absorption process.
Desulfurize paste with caustic soda or soda ash to reduce SO_2 emissions.
Use energy-efficient measures such as waste heat recovery from process gases to reduce fuel usage and associated emissions.
Recover acid, plastics, and other materials when handling battery scrap in secondary lead production.
Recycle condensates, rainwater, and excess process water for washing, for dust control, for gas scrubbing, and for other process applications where water quality is not of particular concern.
Give preference to natural gas over heavy fuel oil for use as fuel and to coke with lower sulfur content.
Use low-NO_x burners.
Use suspension or fluidized bed roasters, where appropriate, to achieve high SO_2 concentrations when roasting zinc sulfides.
Recover and reuse iron-bearing residues from zinc production for use in the steel or construction industries.
Give preference to fabric filters over wet scrubbers or wet electrostatic precipitators (ESPs) for dust control.
Good housekeeping practices are key to minimizing losses and preventing fugitive emissions. Losses and emissions are minimized by enclosed buildings, covered conveyors and transfer points, and dust collection equipment. Yards should be paved and runoff water routed to settling ponds.

Arsenic trioxide or pentoxide is in vapor form because of the high gas temperatures and must be condensed by gas cooling so that it can be removed in fabric filters. Collection and treatment of vent gases by alkali scrubbing may be required when sulfur dioxide is not being recovered in an acid plant.

Effluent treatment of process bleed streams, filter backwash waters, boiler blowdown, and other streams is required to reduce suspended and dissolved solids and heavy metals and to adjust pH. Residues that result from treatment are recycled to other industries such as the construction industry, sent to settling ponds (provided that groundwater and surface water contamination is not a concern), or disposed of in a secure landfill.

Slag should be either landfilled or granulated and sold for use in building materials.

NICKEL ORE PROCESSING AND REFINING

Industry Description and Practices

Primary nickel is produced from two very different ores, lateritic and sulfidic. Lateritic ores are normally found in tropical climates where weathering, with time, extracts and deposits the ore in layers at varying depths below the surface. Lateritic ores are excavated using large earth-moving equipment and are screened to remove boulders. Sulfidic ores, often found in conjunction with copper-bearing ores, are mined from underground. Following is a description of the processing steps used for the two types of ores.

Lateritic Ore Processing

Lateritic ores have a high percentage of free and combined moisture, which must be removed. Drying removes free moisture; chemically bound water is removed by a reduction furnace, which also reduces the nickel oxide. Lateritic ores have no significant fuel value, and an electric furnace is needed to obtain the high temperatures required to accommodate the high magnesia content of the ore. Some laterite smelters add sulfur to the furnace to produce a matte for processing. Most laterite nickel processers run the furnaces so as to reduce the iron content sufficiently to produce ferronickel products. Hydrometallurgical processes based on ammonia or sulfuric acid leach are also used. Ammonia leach is usually applied to the ore after the reduction roast step.

Sulfidic Ore Processing

Flash smelting is the most common process, but electric smelting is used for more complex raw materials when increased flexibility is needed. Both processes use dried concentrates. Electric smelting requires a roasting step before smelting to reduce sulfur content and volatiles. Older nickel-smelting processes, such as blast or reverberatory furnaces, are no longer acceptable because of low energy efficiencies and environmental concerns.

In flash smelting, dry sulfide ore containing less than 1% moisture is fed to the furnace along with preheated air, oxygen-enriched air (30-40% oxygen), or pure oxygen. Iron and sulfur are oxidized. The heat that results from exothermic reactions is adequate to smelt concentrate, producing a liquid matte (up to 45% nickel) and a fluid slag. Furnace matte still contains iron and sulfur, and these are oxidized in the converting step to sulfur dioxide and iron oxide by injecting air or oxygen into the molten bath. Oxides form a slag, which is skimmed off. Slags are processed in an electric furnace prior to discard to recover nickel. Process gases are cooled, and particulates are then removed by gas-cleaning devices.

Nickel Refining

Various processes are used to refine nickel matte. Fluid bed roasting and chlorine-hydrogen reduction produce high-grade nickel oxides (more than 95% nickel). Vapor processes such as the carbonyl process can be used to produce high-purity nickel pellets. In this process, copper and precious metals remain as a pyrophoric residue that requires separate treatment. Use of electrical cells equipped with inert cathodes is the most common technology for nickel refining. Electrowinning, in which nickel is removed from solution in cells equipped with inert anodes, is the more common refining process. Sulfuric acid solutions or, less commonly, chloride electrolytes are used.

Pollution Prevention Practices and Opportunities

Sulfur dioxide is a major air pollutant emitted in the roasting, smelting, and converting of sulfide ores. (Nickel sulfide concentrates contain 6-20% nickel and up to 30% sulfur.) SO_2 releases can be as high as 4 metric tons (t) of sulfur dioxide per metric ton of nickel produced, before controls. Reverberatory furnaces and electric furnaces produce SO_2 concentrations of 0.5-2.0%, while flash furnaces produce SO_2 concentrations of over 10% - a distinct advantage for the conversion of the sulfur dioxide to sulfuric acid. Particulate emission loads for various process steps include 2.0-5.0 kilograms per metric ton (kg/t) for the multiple hearth roaster; 0.5-2.0 kg/t for the fluid bed roaster; 0.2-1.0 kg/t for the electric furnace; 1.0-2.0 kg/t for the Pierce-Smith converter; and 0.4 kg/t for the dryer upstream of the flash furnace. Ammonia and hydrogen sulfide are pollutants associated with the ammonia leach process; hydrogen sulfide emissions are associated with acid leaching processes. Highly toxic nickel carbonyl is a contaminant of concern in the carbonyl refining process. Various process offgases contain fine dust particles and volatilized impurities. Fugitive emissions occur at furnace openings, launders, casting molds, and ladles that carry molten product. The transport and handling of ores and concentrates produce windborne dust.

Pyrometallurgical processes for processing sulfidic ores are generally dry, and effluents are of minor importance, although wet ESPs are often used for gas treatment, and the resulting wastewater could have high metal concentrations. Process bleed streams may contain antimony, arsenic, or mercury. Large quantities of water are used for slag granulation, but most of this water should be recycled.

The smelter contributes a slag that is a dense silicate. Sludges that require disposal will result when neutralized process effluents produce a precipitate.

Pollution prevention is always preferred to the use of end-of-pipe pollution control facilities. Therefore, every attempt should be made to incorporate cleaner production processes and facilities to limit, at source, the quantity of pollutants generated. The choice of flash smelting over older technologies is the most significant means of reducing pollution at source. Sulfur dioxide emissions can be controlled by:

- Recovery as sulfuric acid,
- Recovery as liquid sulfur dioxide (absorption of clean dry off-gas in water or chemical absorption by ammonium bisulfite or dimethyl aniline),
- Recovery as elemental sulfur, using reductants, such as hydrocarbons, carbon, or hydrogen sulfide.

Toxic nickel carbonyl gas is normally not emitted from the refining process because it is broken down in decomposer towers. However, very strict precautions throughout the refining process are required to prevent the escape of the nickel carbonyl into the workplace. Continuous monitoring for the gas, with automatic isolation of any area of the plant where the gas is detected, is required. Impervious clothing is used to protect workers against contact of liquid nickel carbonyl with skin.

Preventive measures for reducing emissions of particulate matter include encapsulation of furnaces and conveyors to avoid fugitive emissions. Covered storage of raw materials should be considered.

Wet scrubbing should be avoided, and cooling waters should be recirculated. Stormwaters should be collected and used in the process. Process water used to transport granulated slag should be recycled. To the extent possible, all process effluents should be returned to the process.

The discharge of particulate matter emitted during drying, screening, roasting, smelting, and converting is controlled by using cyclones followed by wet scrubbers, ESPs, or bag filters. Fabric filters may require reduction of gas temperatures by, for example, dilution with low temperature gases from hoods used for fugitive dust control. Preference should be given to the use of fabric filters over wet scrubbers.

Liquid effluents are used to slurry tailings to the tailings ponds, which act as a reservoir for the storage and recycle of plant process water. However, there may be a need to treat bleed streams of some process effluents to prevent a buildup of various impurities. Solid wastes from nickel sulfide ores often contain other metals such as copper and precious metals, and consideration should be given to further processing for their recovery. Slag can be used as construction material after nickel recovery, as appropriate (e.g., return of converter slag to the furnace). Sanitary sewage effluents require treatment in a separate facility or discharge to a municipal sewer.

ALUMINUM MANUFACTURING

Industry Description and Practices

The production of aluminum begins with the mining and beneficiation of bauxite. At the mine (usually of the surface type), bauxite ore is removed to a crusher. The crushed ore is then screened and stockpiled, ready for delivery to an alumina plant. In some cases, ore is upgraded by beneficiation (washing, size classification, and separation of liquids and solids) to remove unwanted materials such as clay and silica.

At the alumina plant, the bauxite ore is further crushed to the correct particle size for efficient extraction of the alumina through digestion by hot sodium hydroxide liquor. After removal of "red mud" (the insoluble part of the bauxite) and fine solids from the process liquor, aluminum trihydrate crystals are precipitated and calcined in rotary kilns or fluidized bed calciners to produce alumina (Al_2O_3). Some alumina processes include a liquor purification step.

Primary aluminum is produced by the electrolytic reduction of the alumina. The alumina is dissolved in a molten bath of fluoride compounds (the electrolyte), and an electric current is passed through the bath, causing the alumina to dissociate to form liquid aluminum and oxygen. The oxygen reacts with carbon in the electrode to produce carbon dioxide and carbon monoxide. Molten aluminum collects in the bottom of the individual cells or pots and is removed under vacuum into tapping crucibles. There are two prominent technologies for aluminum smelting: prebake and Soderberg. The following discussions focus on the prebake technology, with its associated reduced air emissions and energy efficiencies.

Raw materials for secondary aluminum production are scrap, chips, and dross. Pretreatment of scrap by shredding, sieving, magnetic separation, drying, and so on is designed to remove undesirable substances that affect both aluminum quality and air emissions. The prevailing process for secondary aluminum production is smelting in rotary kilns under a salt cover. Salt slag can be processed and reutilized. Other processes (smelting in induction furnaces and hearth furnaces) need no or substantially less salt and are associated with lower energy demand, but they are only suitable for high-grade scrap.

Depending on the desired application, additional refining may be necessary. For demagging (removal of magnesium from the melt), hazardous substances such as chlorine and hexachloroethane are often used, which may produce dioxins and dibenzofurans. Other, less hazardous methods, such as adding chlorine salts, are available. Because it is difficult to remove alloying elements such as copper and zinc from an aluminum melt, separate collection and separate reutilization of different grades of aluminum scrap are necessary. Note that secondary aluminum production uses substantially less energy than primary production (less than 10-20 gigajoules per metric ton (GJ/t) of aluminum produced, compared with 164 GJ/t for primary production).

Pollution Prevention Practices and Opportunities

At the bauxite production facilities, dust is emitted to the atmosphere from dryers and materials-handling equipment, through vehicular movement, and from blasting. Although the dust is not hazardous, it can be a nuisance if containment systems are not in place, especially on the dryers and handling equipment. Other air emissions could include NO_x, SO_x, and other products of combustion from the bauxite dryers. Ore washing and beneficiation yield process wastewaters containing suspended solids. Runoff from precipitation may also contain suspended solids. At the alumina plant, air emissions can include bauxite dust from handling and processing, limestone dust from limestone handling, burnt lime dust from conveyors and bins, alumina dust from materials handling, red mud dust and sodium salts from red mud stacks (impoundments), caustic aerosols from cooling towers, and products of combustion such as sulfur dioxide and nitrogen oxides from boilers, calciners, various mobile equipment, and kilns. The calciners may also emit alumina dust and the kilns, and burnt lime dust.

Although alumina plants do not normally discharge effluents, heavy rainfalls can result in surface runoff that exceeds what the plant can use in the process. The excess may require treatment.

The main solid waste from the alumina plant is red mud (as much as 2 tons of mud per ton of alumina produced), which contains oxides of alumina, silicon, iron, titanium, sodium, calcium, and other elements. The pH is typically between 10 and 12. Disposal is to an impoundment.

Hazardous wastes from the alumina plant include spent sulfuric acid from descaling in tanks and pipes. Salt cake may be produced from liquor purification if this is practiced.

In the aluminum smelter, air emissions include alumina dust from handling facilities; coke dust from coke handling; gaseous and particulate fluorides; sulfur and carbon dioxides and various dusts from the electrolytic reduction cells; gaseous and particulate fluorides; sulfur dioxide; tar vapor and carbon particulates from the baking furnace; coke dust, tars, and polynuclear aromatic hydrocarbons (PAHs) from the green carbon and anode-forming plant; carbon dust from the rodding room; and fluxing emissions and carbon oxides from smelting, anode production, casting, and finishing operations. The electrolytic reduction cells (pot line) are the major source of the air emissions, with the gaseous and particulate fluorides being of prime concern. The anode effect associated with electrolysis also results in emissions of carbon tetrafluoride (CF_4) and carbon hexafluoride (C_2F_6), which are greenhouse gases of concern because of their potential for global warming. Emissions numbers that have been reported for uncontrolled gases from smelters are 20-80 kg/t for particulates, 6-12 kg/t for hydrogen fluoride, and 6-10 kg/t for fluoride particulates. Corresponding concentrations are 200-800 mg/m^3; 60-120 mg/m^3; and 60-100 mg/m^3. An aluminum smelter produces 40-60 kg of mixed solid wastes per ton of product, with spent cathodes (spent pot and cell linings) being the major fraction. The linings consist of 50% refractory material and 50% carbon. Over the useful life of the linings, the carbon becomes impregnated with aluminum and silicon oxides (averaging 16% of the carbon lining), fluorides (34% of the lining), and cyanide compounds (about 400 parts per million). Contaminant levels in the refractories portion of linings that have failed are generally low. Other by-products for disposal include skim, dross, fluxing slags, and road sweepings.

Atmospheric emissions from secondary aluminum melting include hydrogen chloride and fluorine compounds. Demagging may lead to emissions of chlorine, hexachloroethane, chlorinated benzenes, and dioxins and furans. Chlorinated compounds may also result from the melting of aluminum scrap that is coated with plastic. Salt slag processing emits hydrogen and methane. Solid wastes from the production of secondary aluminum include particulates, pot lining refractory material, and salt slag. Particulate emissions containing heavy metals are also associated with secondary aluminum production.

Pollution prevention is always preferred to the use of end-of-pipe pollution control facilities. Therefore every attempt should be made to incorporate cleaner production processes and facilities to limit, at source, the quantity of pollutants generated.

In the bauxite mine, where beneficiation and ore washing are practiced, a tailings slurry of 79% solids is produced for disposal. The preferred technology is to concentrate these tailings and dispose of them in the mined-out area. A concentration of 25-30% can be achieved through gravity settling in a tailings pond. The tailings can be further concentrated, using a thickener, to 30-50%, yielding a substantially volume reduced slurry.

The alumina plant discharges red mud in a slurry of 25-30% solids, and this also presents an opportunity to reduce disposal volumes. Modern technology, in the form of high-efficiency deep thickeners, and large-diameter conventional thickeners, can produce a mud of 50-60% solids concentration. The lime used in the process forms insoluble solids that leave the plant along with the red mud. These lime-based solids can be minimized by recycling the lime used as a filtering aid to digestion to displace the fresh lime that is normally added at this point. Also, effluent volume from the alumina plant can be minimized or eliminated by good design and operating practices: reducing the water added to the process, segregating condensates and recycling to the process, and using rainwater in the process. Using the prebake technology rather than the Soderberg technology for aluminum smelting is a significant pollution prevention measure. In the smelter, computer controls and point feeding of aluminum oxide to the centerline of the cell help reduce emissions, including emissions of organic fluorides such as CF_4, which can be held at less than 0.1 kg/t aluminum. Energy consumption is typically 14 megawatt hours per ton (MWh/t) of aluminum, with prebake technology. Soderberg technology uses 17.5 MWh/t. Gas collection efficiencies for the prebake process is better than for the Soderberg process: 98% vs. 90%. Dry scrubber systems using aluminum oxide as the adsorbent for the cell gas permits the recycling of fluorides. The use of low-sulfur tars for baking anodes helps control SO_2 emissions. Spent pot linings are removed after they fail, typically because of cracking or heaving of the lining. The age of the pot linings can vary from 3 to 10 years. By improving the life of the lining through better construction and operating techniques, discharge of pollutants can be reduced. Note that part of the pot lining carbon can be recycled when the pots are relined. Emissions of organic compounds from secondary aluminum production can be reduced by thoroughly removing coatings, paint, oils, greases, and the like from raw feed materials before they enter the melt process. European experience has shown that red mud produced at the alumina plant can be reduced from 2 t/t alumina to about 1 t/t alumina through implementation of good industrial practices.

At bauxite facilities, the major sources of dust emissions are the dryers, and emissions are controlled with electrostatic precipitators (ESPs) or baghouses. Removal efficiencies of 99% are achievable. Dust from conveyors and material transfer points is controlled by hoods and enclosures. Dust from truck movement can be minimized by treating road surfaces and by ensuring that vehicles do not drop material as they travel. Dusting from stockpiled material can be minimized by the use of water sprays or by enclosure in a building.

At the alumina plant, pollution control for the various production and service areas is implemented as follows:

- *Bauxite and limestone handling and storage:* dust emissions are controlled by baghouses.
- *Lime kilns*: dust emissions are controlled by baghouse systems. Kiln fuels can be selected to reduce SO_x emissions; however, this is not normally a problem, since most of the sulfur dioxide that is formed is absorbed in the kiln.
- *Calciners:* alumina dust losses are controlled by ESPs; SO_2 and NO_x emissions are reduced to acceptable levels by contact with the alumina.
- *Red mud disposal*: the mud impoundment area must be lined with impervious clay prior to use to prevent leakage. Water spraying of the mud stack may be required to prevent fine dust from being blown off the stack. Longer-term treatment of the mud may include reclamation of the mud, neutralization, covering with topsoil, and planting with vegetation.

In the smelter, primary emissions from the reduction cells are controlled by collection and treatment using dry sorbent injection; fabric filters or ESPs are used for controlling particulate matter. Primary emissions comprise 97.5% of total cell emissions; the balance consists of secondary emissions that escape into the potroom and leave the building through roof ventilators. Wet scrubbing of the primary emissions can also be used, but large volumes of toxic waste liquors will need to be treated or disposed of. Secondary emissions result from the periodic replacement of anodes and other operations; the fumes escape when the cell hood panels have been temporarily removed. While wet scrubbing can be used to control the release of secondary fumes, the high-volume, low-concentration gases offer low scrubbing efficiencies, have high capital and operating costs, and produce large volumes of liquid effluents for treatment. Wet scrubbing is seldom used for secondary fume control in the prebake process.

When anodes are baked on site, the dry scrubbing system using aluminum oxide as the adsorbent is used. It has the advantage of being free of waste products, and all enriched alumina and absorbed material are recycled directly to the reduction cells. Dry scrubbing may be combined with incineration for controlling emissions of tar and volatile organic compounds (VOCs) and to recover energy. Wet scrubbing can also be used but is not recommended, since a liquid effluent, high in fluorides and hydrocarbons, will require treatment and disposal.

Dry scrubber systems applied to the pot fumes and to the anode baking furnace result in the capture of 97% of all fluorides from the process.

The aluminum smelter solid wastes, in the form of spent pot lining, are disposed of in engineered landfills that feature clay or synthetic lining of disposal pits, provision of soil layers for covering and sealing, and control and treatment of any leachate. Treatment processes are available to reduce hazards associated with spent pot lining prior to disposal of the lining in a landfill. Other solid wastes such as bath skimmings are sold for recycling, while spalled refractories and other chemically stable materials are disposed of in landfill sites.

Modern smelters using good industrial practices are able to achieve the following in terms of pollutant loads (all values are expressed on an annualized basis): hydrogen fluoride, 0.2-0.4 kg/t; total fluoride, 0.3-0.6 kg/t; particulates, 1 kg/t; sulfur dioxide, 1 kg/t; and nitrogen oxides, 0.5 kg/t. CF_4 emissions should be less than 0.1 kg/t.

For secondary aluminum production, the principal treatment technology downstream of the melting furnace is dry sorbent injection using lime, followed by fabric filters. Waste gases from salt slag processing should be filtered as well. Waste gases from aluminum scrap pretreatment that contain organic compounds of concern may be treated by post-combustion practices.

Air emissions should be monitored regularly for particulate matter and fluorides. Hydrocarbon emissions should be monitored annually on the anode plant and baking furnaces. Liquid effluents should be monitored weekly for pH, total suspended solids, fluoride, and aluminum and at least monthly for other parameters. Monitoring data should be analyzed and reviewed at regular intervals and compared with the operating standards so that any necessary corrective actions can be taken.

COPPER SMELTING

Industry Description and Practices

Copper can be produced either pyrometallurgically or hydrometallurgically. The hydrometallurgical route is used only for a very limited amount of the world's copper production and is normally only considered in connection with in situ leaching of copper ores. From an environmental point of view, this is a questionable production route. Several different processes can be used for copper production. The traditional process is based on roasting, smelting in reverbatory furnaces (or electric furnaces for more complex ores), producing matte (copper-iron sulfide), and converting for production of blister copper, which is further refined to cathode copper. This route for production of cathode copper requires large amounts of energy per ton of copper: 30-40 million British thermal units (Btu) per ton cathode copper. It also produces furnace gases with low sulfur dioxide concentrations from which the production of sulfuric acid or other products is less efficient. The sulfur dioxide concentration in the exhaust gas from a reverbatory furnace is about 0.5-1.5%; that from an electric furnace is about 2-4%. So-called flash smelting techniques have therefore been developed that utilize the energy released during oxidation of the sulfur in the ore. The flash techniques reduce the energy demand to about 20 million Btu/ton of produced cathode copper. The SO_2 concentration in the off gases from flash furnaces is also higher, over 30%, and is less expensive to convert to sulfuric acid. The INCO process results in 80% sulfur dioxide in the off gas. Flash processes have been in use since the early 1950s.

In addition to the above processes, there are a number of newer processes such as Noranda, Mitsubishi, and Contop, which replace roasting, smelting, and converting, or processes such as ISA-SMELT and KIVCET, which replace roasting and smelting. For converting, the Pierce-Smith and Hoboken converters are the most common processes.

The matte from the furnace is charged to converters, where the molten material is oxidized in the presence of air to remove the iron and sulfur impurities (as converter slag) and to form blister copper. Blister copper is further refined as either fire-refined copper or anode copper (99.5% pure copper), which is used in subsequent electrolytic refining. In fire refining, molten blister copper is placed in a fire-refining furnace, a flux may be added, and air is blown through the molten mixture to remove residual sulfur. Air blowing results in residual oxygen, which is removed by the addition of natural gas, propane, ammonia, or wood. The fire-refined copper is then cast into anodes for further refining by electrolytic processes or is cast into shapes for sale.

In the most common hydrometallurgical process, the ore is leached with ammonia or sulfuric acid to extract the copper. These processes can operate at atmospheric pressure or as pressure leach circuits. Copper is recovered from solution by electrowinning, a process similar to electrolytic refining. The process is most commonly used for leaching low-grade deposits in situ or as heaps.

Recovery of copper metal and alloys from copper-bearing scrap metal and smelting residues requires preparation of the scrap (e.g., removal of insulation) prior to feeding into the primary process. Electric arc furnaces using scrap as feed are also common.

Pollution Prevention Practices and Opportunities

The principal air pollutants emitted from the processes are sulfur dioxide and particulate matter. The amount of sulfur dioxide released depends on the characteristics of the ore-complex ores which may contain lead, zinc, nickel, and other metals, and on whether facilities are in place for capturing and converting the sulfur dioxide. SO_2 emissions may range from less than 4 kilograms per metric ton (kg/t) of copper to 2,000 kg/t of copper. Particulate emissions can range from 0.1 kg/t of copper to as high as 20 kg/t of copper. Fugitive emissions occur at furnace openings and from launders, casting molds, and ladles carrying molten materials. Additional fugitive particulate emissions occur from materials handling and transport of ores and concentrates. Some vapors, such as arsine, are produced in hydrometallurgy and various refining processes. Dioxins can be formed from plastic and other organic material when scrap is melted. The principal constituents of the particulate matter are copper and iron oxides. Other copper and iron compounds, as well as sulfides, sulfates, oxides, chlorides, and fluorides of arsenic, antimony, cadmium, lead, mercury, and zinc, may also be present. Mercury can also be present in metallic form. At higher temperatures, mercury and arsenic could be present in vapor form. Leaching processes will generate acid vapors, while fire-refining processes result in copper and SO_2 emissions. Emissions of arsine, hydrogen vapors, and acid mists are associated with electrorefining. Wastewater from primary copper production contains dissolved and suspended solids that may include concentrations of copper, lead, cadmium, zinc, arsenic, and mercury and residues from mold release agents (lime or aluminum oxides). Fluoride may also be present, and the effluent may have a low pH. Normally there is no liquid effluent from the smelter other than cooling water; wastewaters do originate in scrubbers (if used), wet electrostatic precipitators, cooling of copper cathodes, and so on. In the electrolytic refining process, by-products such as gold and silver are collected as slimes that are subsequently recovered. Sources of wastewater include spent electrolytic baths, slimes recovery, spent acid from hydrometallurgy processes, cooling water, air scrubbers, washdowns, stormwater, and sludges from wastewater treatment processes that require reuse/recovery or appropriate disposal. The main portion of the solid waste is discarded slag from the smelter. Discard slag may contain 0.5-0.7% copper and is frequently used as construction material or for sandblasting. Leaching processes produce residues, while effluent treatment results in sludges, which can be sent for metals recovery. The smelting process typically produces less than 3 tons of solid waste per ton of copper produced.

Process gas streams containing sulfur dioxide are processed to produce sulfuric acid, liquid sulfur dioxide, or sulfur. The smelting furnace will generate process gas streams with SO_2 concentrations ranging from 0.5% to 80%, depending on the process used. It is important, therefore, that a process be selected that uses oxygen-enriched air (or pure oxygen) to raise the SO_2 content of the process gas stream and reduce the total volume of the stream, thus permitting efficient fixation of sulfur dioxide. Processes should be operated to maximize the concentration of the sulfur dioxide. An added benefit is the reduction of NO_x. Some pollution prevention practices for this industry include the following:

- Closed-loop electrolysis plants will contribute to prevention of pollution.
- Continuous casting machines should be used for cathode production to avoid the need for mold release agents.
- Furnaces should be enclosed to reduce fugitive emissions, and dust from dust control equipment should be returned to the process.
- Energy efficiency measures (such as waste heat recovery from process gases) should be applied to reduce fuel usage and associated emissions.
- Recycling should be practiced for cooling water, condensates, rainwater, and excess process water used for washing, dust control, gas scrubbing, and other process applications where water quality is not a concern.
- Good housekeeping practices are key to minimizing losses and preventing fugitive emissions. Such losses and emissions are minimized by enclosed buildings, covered or enclosed conveyors and transfer points, and dust collection equipment. Yards should be paved and runoff water routed to settling ponds. Regular sweeping of yards and indoor storage or coverage of concentrates and other raw materials also reduces materials losses and emissions.

Pollution control technologies acceptable for this industry are as follows. Fabric filters are used to control particulate emissions. Dust that is captured but not recycled will need to be disposed of in a secure landfill or other acceptable manner. Vapors of arsenic and mercury present at high gas temperatures are condensed by gas cooling and removed. Additional scrubbing may be required. Effluent treatment by precipitation, filtration, and so on, of process bleed streams, filter backwash waters, boiler blowdown, and other streams may be required to reduce suspended and dissolved solids and heavy metals. Residues that result from treatment are sent for metals recovery or to sedimentation basins. Stormwaters should be treated for suspended solids and heavy metals reduction. Slag should be landfilled or granulated and sold. Modern plants using good industrial practices should set as targets total dust releases of 0.5-1.0 kg/t of copper and SO_2 discharges of 25 kg/t of copper. A double-contact, double-absorption plant should emit no more than 0.2 kg of sulfur dioxide per ton of sulfuric acid produced (based on a conversion efficiency of 99.7%).

Chapter 8
Miscellaneous Industry Practices and Case Studies

INTRODUCTION

This chapter provides industry profiles for several sectors. In each profile, a concise description of the manufacturing operations is provided, an identification of the major environmental problems, and fate of the pollutants. The pollution prevention practices and potential opportunities are then discussed for each industry profile. Case studies are included where appropriate.

PULP AND PAPER INDUSTRY

Industry Description and Practices

Pulp and paper are manufactured from raw materials containing cellulose fibers, generally wood, recycled paper, and agricultural residues. In developing countries, nearly 60% of cellulose fibers originate from nonwood raw materials such as bagasse (sugar cane fibers), cereal straw, bamboo, reeds, esparto grass, jute, flax, and sisal. This section focuses on the environmental issues and pollution prevention opportunities and practices in pulp and paper manufacturing with unit production capacities greater than 100 metric tons per day (tpd).

The main operations in pulp and paper manufacturing are raw material preparation, such as wood debarking and chip making; pulp manufacturing; pulp bleaching; paper manufacturing; and fiber recycling. Pulp mills and paper mills may exist separately or as integrated operations. Manufactured pulp is used as a source of cellulose for fiber manufacture and for conversion into paper or cardboard.

Pulp manufacturing starts with raw material preparation, which includes debarking (when wood is used as raw material), chipping, and other processes such as depithing (for example, when bagasse is used as the raw material). Cellulosic pulp is manufactured from the raw materials, using chemical and mechanical means. The manufacture of pulp for paper and cardboard employs mechanical (including thermomechanical), chemimechanical, and chemical methods. Mechanical pulping separates fibers by such methods as disk abrasion and billeting. Chemimechanical processes involve mechanical abrasion and the use of chemicals. Thermomechanical pulps, which are used for making products such as newsprint, are manufactured from raw materials by the application of heat, in addition to mechanical operations. Chemimechanical pulping and chemithermomechanical pulping (CTMP) are similar but use less mechanical energy, softening the pulp with sodium sulfite, carbonate, or hydroxide.

Chemical pulps are made by cooking (known as *digesting*) the raw materials, using the kraft (sulfate) and sulfite processes. Kraft processes produce a variety of pulps used mainly for packaging and high-strength papers and board. Wood chips are cooked with caustic soda to produce brownstock, which is then washed with water to remove cooking (black) liquor for the recovery of chemicals and energy. Pulp is also manufactured from recycled paper. Mechanical pulp can be used without bleaching to make printing papers for applications in which low brightness is acceptable - primarily, newsprint. However, for most printing, for copying, and for some packaging grades, the pulp must be bleached. For mechanical pulps, most of the original lignin in the raw pulp is retained but is bleached with peroxides and hydrosulfites. In the case of chemical pulps (kraft and sulfite), the objective of bleaching is to remove the small fraction of the lignin remaining after cooking. Oxygen, hydrogen peroxide, ozone, peracetic acid, sodium hypochlorite, chlorine dioxide, chlorine, and other chemicals are used to transform lignin into an alkali-soluble form. An alkali, such as sodium hydroxide, is necessary in the bleaching process to extract the alkali-soluble form of lignin. Pulp is washed with water in the bleaching process.

In modern pulp and paper mills, oxygen is normally used in the first stage of bleaching. The trend is to avoid the use of any kind of chlorine chemicals and employ "total chlorine-free" (TCF) bleaching. TCF processes allow the bleaching effluents to be fed to the recovery boiler for steam generation; the steam is then used to generate electricity, thereby reducing the amount of pollutants discharged. Elemental chlorine-free (ECF) processes, which use chlorine dioxide, are required for bleaching certain grades of pulp. Not that the use of elemental chlorine for bleaching is not recommended. Only ECF processes are acceptable, and, from an environmental perspective, TCF processes are preferred.

The soluble organic substances removed from the pulp in bleaching stages that use chlorine or chlorine compounds, as well as the substances removed in the subsequent alkaline stages, are then chlorinated. Some of these chlorinated organic substances are toxic; they include dioxins, chlorinated phenols, and many other chemicals. It is generally not practical to recover chlorinated organics in effluents, since the chloride content causes excessive corrosion. The finished pulp may be dried for shipment (market pulp) or may be used to manufacture paper on-site (in an "integrated" mill).

Paper and cardboard are made from pulp by deposition of fibers and fillers from a fluid suspension onto a moving forming device that also removes water from the pulp. The water remaining in the wet web is removed by pressing and then by drying, on a series of hollow-heated cylinders (for example, calender rolls). Chemical additives are added to impart specific properties to the paper, and pigments may be added for coloring purposes.

Pollution Prevention Practices and Opportunities

The significant environmental impacts of the manufacture of pulp and paper result from the pulping and bleaching processes. In some processes, sulfur compounds and nitrogen oxides are emitted to the air, and chlorinated and organic compounds, nutrients, and metals are discharged to the wastewaters. In the kraft pulping process which is the most widely used, highly malodorous emissions of reduced sulfur compounds, measured as total reduced sulfur (TRS) and including hydrogen sulfide, methyl mercaptan, dimethyl sulfide, and dimethyl disulfide, are emitted, typically at a rate of 0.3-3 kilograms per metric ton (kg/t) of air-dried pulp (ADP). Note that the definition of ADP is 90% bone-dry fiber and 10% moisture. Other typical generation rates are: particulate matter, 75-150 kg/t; sulfur oxides, 0.5-30 kg/t; nitrogen oxides, 1-3 kg/t; and VOCs, 15 kg/t from black liquor oxidation. In the sulfite pulping process, sulfur oxides are emitted at rates ranging from 15 kg/t to over 30 kg/t. Other pulping processes, such as the mechanical and thermomechanical methods, generate significantly lower quantities of air emissions.

Steam- and electricity-generating units using coal or fuel oil emit fly ash, sulfur oxides, and nitrogen oxides. Coal burning can emit fly ash at the rate of 100 kg/t of ADP.

Wastewaters are discharged at a rate of 20-250 cubic meters per metric ton (m^3/t) of ADP. They are high in biochemical oxygen demand (BOD), at 10-40 kg/t of ADP; total suspended solids, 10-50 kg/t of ADP; chemical oxygen demand (COD), 20-200 kg/t of ADP; and chlorinated organic compounds, which may include dioxins, furans, and other adsorbable organic halides, AOH, at 0-4 kg/t of ADP.

Wastewater from chemical pulping contains 12-20 kg of BOD/t of ADP, with values of up to 350 kg/t. The corresponding values for mechanical pulping wastewater are 15-25 kg BOD/t of ADP. For chemimechanical pulping, BOD discharges are 3 to 10 times higher than those for mechanical pulping. Pollution loads for some processes, such as those using nonwood raw materials, could be significantly different. Phosphorus and nitrogen are also released into wastewaters. The main source of nutrients, nitrogen, and phosphorus compounds is raw material such as wood. The use of peroxide, ozone, and other chemicals in bleaching makes it necessary to use a complexing agent for heavy metals such as manganese.

The principal solid wastes of concern include wastewater treatment sludges (50-150 kg/t of ADP). Solid materials that can be reused include waste paper, which can be recycled, and bark, which can be used as fuel. Lime sludge and ash is usually disposed of in an appropriate landfill.

The most significant environmental issues are related to the discharge of chlorine-based organic compounds (from bleaching) and of other toxic organics. The unchlorinated material is essentially black liquor that has escaped the mill recovery process. Some mills today are approaching 100% recovery. Industry developments demonstrate that total chlorine-free bleaching is feasible for many pulp and paper products but cannot produce certain grades of paper. The adoption of these modern process developments, wherever feasible, is encouraged. Pollution prevention programs should focus on reducing wastewater discharges and on minimizing air emissions. Process recommendations are listed in Table 1.

Table 1. *P2 Practices and Recommendations in Pulp and Paper Manufacturing.*

Use energy-efficient pulping processes wherever feasible. Acceptability of less bright products should be promoted. For less bright products such as newsprint, thermomechanical processes and recycled fiber may be considered.
Minimize the generation of effluents through process modifications and recycle wastewaters, aiming for total recycling.
Reduce effluent volume and treatment requirements by using dry instead of wet debarking; recovering pulping chemicals by concentrating black liquor and burning the concentrate in a recovery furnace; recovering cooking chemicals by recausticizing the smelt from the recovery furnace; and using high-efficiency washing and bleaching equipment.
Minimize unplanned or nonroutine discharges of wastewater and black liquor, caused by equipment failures, human error, and faulty maintenance procedures, by training operators, establishing good operating practices, and providing sumps and other facilities to recover liquor losses from the process.
Minimize sulfur emissions to the atmosphere by using a low-odor design black liquor recovery furnace.
Prevent and control spills of black liquor.
Aim for zero-effluent discharge where feasible. Reduce wastewater discharges to the extent feasible. Incinerate liquid effluents from the pulping and bleaching processes.
Reduce the odor from reduced sulfur emissions by collection and incineration and by using modem, low-odor recovery boilers fired at over 75% concentration of black liquor.

Table 1. continued

Dewater and properly manage sludges.
Where wood is used as a raw material to the process, encourage plantation of trees to ensure sustainability of forests.
Use energy-efficient processes for black liquor chemical recovery, preferably aiming for a high solid content (say, 70%).
Reduce bleaching requirements by process design and operation. Use the following measures to reduce emissions of chlorinated compounds to the environment: before bleaching, reduce the lignin content in the pulp (Kappa number of 10) for hardwood by extended cooking and by oxygen delignification under elevated pressure; optimize pulp washing prior to bleaching; use TCF or at a minimum, ECF bleaching systems; use oxygen, ozone, peroxides (hydrogen peroxide), peracetic acid, or enzymes (cellulose-free xylanase) as substitutes for chlorine-based bleaching chemicals; recover and incinerate maximum material removed from pulp bleach grade where chlorine bleaching is used, reduce the chlorine charge on the lignin by controlling pH and by splitting the addition of chlorine.

For air emissions, the World bank Organization recommends the target of 1.5 kg NO_x per ton for both kraft and sulfite processes; for mechanical and chemimechanical processes used in newsprint manufacture, 260 nanograms per joule (ng/J) of NO_x, for coal; 130 ng/J for oil; and 86 ng/J for gas used as fuel. Wastewater generation rates should not exceed 50 m^3/tof ADP, and levels of 20 m^3/t of ADP (or product) should be targeted. For paper mills, effluent discharges should be less than 5 m^3/t of ADP. Wherever feasible, use a total wastewater recycling system, along with a TCF pulp-bleaching system, and incinerate bleaching effluents in the recovery boiler. As a minimum, use chlorine dioxide as a substitute for elemental chlorine in pulp bleaching.

Standard pollution control systems applicable to this industry are as follows. Sulfur oxide emissions are scrubbed with slightly alkaline solutions. The reduced sulfur-compounds gases are collected using headers, hoods, and venting equipment. Condensates from the digester relief condenser and evaporation of black liquor are stripped of reduced sulfur compounds. The stripper overhead and noncondensable are incinerated in a lime kiln or a dedicated combustion unit. Approximately 0.5 kg sulfur per ton of pulp for the kraft process and 1.5 kg sulfur per ton for the sulfite process are considered acceptable emissions levels. Electrostatic precipitators are used to control the release of particulate matter into the atmosphere.

Wastewater treatment typically includes the following unit operations: (a) neutralization, screening, sedimentation, and floatation/hydrocycloning to remove suspended solids and (b) biological/secondary treatment to reduce the organic content in wastewater and destroy toxic organics. Chemical precipitation is also used to remove certain cations. Fibers collected in primary treatment should be recovered and recycled. A mechanical clarifier or a settling pond is used in primary treatment. Flocculation to assist in the removal of suspended solids is also sometimes necessary. Biological treatment systems, such as activated sludge, aerated lagoons, and anaerobic fermentation, can reduce BOD by over 99% and achieve a COD reduction of 50% to 90%.

Tertiary treatment may be performed to reduce toxicity, suspended solids, and color. Solid waste treatment steps include dewatering of sludge and combustion in an incinerator, bark boiler, or fossil-fuel-fired boiler. Sludges from a clarifier are dewatered and may be incinerated; otherwise, they are landfilled. Solid wastes should be sent to combustion devices or disposed of in a manner that avoids odor generation and the release of toxic organics to the environment.

SUGAR MANUFACTURING

Industry Description and Practices

The sugar industry processes sugar cane and sugar beet to manufacture edible sugar. More than 60% of the world's sugar production is from sugar cane; the balance is from sugar beet. Sugar manufacturing is a highly seasonal industry, with season lengths of about 6 to 18 weeks for beets and 20 to 32 weeks for cane. Approximately 10% of the sugar cane can be processed to commercial sugar, using approximately 20 cubic meters of water per metric ton of cane processed. Sugar cane contains 70% water; 14% fiber; 13.3% saccharose (about 10 to 15% sucrose), and 2.7% soluble impurities.

Sugar canes are generally washed, after which juice is extracted from them. The juice is clarified to remove mud, evaporated to prepare syrup, crystallized to separate out the liquor, and centrifuged to separate molasses from the crystals. Sugar crystals are then dried and may be further refined before bagging for shipment. In some places (for example, in South Africa), juicc is extracted by a diffusion process that can give higher rates of extraction with lower energy consumption and reduced operating and maintenance costs. For processing sugar beet (water, 75%; sugar, 17%), only the washing, preparation, and extraction processes are different.

After washing, the beet is sliced, and the slices are drawn into a slowly rotating diffuser where a countercurrent flow of water is used to remove sugar from the beet slices. About 15 m^3 of water and 28 kilowatt-hours (kWh) of energy are consumed per metric ton of beet processed. Sugar refining involves removal of impurities and decolorization. The unit operations that follow include affination (mingling and centrifugation), melting, clarification, decolorization, evaporation, crystallization, and finishing. Decolorization methods use granular activated carbon, powdered activated carbon, and ion exchange resins.

Pollution Prevention Practices and Opportunities

The main air emissions from sugar processing and refining result primarily from the combustion of bagasse (the fiber residue of sugar cane), fuel oil, or coal. Other air emission sources include juice fermentation units, evaporators, and sulfitation units. Approximately 5.5 kilograms of fly ash per metric ton (kg/t) of cane processed (or 4,500 mg/m^3 of fly ash) are present in the flue gases from the combustion of bagasse. Sugar manufacturing effluents typically have biochemical oxygen demand (BOD) of 1,7006,600 mg/ℓ in untreated effluent from cane processing and 4,000-7,000 mg/ℓ from beet processing; chemical oxygen demand (COD) of 2,300-8,000 mg/ℓ from cane processing and up to 10,000 mg/l from beet processing; total suspended solids of up to 5,000 mg/ℓ; and high ammonium content. The wastewater may contain pathogens from contaminated materials or production processes. A sugar mill often generates odor and dust, which need to be controlled. Most of the solid wastes can be processed into other products and by-products. In some cases, pesticides may be present in the sugar cane rinse liquids. Pollution prevention practices in sugar manufacturing focus on the areas listed in Table 2.

Table 2. *P2 Practices and Recommendations in Sugar Manufacturing.*

Reduce product losses to less than 10% by better production control. Perform sugar auditing.
Discourage spraying of molasses on the ground for disposal.
Minimize storage time for juice and other intermediate products to reduce product losses and discharge of product into the wastewater stream.
Give preference to less polluting clarification processes such as those using bentonite instead of sulfite for the manufacture of white sugar.
Collect waste product for use in other industries for example, bagasse for use in paper mills and as fuel. Cogeneration systems for large sugar mills generate electricity for sale. Beet chips can be used as animal feed.
Optimize the use of water and cleaning chemicals. Procure cane washed in the field. Prefer the use of dry cleaning methods.
Recirculate cooling waters.
Continuous sampling and measurement of key production parameters allow production losses to be identified and reduced, thus reducing the waste load. Fermentation processes and juice handling are the main sources of leakage.
Odor problems can usually be prevented with good hygiene and storage practices.

Since the pollutants generated are largely losses in production, improvements in production efficiency are recommended to reduce pollutant loads. Approximately 90% of the saccharose should be accounted for, and 85% of the sucrose can be recovered. Recirculation of water should be maximized. Wastewater loads can be reduced to at least 1.3 m^3/t of cane processed, and plant operators should aim at rates of 0.9 m^3/t or less through recirculation of wastewater. Wastewater loads from beet processing should be less than 4 m^3/t of sugar produced or 0.75 m^3/t of beet processed, with a target of 0.3 to 0.6 m^3/t of beet processed. Standard pollution control technologies applicable to the industry are as follows. Pretreatment of effluents consists of screening and aeration, normally followed by biological treatment. If space is available, land treatment or pond systems are potential treatment methods. Other possible biological treatment systems include activated sludge and anaerobic systems which can achieve a reduction in the BOD level of over 95%. Odor control by ventilation and sanitation may be required for fermentation and juice-processing areas. Biofilters may be used for controlling odor. Cyclones, scrubbers, and electrostatic precipitators are used for dust control.

TANNING AND LEATHER FINISHING

Industry Description and Practices

Hides and skins are preserved by drying, salting, or chilling, so that raw hides and skins will reach leather tanneries in an acceptable condition. The use of environmentally persistent toxics for preservation of raw hides and skins is to be avoided. In the tanning process, animal hides and skins are treated to remove hair and nonstructured proteins and fats, leaving an essentially pure collagen matrix. The hides are then preserved by impregnation with tanning agents. Leather production usually involves three distinct phases: preparation (in the beamhouse); tanning (in the tanyard); and finishing, including dyeing and surface treatment. A range of processes and chemicals, including chrome salts, is used in the tanning and finishing processes. The tanning and finishing process generally consists of the following operations:

1. Soaking and washing to remove salt, restore the moisture content of the hides, and remove any foreign material such as dirt and manure;
2. Liming to open up the collagen structure by removing interstitial material;
3. Fleshing to remove excess tissue from the interior of the hide;
4. Dehairing or dewooling to remove hair or wool by mechanical or chemical means;
5. Bating and pickling to delime the skins and condition the hides to receive

the tanning agents;

6. Tanning to stabilize the hide material and impart basic properties to the hides;
7. Retanning, dyeing, and fat-liquoring to impart special properties to the leather, increase penetration of tanning solution, replenish oils in the hides, and impart color to the leather;
8. Finishing to attain final product specifications.

Pollution Prevention Practices and Opportunities

The environmental impacts of tanning operations are significant. Composite untreated wastewater, amounting to 20-80 m^3/t of hide or skin, is turbid, colored, and foul smelling. It consists of acidic and alkaline liquors, with chromium levels of 100-400 mg/ℓ; sulfide levels of 200-800 mg/ℓ; nitrogen levels of 200-1,000 mg/l; biochemical oxygen demand (BOD) levels of 900-6,000 mg/ℓ, usually ranging from 160 to 24,000 mg/ℓ; chemical oxygen demand (COD) ranging from 800 to 43,000 mg/ℓ in separate streams, with combined wastewater levels of 2,400 to 14,000 mg/ℓ; chloride ranging from 200 to 70,000 mg/ℓ in individual streams and 5,600 to 27,000 mg/ℓ in the combined stream; and high levels of fat. Suspended solids are usually half of chloride levels.

Wastewater may also contain residues of pesticides used to preserve hides during transport, as well as significant levels of pathogens. Significant volumes of solid wastes are produced, including trimmings, degraded hide, and hair from the beamhouse processes. The solid wastes can represent up to 70% of the wet weight of the original hides.

In addition, large quantities of sludges are generated. Decaying organic material produces strong odors. Hydrogen sulfide is released during dehairing, and ammonia is released in deliming. Air quality may be further degraded by release of solvent vapors from spray application, degreasing, and finishing (for example, dye application). The pollution prevention practices that can be implemented in this industry are summarized in Table 3.

Use of techniques such as water-based paint and roller coating can help achieve emissions of VOCs from finishing of less than 4 kg/t (aim for 2 kg/t). Treatment of tannery wastewaters is always required. Some streams, such as soaking liquor (which has high salinity), sulfide-rich lime liquor, and chrome wastewaters should be segregated. Preliminary screening of wastewaters is required because of the large quantities of solids present. Recovery of hair from the dehairing and liming process reduces the BOD of the process effluent. Physical-chemical treatment precipitates metals and removes a large portion of solids, BOD, and COD.

Table 3. *P2 Practices and Recommendations in Tanning and Leather Finishing.*

AREA OF OPPORTUNITY	POLLUTION PREVENTION RECOMMENDATION
General	1. Process fresh hides or skins to reduce the quantity of salt in wastewater, where feasible. 2. Reduce the quantities of salt used for preservation. When salted skins are used as raw material, pretreat the skins with salt elimination methods. 3. Use salt or chilling methods to preserve hides, instead of persistent insecticides and fungicides. 4. When antiseptics or biocides are necessary, avoid toxic and less degradable ones, especially those containing arsenic, mercury, lindane, or pentachlorophenol or other chlorinated substances.
	5. Flesh green hides instead of limed hides. 6. Use sulfide and lime as a 20-50% solution to reduce sulfide levels in wastewater. 7. Split limed hides to reduce the amount of chrome needed for tanning. 8. Consider the use of carbon dioxide in deliming to reduce ammonia in wastewater. 9. Use only trivalent chrome when required for tanning. 10. Inject tanning solution in the skin using high pressure nozzles; recover chrome from chrome-containing wastewaters, which should be kept segregated from other wastewaters. Recycle, chrome after precipitation and acidification. Improve fixation of chrome by addition of dicarboxylic acids. 11. Recycle spent chrome liquor to the tanning process or to the pickling vat. 12. Examine alternatives to chrome in tanning, such as titanium, aluminum, iron, zirconium, and vegetable tanning agents. 13. Use nonorganic solvents for dyeing and finishing. 14. Use photocell-assisted paint-spraying techniques to avoid overspraying. 15. Recover hair by using hair-saving methods to reduce pollution loads. For example, avoid dissolving hair in chemicals by making a proper choice of chemicals and using screens to remove hair from wastewater. 16. Precondition hides before vegetable tanning.
Water Conservation	17. Monitor and control process waters; reductions of up to 50% can be achieved. 18. Use batch washing instead of continuous washing, for reductions of up to 50%.

	19. Use low-float methods (for example, use 40-80% floats). Recycle liming, pickling, and tanning floats. Recycle sulfide in spent liming liquor after screening to reduce sulfide losses (by, say, 20-50%) and lime loss (by about 40 - 60%). 20. Use drums instead of pits for immersion of hides. 21. Reuse wastewaters for washing-for example, by recycling lime wash water to the soaking stage. Reuse treated wastewaters in the process to the extent feasible (for example, in soaking and pickling).
Waste Reduction	22. Recover hide trimmings for use in the manufacture of glue, gelatin, and similar products. 23. Recover grease for rendering. Use aqueous degreasing methods. 24. Recycle wastes to the extent feasible in the manufacture of fertilizer, animal feed, and tallow, provided the quality of these products is not compromised.
	25. Use tanned shavings in leather board manufacture. 26. Control odor problems by good housekeeping methods such as minimal storage of flesh trimmings and organic material. 27. Recover energy from the drying process to heat process water.

Biological treatment is usually required to reduce the remaining organic loads to acceptable levels (0.3 kg BOD, 2 kg COD, and 0.004 kg chromium per metric ton of raw hide). Good ventilation and minimization of solvent release can avoid the need to collect and treat vapors in carbon adsorption beds. VOC emissions from finishing are approximately 30 kg/t if pollution prevention measures are not adopted. Maximum upstream pollutant reduction is essential for tanneries, but treatment is also required.

BREWERIES

Industry Description and Practices

Beer is a fermented beverage with low alcohol content made from various types of grain. Barley predominates, but wheat, maize, and other grains can be used. The production steps include:

Malt production and handling: grain delivery and cleaning; steeping of the grain in water to start germination; growth of rootlets and development of enzymes (which convert starch into maltose); kilning and polishing of the malt to remove rootlets; storage of the cleaned malt.

Wort production: grinding the malt to grist; mixing grist with water to produce a mash; heating of the mash to activate enzymes; separation of grist residues to leave a liquid wort; boiling of the wort with hops; separation of the wort from the trub/hot break (precipitated residues), with the liquid part of the trub being returned to the lauter tub and the spent hops going to a collection vessel; and cooling of the wort.

Beer production: addition of yeast to cooled wort; fermentation; separation of spent yeast by filtration, centrifugation or settling; bottling or kegging.

Water consumption for breweries generally ranges 4-8 cubic meter per cubic meter of beer produced. Water consumption for individual process stages, as reported for the German brewing industry, is shown in Table 4.

Table 4. *Water Consumption Reported for the German Brewing Industry.*

PROCESS STEP	Water Consumption m^3/m^3 of sold beer	
	avg.	range
Gyle (unfermented wort) to whirlpool	2.0	1.8-2.2
Wort cooling	1.2	0.0-2.4
Fermentation cellar and yeast treatment	0.6	0.5-0.8
Filter and pressure tank room	0.3	0.1-0-5
Storage cellar	0.5	0.3-0.6
Bottling (70% of beer produced)	1.1	0.9-2.1
Barrel filling (30% of beer produced)	0.1	0.1-0.2
Wastewater from cleaning of vehicles, sanitary use, etc.	1.5	1.0-3.0
Steam boiler	0.2	0.1-0.3
Air compressor	0.3	0.1-0.5
Total	6.6	4.9-12.6

Pollution Prevention Practices and Opportunities

Untreated effluents typically contain suspended solids in the range 10-60 mg/ℓ, biochemical oxygen demand (BOD) in the range 1,000-1,500 mg/ℓ, chemical oxygen demand (COD) in the range 1,800-3,000 mg/ℓ, and nitrogen in the range 30-100 mg/ℓ. Phosphorus can also be present at concentrations of the order of 10-30 mg/ℓ. Effluents from individual process steps are variable. For example, bottle washing produces a large volume of effluent that, however, contains only a minor part of the total organics discharged from the brewery. Effluents from fermentation and filtering are high in organics and BOD but low in volume, accounting for about 3% of total wastewater volume but 97% of BOD. Effluent pH averages about 7 for the combined effluent but can fluctuate from 3 to 12 depending on the use of acid and alkaline cleaning agents. Solid wastes include grit, seed, and grain of less than 2.2 mm in diameter, removed when grain is cleaned; spent grain and yeast; spent hops; broken and reject bottles; and cardboard and other solid wastes associated with the process, such as kieselguhr (diatomaceous earth used for clarifying). Breweries do not discharge air pollutants, other than some odors. Some P2 options that may be considered are listed in Table 5.

Table 5. *P2 Practices and Recommendations for Breweries.*

Breweries have a favorable steam-to-electricity ratio. Planning for cogeneration of electricity may be advantageous.
Reduction of energy consumption through reuse of wort-cooling water as the process water for the next mash.
Collection of broken glass, bottles that can be reused, and waste cardboard for recycling.
Consideration should be given to the use of non-phosphate-containing cleaning agents.
Filtration of bottom sediments from final fermentation tanks for use as animal feed.
Recovery of spilled beer, adding it to spent grain that is being dried through evaporation.
Disposal of trub by adding it to spent grain.
Use for livestock feed of spent yeast that is not reused.
Disposal of wet hops by adding them to the spent grain.
Disposal of spent hop liquor by mixing with spent grain.
Use of spent grain as animal feed, either 80% wet, or dry after evaporation.
Use of grit, weed seed, and discarded grain as chicken feed.
Recirculating systems on cooling water circuits.
High-pressure, low-volume hoses for equipment cleaning.
Clean-in-place methods for decontaminating equipment.

Pollution prevention and control are best practiced through effective management, maintenance, and housekeeping in a process that incorporates water conservation and recycling, energy conservation, and disposal of solid wastes as by-products. If the brewery does not discharge to a municipal sewer, primary and secondary treatment of the effluent is required. Primary treatment facilities may include pH adjustment, roughing screens, grit-settling chambers, and a clarifier. Choices of processes for removing BOD in a secondary treatment stage include anaerobic treatment followed by aerobic treatment and activated sludge systems. Sludges from the clarifier are dewatered and disposed of through incineration or to an approved landfill. Where the brewery is permitted to discharge to a municipal sewer, pretreatment may be required to meet municipal by-laws and to lessen the load on the municipal treatment plant. In some cases, sewer discharge fees imposed by the municipality on effluent volume and on the suspended and BOD loads may encourage the brewery to install its own treatment facility. Plants using good industrial practices are able to achieve the following performance in terms of pollutant loads. Water conservation and recycling will allow water consumption to be kept to a minimum. A new brewery should target on achieving an effluent range of 3-5 m^3/m^3 beer produced. Provision for recycling liquors and reusing wash waters will help reduce the total volume of liquid effluent. A new brewery should set as a target the achievement of a treated effluent that has less than 0.3 kg of BOD/m^3 beer produced and 0.3 kg of suspended solids/m^3 beer produced (assuming discharge to receiving waters). Odor emissions can be minimized if exhaust vapors are condensed before they are released to the atmosphere or if vapors are sent to the boiler and burned.

CEMENT MANUFACTURING

Industry Description and Practices

The preparation of cement involves mining; crushing, and grinding of raw materials (principally limestone and clay); calcining the materials in a rotary kiln; cooling the resulting clinker; mixing the clinker with gypsum; and milling, storing, and bagging the finished cement. The process generates a variety of wastes, including dust, which is captured and recycled to the process. The process is energy-intensive, and there are strong incentives for energy conservation. Gases from the clinker cooler are used as secondary combustion air. The dry process, using preheaters and precalciners, is both economically and environmentally preferable to the wet process because the energy consumption, which is around 200 joules per kilogram J/kg, which is approximately half that for the wet process. Certain solid waste products from other industries, such as pulverized fly ash (PFA) from power stations, slag, roasted pyrite residues, and foundry sand, can be used as additives or fillers in cement production.

Pollution Prevention Practices and Opportunities

The generation of fine particulates is inherent in the process, but most are recovered and recycled. Approximately 10-20% of the kiln feed can be suspended in the kiln exhaust gases, captured, and returned to the feed. Other sources of dust emissions include the clinker cooler, crushers, grinders, and materials-handling equipment. When the raw materials have high alkali or chloride content, a portion of the collected dust must be disposed of as solid waste, to avoid alkali build-up. Leaching of the dust to remove the alkali is rarely practiced. Grinding mill operations also result in particulate emissions. Other materials-handling operations, such as conveyors, result in fugitive emissions.

Ambient particulate levels (especially at sizes less than 10 microns) have been clearly demonstrated to be related to health impacts. Gases such as NO_x and SO_x are formed from the combustion of the fuel (oil and coal) and oxidation of sulfur present in the raw materials, but the highly alkaline conditions in the kiln can absorb up to 90% of the sulfur oxides. Heavy metals may also be present in the raw materials and fuel used and are released in kiln gases. The principal aim of pollution control in this industry is to avoid increasing ambient levels of particulates by minimizing the loads emitted.

Cement kilns, with their high flame temperatures, are sometimes used to burn waste oils, solvents, and other organic wastes. These practices can result in the release of toxic metals and organics. Cement plants are not normally designed to burn wastes, but if such burning is contemplated, technical and environmental acceptability needs to be demonstrated. To avoid the formation of toxic chlorinated organics from the burning of organic wastes, air pollution control devices for such plants should not be operated in the temperature range of 230-400° C.

The priority in the cement industry is to minimize the increases in ambient particulate levels by reducing the mass load emitted from the stacks, from fugitive emissions, and from other sources. Collection and recycling of dust in kiln gases is required to improve the efficiency of the operation and to reduce atmospheric emissions. Units that are well designed, well operated, and well maintained can normally achieve generation of less than 0.2 kilograms of dust per metric ton of clinker, using dust recovery systems. NO_x emissions should be controlled by using proper kiln design, low-NO_x burners, and an optimum level of excess air. NO_x emissions from a dry kiln with preheater and precalciner are typically 1.5 kg/t of clinker, as opposed to 4.5 kg/t for the wet process. The nitrogen oxide emissions can be reduced further to 0.5 kg/t of clinker, by afterburning in a reducing atmosphere, and the energy of the gases can be recovered in a preheater/precalciner.

For control of fugitive particulate emissions, ventilation systems should be used in conjunction with hoods and enclosures covering transfer points and conveyors. Drop distances should be minimized by the use of adjustable conveyors. Dusty areas such as roads should be wetted down to reduce dust generation. Appropriate stormwater and runoff control systems should be provided to minimize the quantities of suspended material carried off-site.

SOx emissions are best controlled by using low sulfur fuels and raw materials. The absorption capacity of the cement must be assessed to determine the quantity of sulfur dioxide emitted, which may be up to about half the sulfur load on the kiln. Precalcining with low-NO_x, secondary firing can reduce nitrogen oxide emissions.

Alkaline dust removed from the kiln gases is normally disposed of as solid waste. When solid wastes such as pulverized fly ash are used with feedstock, appropriate steps must be taken to avoid environmental problems from contaminants or trace elements.

Mechanical systems such as cyclones trap the larger particulates in kiln gases and act as preconditioners for downstream collection devices. Electrostatic precipitators and baghouses are the principal options for collection and control (achieving over 99% removal efficiency) of fine particulates. ESPs are sensitive to gas characteristics, such as temperature, and to variation in voltage; baghouses are generally regarded as more reliable. The overall costs of the two systems are similar. The choice of system will depend on flue gas characteristics and local considerations.

Both ESPs and baghouses can achieve high levels of particulate removal from the kiln gas stream, but good operation and maintenance are essential for achieving design specifications. Two significant types of control problem can occur:

- complete failure (or automatic shutoff) of systems related to plant shutdown and start-up, power failures, and the like, leading to the emission of very high levels of particulates for short periods of time; and
- a gradual decrease in the removal efficiency of the system over time because of poor maintenance or improper operation. The lime content of raw materials can be used to control sulfur oxides.

Stormwater systems and storage areas should be designed to minimize washoff of solids. Normally, effluents requiring treatment originate from cooling operations or as stormwater. Treated effluent discharges should have a pH in the range of 6-9. Cooling water should preferably be recycled. If this is not economical, the effluent should not increase the temperature of the receiving waters at the edge of the mixing zone (or 100 meters, where the mixing zone is not defined) by more than 3° C.

If quantities of suspended solids in the effluent are high in relation to receiving waters, treatment may be required to reduce levels in the effluent to a maximum of 50 mg/l. Note that the effluent requirements are for direct discharge to surface waters.

Table 6 provides a list of pollution prevention action items.

Table 6. *P2 Practices and Recommendations for Cement Factories.*

Develop a strong unit or division to undertake environmental management responsibilities.
Operate control systems to achieve the required emissions levels.
Use low sulfur fuels in the kiln.
Use low-NO_x burners with the optimum level of excess air.
Wet down intermediate and finished product storage piles.
Use enclosed adjustable conveyors to minimize drop distances.
Install equipment covers and filters for crushing, grinding, and milling operations.

VEGETABLE OIL PROCESSING

Industry Description and Practices

The vegetable oil processing industry involves the extraction and processing of oils and fats from vegetable sources. Vegetable oils and fats are principally used for human consumption but are also used in animal feed, for medicinal purposes, and for certain technical applications. The oils and fats are extracted from a variety of fruits, seeds, and nuts. The preparation of raw materials includes husking, cleaning, crushing, and conditioning. The extraction processes are generally mechanical (e.g., boiling for fruits, pressing for seeds and nuts) or involve the use of a solvent such as hexane. After boiling, the liquid oil is skimmed; after pressing, the oil is filtered; and after solvent extraction, the crude oil is separated and the solvent is evaporated and recovered. Residues are conditioned (for example, dried) and are reprocessed to yield by-products such as animal feed. Crude oil refining includes degumming, neutralization, bleaching, deodorization, and further refining.

Pollution Prevention Practices and Opportunities

Dust is generated in materials handling and in the processing of raw materials, including in the cleaning, screening, and crushing operations. For palm fruit, about 2-3 cubic meters of wastewater is generated per metric ton of crude oil. The wastewater is high in organic content, resulting in a biochemical oxygen demand (BOD) of 20,000-35,000 mg/ℓ and a chemical oxygen demand (COD) of 30,000-60,000 mg/ℓ. In addition, the wastewaters are high in dissolved solids (10,000 mg/ℓ), oil and fat residues (5,000-10,000 mg/1), organic nitrogen (500-800 mg/ℓ), and ash residues (4,000 to 5,000 mg/ℓ). Seed dressing and edible fat and oil processing generate approximately 10-25 m^3 of wastewater per metric ton (t) of product. Most of the solid wastes which are mainly of vegetable origin, can be processed into by-products or used as fuel. Molds may be found on peanut kernels, and aflatoxins may be present. Table 7 provides an action list of pollution prevention opportunities. Since the pollutants generated by the industry are largely losses in production, improvements in production efficiency, as described in Table 7, are recommended to reduce pollutant loads.

Table 7. *P2 Practices and Recommendations for Vegetable Oil Processing.*

Collect waste product for use in by-products such as animal feed, where feasible without exceeding cattle-feed quality limits.
Continuous sampling and measuring of key production parameters allow production losses to be identified and reduced, thus reducing the waste load.
Odor problems can usually be prevented through good hygiene and storage practices.
Chlorinated fluorocarbons should not be used in the refrigeration system.
Recirculate cooling waters.
Optimize the use of water and cleaning chemicals.
Recover solvent vapors to minimize losses.
Provide dust extractors to maintain a clean workplace, recover product, and control air emissions.
Maintain volatile organic compounds (VOCs) well below explosive limits. Hexane should be below 150 mg/m^3 of air (its explosive limit is 42,000 mg/m^3).
Reduce product losses through better production control.
Where appropriate, give preference to physical refining rather than chemical refining of crude oil, as active clay has a lower environmental impact than the chemicals generally used.
Use citric acid instead of phosphoric acid, where feasible, in degumming operations.
Prevent the formation of molds on edible materials by controlling and monitoring air humidity.

Wastewater loads are typically 3-5 m^3/t of feedstock; plant operators should aim to achieve lower rates at the intake of the effluent treatment system. Hexane, if used, should be below 50 mg/ℓ in wastewater. The BOD level should be less than 2.5 kg/t of product, with a target of 1-1.5 kg/t.

Pretreatment of effluents comprises screening and air flotation to remove fats and solids; it is normally followed by biological treatment. if space is available, land treatment or pond systems are potential treatment methods. Other possible biological treatment systems include trickling filters, rotating biological contactors, and activated sludge treatment. Pretreated effluents can be discharged to a municipal sewerage system, if capacity exists, with the approval of the relevant authority. Proper circulation of air, using an extractive and cleaning system, is normally required to maintain dust at acceptable levels. Dust control is provided by fabric filters. Odor control is done by ventilation, but scrubbing may also be required.

WOOD PRESERVING

Industry Description and Practices

Wood preserving involves imparting protective properties to wood to guard against weathering and attack by pests. Three main types of preservatives are used: water based (for example, sodium phenylphenoxide, benzalconium chloride, guazatin, and copper chrome arsenate); organic solvent based (for example, pentachlorophenol and such substitutes as propiconazol, tebuconazol, lindane, permethrin, triazoles, tributyltm compounds, and copper and zinc naphthenates); borates; and tar oils (such as creosote). Note that some of the preservatives mentioned (for example, lindane, tributyltin, and pentachlorophenol) are banned in some countries. The preservatives are applied to the surface of wood by pressure impregnation, with a pressure range of 800 kilopascals (kPa) to 1,400 kPa; by deluging (mechanical application by flooding or spraying), by dipping or immersion; and by thermal processing (immersion in a hot bath of preservative). Application of vacuum helps to improve the effectiveness of the process and to recover some of the chemicals used. Pesticides are applied using appropriate protective clothing, including gloves, aprons, overalls, and inhalation protection.

Pollution Prevention Practices and Opportunities

The substances used in wood preserving, such as preservatives and solvents, can be found in the drips and the surface runoff streams. Air emissions of solvents and other volatile organics result from the surface treatment steps, drying of the treated wood, and storage and transfer of chemicals. Soil contamination may result from the drippage and surface runoff, and this may happen near the process areas and the treated wood storage areas. Some of the major pollutants present in drips, surface runoff, and contaminated soil include polynuclear aromatic hydrocarbons, pentachlorophenol, pesticides, dioxins, chrome, copper, and arsenic. Wood preserving involves different combinations of a wide variety of processes, and there are many opportunities to improve on the traditional practices in the industry. Table 8 provides an action list of pollution prevention opportunities.

Table 8. *P2 Practices and Recommendations for Wood Preserving.*

Proper labels should be applied, and used packaging should be returned to the supplier for reuse or sent for other acceptable uses or destruction.
Preservatives and other hazardous substances should be stored safely, preferably under a roof with a spill collection system.
Sites should be selected that are not prone to flooding or adjacent to water intake points or valuable groundwater resources.
Cover process areas and collect surface runoff for recycling and treatment. Where water-based preservatives are used, prevent freshly treated wood from coming into contact with rainwater.
Minimize surface runon by diversion of stormwater away from the process areas.
Use concrete pads for the wood treatment area and intermediate storage areas to ensure proper collection of drippage. Treated wood should be sent for storage only after drippage has completely stopped.
Heat treated wood when water-based preservatives are used.
Recycle collected drips after treatment, if necessary.
Minimize drippage by effective removal of extra preservative from the wood surface by mechanical shaking until no drippage is noticeable. Provide sufficient holding time after preservative application to minimize free liquid.
Give preference to pressurized treatment processes to minimize both wastage of raw materials and the release of toxics that may be present.
Minimize contamination of surface runoff and soil. Have a closed system for managing liquids to avoid the discharge of liquid effluents.
Exhaust streams should be treated, using carbon filters that allow the reuse of solvents, to reduce volatile organic compounds (VOCs) to acceptable levels before venting to the atmosphere. Where VOC recovery is not feasible, destruction is carried out in combustion devices or bio-oxidation systems.
Do not use pentachlorophenol, lindane, tributyltin, or copper chrome arsenate (or its derivatives).

The main treatment process is recycling of collected drips and surface runoff after evaporation. Other processes include detoxification (using ultraviolet oxidation) and precipitation or stabilization of heavy metals. Contaminated soil may contain heavy metals and toxic organics and should normally be managed as hazardous waste. Treatment methods include incineration of toxic organics and stabilization of heavy metals.

ELECTRONICS MANUFACTURING

Industry Description and Practices

The electronics industry includes the manufacture of *passive components* (resistors, capacitors, inductors); semiconductor components (discretes, integrated circuits); printed circuit boards (single and multilayer boards); and printed wiring assemblies. This section addresses the environmental issues associated with the last three manufacturing processes.

The manufacture of passive components is not included because it is similar to that of semiconductors. A difference is that passive component manufacturing uses less of the toxic chemicals employed in doping semiconductor components and more organic solvents, epoxies, plating metals, coatings, and lead.

Semiconductors

Semiconductors are produced by treating semiconductor substances with dopants such as boron or phosphorus atoms to impart electrical properties. Important semiconductor substances are silicon and gallium arsenide. Manufacturing stages include crystal growth; acid etch and epitaxy formation; doping and oxidation; diffusion and ion implantation; metallization; chemical vapor deposition; die separation; die attachment; postsolder cleaning; wire bonding; encapsulation packaging; and final testing, marking, and packaging. Several of these process steps are repeated several times, so the actual length of the production chain may well exceed 100 processing steps. Between the repetitions, a cleaning step that contributes to the amount of effluent produced by the process is often necessary. Production involves carcinogenic and mutagenic substances and should therefore be carried out in closed systems.

Printed Circuit Board (PCB) Manufacturing

There are three types of boards: single sided (circuits on one side only), double sided (circuits on both sides), and multilayer (three or more circuit layers). Board manufacturing is accomplished by producing patterns of conductive material on a nonconductive substrate by subtractive or additive processes. The conductor is usually copper; the base can be pressed epoxy, Teflon, or glass. In the subtractive process, which is the preferred route, the steps include cleaning and surface preparation of the base, electroless copperplating, pattern printing and masking, electroplating, and etching.

Printed Wiring Assemblies

Printed wiring assemblies consist of components attached to one or both sides of the printed circuit board. The attachment may be by through-hole technology, in which the "legs" of the components are inserted through holes in the board and are soldered in place from underneath, or by surface mount technology (SMT), in which components are attached to the surface by solder or conductive adhesive.

The solder is generally a tin-lead alloy. In printed circuit boards of all types, drilled holes may have to be copper-plated to ensure interconnections between the different copper layers. SMT, which eliminates the drilled holes, allows much denser packing of components, especially when components are mounted on both sides. It also offers higher-speed performance and is gaining over through-hole technology.

Pollution Prevention Practices and Opportunities

Air emissions from semiconductor manufacturing include toxic, reactive, and hazardous gases; organic solvents; and particulates from the process. The changing of gas cylinders may also result in fugitive emissions of gases. Chemicals in use may include hydrogen, silane, arsine, phosphine, diborane, hydrogen chloride, hydrogen fluoride, dichlorosilane, phosphorous oxychloride, and boron tribromide.

Air emissions from the manufacture of printed circuit boards include sulfuric, hydrochloric, phosphoric, nitric, acetic, and other acids; chlorine; ammonia; and organic solvent vapors (isopropanol, acetone, trichloroethylene; n-butyl acetate; xylene; petroleum distillates; and ozone-depleting substances).

In the manufacture of printed wiring assemblies, air emissions may include organic solvent vapors and fumes from the soldering process, including aldehydes, flux vapors, organic acids, and so on.

Throughout the electronics manufacturing sector, chlorofluorocarbons (CFCs) have been a preferred organic solvent for a variety of applications. CFCs are ozone-depleting substances (ODSs). Their production in and import into developing countries will soon be banned. Hydrochlorofluorocarbons (HCFCs) have been developed as a substitute for CFCs, but they too are ODSs and will be phased out. Methyl chloroform, another organic solvent, has also been used by the electronics industry; it too is an ODS and is being eliminated globally on the same schedule as CFCs. Chlorobromomethane and n-propyl bromide are also unacceptable because of their high ozone-depleting potential.

Effluents from the manufacture of semiconductors may have a low pH from hydrofluoric, hydrochloric, and sulfuric acids (the major contributors to low pH) and may contain organic solvents, phosphorous oxychloride (which decomposes in water to form phosphoric and hydrochloric acids), acetate, metals, and fluorides.

Effluents from the manufacture of printed circuit boards may contain organic solvents, vinyl polymers; stannic oxide; metals such as copper, nickel, iron, chromium, tin, lead, palladium, and gold; cyanides (because some metals may be complexed with chelating agents); sulfates; fluorides and fluoroborates; ammonia; and acids.

Effluents from printed wiring assemblies may contain acids, alkalis, fluxes, metals, organic solvents, and, where electroplating is involved, metals, fluorides, cyanides, and sulfates.

Solid and hazardous wastes from semiconductor manufacture may include heavy metals, solder dross (solder pot skimmings), arsenic, spent epoxy, and waste organic solvents (contributing the largest volume of waste). In printed circuit board operations, solid wastes may include scrap board materials, plating and hydroxide sludges, and inks. In the manufacture of printed wiring assemblies, solid wastes may include solder dross, scrap boards, components, organic solvents, and metals.

Boards may also be treated with brominated flame retardants, which may pose some environmental risk when boards are disposed of in landfills. All conventional electronics present additional hazards in landfills because of the presence of lead in cathode-ray tube envelopes and in solder, as well as lead and other metal salts, particularly if they have not been cleaned in a postsoldering operation. All three manufacturing processes may generate sludges containing heavy metals from wastewater treatment plants.

Organic solvent residues also require management and disposal. Measures such as plasma etching of silicon nitride (a dry process) in metal oxide semiconductor (MOS) technology replace the hot corrosive phosphoric acid (H_3PO_4) wet process and offer reductions in generated waste and better safety for workers while reducing the number of processing steps. Because of the reaction of the plasma with the substrate, several substances are formed that are regarded as carcinogenic or mutagenic and that may pose a danger to maintenance personnel. Risks are minimized by sweeping equipment with nitrogen before opening it. A gas mask with breathing equipment should be worn by personnel during repair and maintenance. A number of process alternatives exist for the manufacture of printed circuit boards. These include:

- In board manufacture: SMT rather than plated through-hole technology; injection molded substrate; additive plating;
- In cleaning and surface preparation: use of nonchelating cleaners; extension of bath life; improvement of rinse efficiency; countercurrent cleaning; recycling and reuse of cleaners and rinses;
- In pattern printing and masking: aqueous processable resist; screen printing to replace photolithography; dry photoresist; recycling and reuse of photoresist strippers; segregation of streams; recovery of metals;
- For electroplating and electroless plating: replacement of these processes by mechanical board production; use of noncyanide baths; extension of bath life; recycling and reuse of cleaners and rinses; improvement of rinse efficiency; countercurrent rinsing; segregation of streams; recovery of metals;
- In etching: use of differential plating; use of nonchelated etchants and nonchrome etchant; use of pattern instead of panel plating; use of additive instead of subtractive processes; recycling and reuse of etchants.

Metal recovery by regenerative electrowinning results in a near-zero effluent discharge for segregated metal-bearing streams. Heavy metals are recovered to metal sheets which eliminates 95% of sludge disposal. Metal-bearing sludges that are not treated for recovery of metals should be disposed of in secure landfills.

In the printed wiring assembly process, nonozone-depleting alternatives are readily available for cleaning printed wiring assemblies. These alternatives include other organic solvents, hydrocarbon/surfactant blends, alcohols, and organic solvent blends, as well as aqueous and semiaqueous processes. More important, the industry has shown that even sophisticated printed wiring assemblies intended for military uses (where specifications are very exacting) can be made without cleaning by using low-residue fluxes that leave very little in the way of contamination on the boards. The no-clean concept does away with the use of organic solvents and the need to dispose of organic solvent waste, eliminates a process step and the corresponding equipment, and has been shown to give adequate product quality according to the application.

Organic solvent losses can be reduced by conservation and recycling, using closed-loop delivery systems, hoods, fans, and stills. Installation of activated carbon systems can achieve up to 90% capture and recycle of organic solvents used in the system. All solvents and hazardous chemicals (including wastes) require appropriate safe storage to prevent spills and accidental discharges. All tanks, pipework, and other containers should be situated over spill containment trays with dimensions large enough to contain the total volume of liquid over them. Containment facilities must resist all chemical attack from the products. In lieu of containment facilities, the floor and walls, to a reasonable height, may be treated (e.g., by an epoxy product, where chemically appropriate) to prevent the possibility of leakage of accidental spills into the ground, and there should be doorsills. (Untreated cement or concrete or grouted tile floors are permeable.) It is unacceptable to have a drain in the floor of any shop where chemicals of any description are used or stored, except where such a drain leads to an adequate water-treatment plant capable of rendering used or stored chemicals in its catchment area.

Waste organic solvents should be sent to a solvent recycling operation for reconstitution and reuse. Where recycling facilities are not available, waste solvents may need to be incinerated or destroyed as appropriate for their chemical composition.

The following production-related targets can be achieved by measures such as those described above. Ozone-depleting substances are not to be used in production operations unless no proven alternative exists. Discharges of organic solvents should be minimized, and alternative technologies should be considered where available. Solder dross should not be sent to landfills. (Waste can be sent to suppliers or approved waste recyclers for recovery of the lead and tin content of the dross.) Scrap boards and assemblies having soldered components should have their components and solder connections removed before they are sent to landfills or recycled for other uses.

Wet scrubbers, point-of-use control systems, and volatile organic compound (VOC) control units are used to control toxic and hazardous emissions of the chemicals used in semiconductor manufacturing. It is often appropriate to scrub acid and alkaline waste gases in separate scrubbers because different scrubber liquids can then be used, resulting in higher removal efficiencies. Air emission concentrations of chemicals such as arsine, diborane, phosphine, silane, and other chemicals used in the process should be reduced below worker health levels for plant operations.

Because of the many chemicals used in the electronics industry, wastewater segregation simplifies waste treatment and allows recovery and reuse of materials. Organic wastes are collected separately from wastewater systems. (Note that solvent used in the semiconductor industry cannot be readily recycled because much of it is generated from complex mixtures such as photoresist.) Acids and alkalis are sent to on-site wastewater treatment facilities for neutralization, after segregation of heavy-metal-bearing streams for separate treatment. Fluoride-bearing streams in a semiconductor plant are segregated and treated on-site or sent off-site for treatment or disposal. Treatment steps for effluents from the electronics industry may include precipitation, coagulation, sedimentation, sludge dewatering, ion exchange, filtering, membrane purification and separation, and neutralization, depending on the particular stream. Sanitary wastes are treated separately (primary and secondary treatment followed by disinfection) or discharged to a municipal treatment system. Table 9 provides an action list of pollution prevention opportunities.

Table 9. *P2 Practices and Recommendations for Electronics Manufacturing.*

Where liquid chemicals are employed, the plant, including loading and unloading areas, should be designed to minimize evaporation (other than water) and to eliminate all risk of chemicals entering the ground or any watercourse or sewerage system in the event of an accidental leak or spill.
Cylinders of toxic gases should be well secured and fitted with leak detection devices as appropriate. Well-designed emergency preparedness programs are required. Note that fugitive emissions occurring when gas cylinders are changed do not normally require capture for treatment, but appropriate safety precautions are expected to be in place.
No ozone-depleting chemicals should be used in the process unless no proven alternatives are available.
Equipment, such as refrigeration equipment, containing ozone-depleting chemicals should not be purchased unless no other option is available.
Toxic and hazardous sludges and waste materials must be treated and disposed of or sent to approved waste disposal or recycling operations.

ELECTROPLATING

Industry Description and Practices

Electroplating involves the deposition of a thin protective layer (usually metallic) onto a prepared metal surface, using electrochemical processes. The process involves pretreatment (cleaning, degreasing, and other preparation steps), plating, rinsing, passivating, and drying. The cleaning and pretreatment stages involve a variety of solvents (often chlorinated hydrocarbons, whose use is discouraged) and surface stripping agents, including caustic soda and a range of strong acids, depending on the metal surface to be plated. The use of halogenated hydrocarbons for degreasing is not necessary, as water-based systems are available. In the plating process, the object to be plated is usually used as the cathode in an electrolytic bath. Plating solutions are acid or alkaline and may contain complexing agents such as cyanides.

Pollution Prevention Practices and Opportunities

The substances used in electroplating (such as acidic solutions, toxic metals, solvents, and cyanides) can be found in the wastewater, either via rinsing of the product or from spillage and dumping of process baths. The solvents and vapors from hot plating baths result in elevated levels of VOCs and, in some cases, volatile metal compounds, which may contain chromates. Approximately 30% of the solvents and degreasing agents used can be released as VOCs when baths are not regenerated.

The mixing of cyanide and acidic wastewaters can generate lethal hydrogen cyanide gas, and this must be avoided. The overall wastewater stream is typically extremely variable (1 liter to 500 liters per square meter of surface plated) but is usually high in heavy metals, including cadmium, chrome, lead, copper, zinc, and nickel, and in cyanides, fluorides, and oil and grease, all of which are process dependent. Air emissions may contain toxic organics such as trichloroethylene and trichloroethane.

Cleaning or changing of process tanks and treatment of wastewaters can generate substantial quantities of wet sludges containing high levels of toxic organics or metals. Plating involves different combinations of a wide variety of processes, and there are many opportunities to improve on traditional practices in the industry. Table 10 provides an action list of pollution prevention opportunities. An important parameter is the water use in each process. Systems should be designed to reduce water use. Where electroplating is routinely performed on objects with known surface area in a production unit, water consumption of no more than 1.3 liters per square meter plated (ℓ/m^2) for rack plating and 10 ℓ/m^2 for drum plating should be achieved.

Cadmium plating should be avoided. Where there are no feasible alternatives, a maximum cadmium load in the waste of 0.3 grams for every kilogram of cadmium processed is recommended. At least 90% of the solvent emissions to air must be recovered by the use of an air pollution control system such as a carbon filter. Ozone-depleting solvents such as chlorofluorocarbons and trichloroethane are not to be used in the process. Segregation of waste streams is essential because of the dangerous reactions that can occur. Strong acid and caustic reactions can generate boiling and splashing of corrosive liquids; acids can react with cyanides and generate lethal hydrogen cyanide gas. In addition, segregated streams that are concentrated are easier to treat.

Table 10. *P2 Practices and Recommendations for Electroplating.*

AREA OF OPPORTUNITY	POLLUTION PREVENTION RECOMMENDATION
Changes in Process	1. Replace cadmium with high-quality, corrosion-resistant zinc plating. Use cyanide-free systems for zinc plating where appropriate. Where cadmium plating is necessary, use bright chloride, high-alkaline baths, or other alternatives. Note, however, that use of some alternatives to cyanides may lead to the release of heavy metals and cause problems in wastewater treatment. 2. Use trivalent chrome instead of hexavalent chrome; acceptance of the change in finish needs to be promoted. 3. Give preference to water-based surface-cleaning agents, where feasible, instead of organic cleaning agents, some of which are considered toxic. 4. Regenerate acids and other process ingredients whenever feasible.
Reduction in Dragout and Wastage	5. Minimize dragout through effective draining of bath solutions from the plated part, by, for example, making drain holes in bucket-type pieces, if necessary. 6. Allow dripping time of at least 10 to 20 seconds before rinsing. 7. Use fog spraying of parts while dripping. 8. Maintain the density, viscosity, and temperature of the baths to minimize dragout. 9. Place recovery tanks before the rinse tanks (also yielding make-up for the process tanks). The recovery tank provides for static rinsing with high dragout recovery.
Minimizing Water Consumption in Rinsing Systems	10. Agitation of rinse water or work pieces to increase rinsing efficiency. 11. Apply multiple countercurrent rinses. 12. Spray rinses (especially for barrel loads).
Management of Process Solutions	13. Recycle process baths after concentration and filtration. Spent bath solutions should be sent for recovery and regeneration of plating chemicals, not discharged into wastewater treatment units. 14. Recycle rinse waters (after filtration). 15. Regularly analyze and regenerate process solutions to maximize useful life. 16. Clean racks between baths to minimize contamination. 17. Cover degreasing baths containing chlorinated solvents when not in operation to reduce losses. Spent solvents should be sent to solvent recyclers and the residue from solvent recovery properly managed (e.g., blended with fuel and burned in a combustion unit with proper controls for toxic metals).

Exhaust hoods and good ventilation systems protect the working environment, but the exhaust streams should be treated to reduce VOCs and heavy metals to acceptable levels before venting to the atmosphere. Acid mists and vapors should be scrubbed with water before venting. In some cases, VOC levels of the vapors are reduced by use of carbon filters, which allow the reuse of solvents, or by combustion (and energy recovery) after scrubbing, adsorption, or other treatment methods.

Cyanide destruction, flow equalization and neutralization, and metals removal are required, as a minimum, for electroplating plants. Individual design is necessary to address the characteristics of the specific plant, but there are a number of common treatment steps. For small facilities, the possibility of sharing a common wastewater treatment plant should be considered. Cyanide destruction must be carried out upstream of the other treatment processes. If hexavalent chrome ($Cr+6$) occurs in the wastewater, the wastewater is usually pretreated to reduce the chromium to a trivalent form using a reducing agent, such as a sulfide.

The main treatment processes are equalization, pH adjustment for precipitation, flocculation, and sedimentation/filtration. The optimum pH for metal precipitation is usually in the range 8.5-11, but this depends on the mixture of metals present. The presence of significant levels of oil and grease may affect the effectiveness of the metal precipitation process; hence, the level of oil and grease affects the choice of treatment options and the treatment sequence. It is preferred that the degreasing baths be treated separately.

Flocculating agents are sometimes used to facilitate the filtration of suspended solids. Pilot testing and treatability studies may be necessary and final adjustment of pH and further polishing of the effluent may be required. Modern wastewater treatment systems use ion exchange, membrane filtration, and evaporation to reduce the release of toxics and the quantity of effluent that needs to be discharged. The design can provide for a closed system with a minor bleed stream.

Treatment sludges contain high levels of metals, and these should normally be managed as hazardous waste or sent for metals recovery. Electrolytical methods may be used to recover metals. Sludges are usually thickened, dewatered, and stabilized using chemical agents (such as lime) before disposal, which must be in an approved and controlled landfill. The high costs of proper sludge disposal are likely to become an increasing incentive for waste minimization.

FOUNDRIES

Industry Description and Practices

In foundries, molten metals are cast into objects of desired shapes. Castings of iron, steel, light metals (such as aluminum), and heavy metals (such as copper and zinc) are made in units that may be independent or part of a production line. Auto manufacturing facilities usually have foundries within their production facilities or as ancillaries. The main production steps include: (1) the preparation of raw materials; (2) metal melting; (3) preparation of molds; (4) casting; (5) finishing stages which includes fettling and tumbling.

Electric induction furnaces are used to melt iron and other metals. However, large car-component foundries and some small foundries melt iron in gas or coke-fired cupola furnaces and use induction furnaces for aluminum components of engine blocks. Melting capacities of cupola furnaces generally range from 3 to 25 metric tons per hour (t/hr). Induction furnaces are also used in zinc, copper, and brass foundries. Electric arc furnaces are usually used in stainless steel and sometimes in copper foundries. Flame ovens, which burn fossil fuels, are often used for melting nonferrous metals. The casting process usually employs nonreusable molds of green sand, which consists of sand, soot, and clay (or water glass). The sand in each half of the mold is packed around a model, which is then removed. The two halves of the mold are joined, and the complete mold is filled with molten metal, using ladles or other pouring devices. Large foundries often have pouring furnaces with automatically controlled pouring. The mold contains channels for introducing and distributing the metal - a "gating system." For hollow casting, the mold is fitted with a core. Cores must be extremely durable, and so strong bonding agents are used for the core, as well as for the molds themselves. These bonding agents are usually organic resins, but inorganic ones are also used. Plastic binders are being used for the manufacture of high-quality products. Sand cores and chemically bonded sand molds are often treated with water-based or spirit-based blacking to improve surface characteristics. Aluminum and magnesium, as well as copper and zinc alloys, are frequently die-cast or gravity-cast in reusable steel molds. Die casting involves the injection of metal under high pressure by a plunger into a steel die. Centrifugal casting methods are used for pipes.

Finishing processes such as fettling involves the removal from the casting of the gating system, fins (burrs), and sometimes feeders. This is accomplished by cutting, blasting, grinding, and chiseling. Small items are usually ground by tumbling, carried out in a rotating or vibrating drum, usually with the addition of water, which may have surfactants added to it.

Pollution Prevention Practices and Opportunities

Emissions of PM from the melting and treatment of molten metal, as well as from mold manufacture, shakeout, cleaning and after-treatment, is generally of greatest concern. PM may contain metals that may be toxic. Oil mists are released from the lubrication of metals. Odor and alcohol vapor (from surface treatment of alcohol-based blacking) and emissions of other VOCs are also of concern. Care must be exercised when handling halogenated organics, including aluminum scrap contaminated with chlorinated organics, polyvinyl chloride (PVC) scrap and turnings with chlorinated cutting oil, as dioxins may be emitted during melting operations.

Oil and suspended solids are released into process effluents, and treatment is warranted before their discharge. Wet scrubbers release wastewaters that may contain metals. Wastewater from tumbling may contain metals and surfactants. Cooling waters, used in amounts of up to 20 cubic meters per metric ton, may contain oil and some chemicals for the control of algae and corrosion.

Sand molding creates large quantities of waste sand. Other wastes include slag (300-500 kilograms per metric ton, kg/t, of metal), collected particulate matter, sludges from separators used in wastewater treatment, and spent oils and chemicals. Discarded refractory lining is another waste produced.

The primary hazardous components of collected dust are zinc, lead, and cadmium, but its composition can vary greatly depending on scrap composition and furnace additives. (Nickel and chromium are present when stainless steel scrap is used.) Generally, foundries produce 10 kg of dust per ton of molten metal, with a range of 5-30 kg/t, depending on factors such as scrap quality. However, induction furnaces and flame ovens tend to have lower air emissions than cupolas and electric arc furnaces (EAF). Major pollutants present in the air emissions include particulates of the order of 1,000 milligrams per normal cubic meter (mg/Nm^3).

Foundries can generate up to 20 cubic meters of wastewater per metric ton of molten metal when cooling water, scrubber water, and process water are not regulated. Untreated wastewaters may contain high levels of total suspended solids, copper (0.9 mg/ℓ), lead (2.5 mg/ℓ), total chromium (2.5 mg/ℓ), hexavalent chromium, nickel (0.25 mg/ℓ), and oil and grease. The characteristics of the wastewater will depend on the type of metal and the quality of scrap used as feed to the process.

Solid wastes (excluding dust) are generated at a rate of 300-500 kg/t of molten metal. Sludges and scale may contain heavy metals such as chromium, lead, and nickel. Table 11 provides an action list of pollution prevention opportunities.

Table 11. *P2 Practices and Recommendations for Foundries.*

Reclaim sand after removing binders.
Reduce nitrogen oxide NO_x emissions by use of natural gas as fuel, use low-NO_x burners.
Control water consumption by recirculating cooling water after treatment.
Store chemicals and other materials in such a way that spills, if any, can be collected.
Use closed-loop systems in scrubbers where the latter are necessary.
Use continuous casting for semifinished and finished products wherever feasible.
Use dry dust collection methods such as fabric filters instead of scrubbers.
Provide hoods for cupolas or doghouse enclosures for EAFs and induction furnaces.
Improve feed quality: use selected and clean scrap to reduce the release of pollutants to the environment. Preheat scrap, with afterburriting of exhaust gases. Store scrap under cover to avoid contamination of stormwater.
Prefer induction furnaces to cupola furnaces.
Replace the cold-box method for core manufacture, where feasible.

Dust emission control technologies include cyclones, scrubbers (with recirculating water), baghouses, and electrostatic precipitators (ESPs). Scrubbers are also used to control mists, acidic gases, and amines. Gas flame is used for incineration of gas from core manufacture. Target values for emissions passing through a fabric filter are normally around 10 mg/Nm^3 (dry). Emissions of PM from furnaces (including casting machines used for die casting) should not exceed 0.1-0.3 kg/t of molten metal, depending on the nature of the PM and the melting capacity of the plant. At small iron foundries, a somewhat higher emission factor may be acceptable, while in large heavy-metal foundries, efforts should be made to achieve a target value lower than 0.1 kg PM per metric ton. Odors may be eliminated by using bioscrubbers. For wastewater treatment, common practice is to recirculate tumbling water by sedimentation or centrifuging followed by filtering (using sand filters or ultrafilters); separate oil from surface water. In the very rare cases in which scrubbers are used, recirculate water and adjust its pH to precipitate metals. Precipitate metals in wastewater by using lime or sodium hydroxide. Cooling waters should be recirculated, and polluted stormwater should be treated before discharge. The reader may refer to chapter 2 again for a description of standard pollution control equipment.

FRUIT AND VEGETABLE PROCESSING

Industry Description and Practices

Processing (canning, drying, freezing, and preparation of juices, jams, and jellies) increases the shelf life of fruits and vegetables. Processing steps include preparation of the raw material (cleaning, trimming, and peeling followed by cooking, canning, or freezing. Plant operation is often seasonal.

Pollution Prevention Practices and Opportunities

The fruit and vegetable industry generates large volumes of effluents and solid waste. The effluents contain high organic loads, cleansing and bleaching agents, salt, and suspended solids such as fibers and soil particles. They may also contain pesticide residues washed from the raw materials. The main solid wastes are organic materials, including discarded fruits and vegetables. Odor problems can occur with poor management of solid wastes and effluents; when onions are processed; and when ready-to-serve meals are prepared. Reductions in wastewater volumes of up to 95% have been reported through implementation of good practices. Where possible, measures such as those listed in Table 12 should be implemented.

Table 12. *P2 Practices and Recommendations for Fruit/Vegetable Processing.*

Reuse concentrated wastewaters and solid wastes for production of by-products.
Remove solid wastes without the use of water.
Minimize the use of water for cleaning floors and machines.
Use steam instead of hot water to reduce the quantity of wastewater going for treatment (taking into consideration, however, the tradeoff with increased use of energy).
Use countercurrent systems where washing is necessary.
Recirculation of process water from onion preparation reduces the organic load by 75% and water consumption by 95%. Similarly, the liquid waste load (in terms of BOD) from apple juice and carrot processing can be reduced by 80%.
Separate and recirculate process wastewaters.
Use dry methods such as vibration or air jets to clean raw fruit and vegetables. Dry peeling methods reduce the effluent volume (by up to 35%) and pollutant concentration (organic load reduced by up to 25%).
Solid wastes, particularly from processes such as peeling and coring, typically have a high nutritional value and may be used as animal feed.
Procure clean raw fruit and vegetables, thus reducing the concentration of dirt and organics (including pesticides) in the effluent.

Good water management should be adopted, where feasible, to achieve the levels of consumption cited in Table 13.

Table 13. *Water Usage in the Fruit and Vegetable Processing Industry.*

PRODUCT CATEGORY	Water Use (cubic meters per metric ton of product)
Canned fruit	2.5-4.0
Canned vegetables	3.5-6.0
Frozen vegetables	5.0-8.5
Fruit juices	6.5
Jams	6.0
Baby food	6.0-9.0

Preliminary treatment of wastewaters should include screening (or sieving to recover pulp) and grit removal, if necessary. This is followed by pH adjustment and biological treatment of the organic load. The flows are frequently seasonal, and robust treatment systems are preferred for on-site treatment. Pond systems are used successfully to treat fruit and vegetable wastes, but odor nuisance, soil deterioration, and groundwater pollution are to be avoided. The quality of the effluent is normally suitable for discharge to municipal systems, although peak hydraulic loads may cause a problem. Odor problems can be avoided by using gas scrubbers or biofilters. Pesticides may be present in significant levels; testing should therefore be performed, and, if pesticides are present at levels above 0.05 milligrams per liter (mg/l), corrective action should be taken. The best course may be to switch to a supplier that provides raw materials without pesticide residues.

Whenever possible, organic wastes should be used in the production of animal feed or organic fertilizers. Other solid wastes should be disposed of in a secure landfill to avoid contamination of surface and groundwater.

GLASS MANUFACTURING

Industry Description and Practices

This section describes the manufacture of flat glass and pressed and blown glass. Flat glass includes plate and architectural glass, automotive windscreens, and mirrors. Pressed and blown glass includes containers, machine-blown and hand-blown glassware, lamps, and television tubing. In both categories, a glass melt is prepared from silica sand, other raw materials such as lime, dolomite, and soda, and cullet (broken glass).

The use of recycled glass is increasing. It reduces the consumption of both raw materials and energy but necessitates extensive sorting and cleaning prior to batch treatment to remove impurities. For the manufacture of special and technical glass, lead oxide, potash, zinc oxide, and other metal oxides are added. Refining agents include arsenic trioxide, antimony oxide, nitrates, and sulfates. Metal oxides and sulfides are used as coloring or decoloring agents. The most common furnace used for manufacturing glass melt is the continuous regenerative type, with either the side or the end ports connecting brick checkers to the inside of the melter. Checkers conserve fuel by acting as heat exchangers; the fuel combustion products heat incoming combustion air. The molten glass is refined (heat conditioning) and is then pressed, blown, drawn, rolled, or floated, depending on the final product. Damaged and broken product (cullet) is returned to the process. The most important fuels for glass-melting furnaces are natural gas, light and heavy fuel oil, and liquefied petroleum gas. Electricity (frequently installed as supplementary heating) is also used. Energy requirements range from 3.7 to 6.0 kilojoules per metric ton (kJ/t) glass produced.

Pollution Prevention Practices and Opportunities

Two types of air emissions are generated: those from the combustion of fuel for operating the glass-melting furnaces, and fine particulates from the vaporization and recrystallization of materials in the melt. The main emissions are sulfur oxides SO_x nitrogen oxides NO_x and particulates, which can contain heavy metals such as arsenic and lead. Particulates from lead crystal manufacture can have a lead content of 20-60% and an arsenic content of 0.5-2%. Certain specialty glasses can produce releases of hydrogen chloride (HCl), hydrogen fluoride (HF), arsenic, boron, and lead from raw materials. Container, pressing, and blowing operations produce a periodic mist when the hot gob comes into contact with the release agent used on the molds. Cold-top electric furnaces, in which the melt surface is covered by raw material feed, release very little particulate matter, as the blanket acts as a filter to prevent the release of particulate matter. Some releases of particulates will take place in tapping, but furnace releases should be of the order of 0.1 kilogram per ton (kg/t) when operated this way. Lead glass manufacture may result in lead emissions of about 2-5 kg/t.

In all cases, the concentration of heavy metals and other pollutants in the raw flue gas mainly depends on the type of fuel used, the composition of the feed material, and the portion of recycled glass. High input of sulfates or potassium nitrate may increase emissions of sulfur dioxide and nitrogen oxides, respectively. Where nitrate is used, more than two thirds of the introduced nitrogen may be emitted as nitrogen oxides. The use of heavy metals as coloring or decoloring agents will increase emissions of these metals.

The grinding and polishing of flat glass to produce plate glass have become obsolete since the development of the float glass process. The chemical make-up of detergents that may be used in float glass manufacturing can vary significantly - some may contain phosphorus. In blowing and pressing, pollutants in effluents are generated by finishing processes such as cutting, grinding, polishing, and etching. The pollutants include suspended solids, fluorides, lead, and variations in pH.

Liquid effluents also result from forming, finishing, coating, and electroplating operations. Heavy metal concentrations in effluents occur where silvering and copperplating processes are in use.

Oxygen-enriched and oxyfuel furnaces are used in specialty glass operations to reduce emissions or to make possible higher production rates with the same size furnace. Although oxyfuel furnaces may produce higher NO_x emissions on a concentration basis, they are expected to yield very low levels of nitrogen oxides on a mass basis (kg/t of product). Low-NO_x furnaces, staged firing, and flue gas recirculation are available to reduce both concentration and the mass of nitrogen oxide emissions. These techniques are also available for air-fuel-fired furnaces. Nitrogen oxide levels can be controlled to 500-800 mg/m^3.

The type of combustion fuel used affects the amount of sulfur oxides and nitrogen oxides emitted. Use of natural gas results in negligible sulfur dioxide emissions from the fuel compared with high-sulfur fuel oils. Fuel oil with a low sulfur content is preferable to fuel oil with a high sulfur content if natural gas is not available. An efficient furnace design will reduce gaseous emissions and energy consumption. Examples of improvements include modifications to the burner design and firing patterns, higher preheater temperatures, preheating of raw material, and electric melting.

Changing the composition of the raw materials can, for example, reduce chlorides, fluorides, and sulfates used in certain specialty glasses. The use of outside-sourced cullet and recycled glass will reduce energy requirements (for an estimated 2% savings for each 10% of cullet used in the manufacture of melt) and thus air emissions (up to 10% for 50% cullet in the mix). Typical recycling rates are 10-20% in the flat glass industry and over 50% for the blown and pressed glass industries.

The amount of heavy metals used as refining and coloring or decoloring agents, as well as use of potassium nitrate, should be minimized to the extent possible.

In the furnace, particulates are formed through the volatilization of materials, leading to formation of condensates and of slag that clogs the furnace checkers. Disposal of the slag requires testing to determine the most suitable disposal method. It is important to inspect the checkers regularly to determine whether cleaning is required.

Particulate matter is also reduced, for example, by enclosing conveyors, pelletizing raw material, reducing melt temperatures, and blanketing the furnace melt with raw material.

Reductions in wastewater volumes are possible through closed cooling water loops and improved blowoff techniques.

Modern plants using good industrial practices are able to achieve the pollutant loads given here. Because of the lack of nitrogen in the oxidant, using oxyfuel-fired furnaces produces four to five times less flue gas volume than regenerative furnaces. As a result nitrogen oxides are reduced by 80%, and particulates are reduced by 20-80%.

For furnaces that operate with a cover of raw material, a target of 0.1 kg/t for particulates is realistic. Reductions in sulfur dioxide are achieved by choosing natural gas over fuel oil where possible.

ESPs are the preferred choice for removing particulates, although fabric filters are also used. Dry scrubbing using calcium hydroxide is used to reduce sulfur dioxide, hydrogen fluoride, and hydrogen chloride. Secondary measures for NO_x control include selective catalytic reduction (SCR), selective noncatalytic reduction (SNCR), and certain proprietary processes such as the Pilkington 3R process.

MEAT PROCESSING AND RENDERING

Industry Description and Practices

The meat processing and rendering industry includes the slaughter of animals and fowl, processing of the carcasses into cured, canned, and other meat products, and the rendering of inedible and discarded remains into useful by-products such as lards and oils. A wide range of processes is used.

Pollution Prevention Practices and Opportunities

The meat industry generates large quantities of solid wastes and wastewater with a biochemical oxygen demand (BOD) of 600 mg/ℓ. BOD can be as high as 8,000 mg/l, or 10-20 kg/t of slaughtered animal; and suspended solids levels can be 800 mg/l, and higher. In some cases, offensive odors may occur. The amounts of wastewater generated and the pollutant load depend on the kind of meat being processed. For example, the processing of gut has a significant impact on the quantity and quality (as measured by levels of BOD and of chemical oxygen demand, COD) of wastewater generated. The wastewater from a slaughterhouse can contain blood, manure, hair, fat, feathers, and bones. The wastewater may be at a high temperature and may contain organic material and nitrogen, as well as such pathogens as salmonella and shigella bacteria, parasite eggs, and amoebic cysts. Pesticide residues may be present from treatment of animals or their feed. Chloride levels from curing and pickling may be very high-up to 77,000 mg/l. Smoking operations can release toxic organics into the air. Rendering is an evaporative process that produces a condensate stream with a foul odor.

All slaughtering wastes (generally, 35% of the animal weight) can be used as by-products or for rendering. The only significant solid waste going for disposal is the manure from animal transport and handling areas.

Separation of product from wastes at each stage is essential for maximizing product recovery and reducing waste loads. The materials being handled are all putrescible; hence, cleanliness is essential. Water management should achieve the necessary cleanliness without waste. The amounts and strength of wastes can be reduced by good practices such as dry removal of solid wastes and installation of screens on wastewater collection channels. In-plant measures that can be used to reduce the odor nuisance and the generation of solid and liquid wastes from the production processes include those listed in Table 14.

Table 14. *P2 Practices and Recommendations for Meat Processing/Rendering.*

Recover and process blood into useful by-products. Allow enough time for blood draining (at least seven minutes).
Process paunches and intestines and utilize fat and slime.
Minimize water consumed in production by, for example, using taps with automatic shutoff, using high water pressure, and improving the process layout.
Eliminate wet transport (pumping) of wastes (for example, intestines and feathers) to minimize water consumption.
Reduce the liquid waste load by preventing any solid wastes or concentrated liquids from entering the wastewater stream.
Cover collection channels in the production area with grids to reduce the amount of solids entering the wastewater.
Remove manure (from the stockyard and from intestine processing) in solid form.
Dispose of hair and bones to the rendering plants.

Reduce air emissions from ham processing through some degree of air recirculation, after filtering.
Isolate and ventilate all sources of odorous emissions. Oxidants such as nitrates can be added to wastes to reduce odor.
Equip the outlets of wastewater channels with screens and fat traps to recover and reduce the concentration of coarse material and fat in the combined wastewater stream.
Implement dry precleaning of equipment and production areas prior to wet cleaning.
Separate cooling water from process water and wastewaters, and recirculate cooling water.
Optimize the use of detergents and disinfectants in washing water.
In *rendering plants,* odor is the most important air pollution issue. To reduce odor: (1) Minimize the stock of raw material and store it in a cold, closed, well-ventilated place; (2) Pasteurize the raw material before processing it in order to halt biological processes that generate odor; (3) Install all equipment in closed spaces and operate under partial or total vacuum; (4) Keep all working and storage areas clean.

Wastewaters from meat processing are suitable for biological treatment and (except for the very odorous rendering wastewater) could be discharged to a municipal sewer system after flow equalization, if the capacity exists. Sewer authorities usually require pretreatment of the wastewater before it is discharged into the sewer. Screens and fat traps are the minimum means of pretreatment in any system. Flotation, in some cases aided by chemical addition, may also be carried out to remove suspended solids and emulsified fats, which can be returned to the rendering plant. The choice of an appropriate biological treatment system will be influenced by a number of factors, including wastewater load and the need to minimize odors. Rendering wastewater typically has a very high organic and nitrogen load. Extended aeration is an effective form of treatment, but care must be taken to minimize odors. Disinfection of the final effluent may be required if high levels of bacteria are detected. Ponding is a simple solution but requires considerable space. Chemical methods, usually based on chlorine compounds, are an alternative. Biofilters, carbon filters, and scrubbers are used to control odors and air emissions from several processes, including ham processing and rendering. Recycling exhaust gases from smoking may be feasible in cases where operations are not carried out manually and smoke inhalation by workers is not of concern.

PRINTING INDUSTRY

The printing industry is very diverse, as can be seen in the multitude of different products that bear some form of printing-books, daily newspapers, periodicals, packaging, cartons, carrier bags, drink containers, signs, forms, brochures, advertisements, wallpaper, textiles, sheeting, metal foil, and so on. Text, diagrams, pictures, and so on are designed and composed on, for example, a newspaper page. If pictures and/or text are to be printed in several colors, these must be separated. The pictures are also often screened, producing an image that consists of a large number of very small dots instead of a solid field. Photographic techniques are used for setting and working on pictures. The page is then transferred to a printing form, a printing block (high-intensity, flexography), plate (offset), roller (rotogravure), or stencil (screen printing). This is done by means of exposure to a light-sensitive coating. In the case of offset and screen printing, the printing form is developed by washing away part of the coating; the form may then, in theory, be used immediately. The offset plate is coated with rubber to protect it from oxidation. The screen sheet's sides are masked with protective paint. Other printing methods require further stages. The small grooves in the gravure roller are etched or, increasingly, engraved, and the surface is chromed for better durability. The rubber printing block for flexographic printing is cast or engraved by laser.Printing is done on single sheets or paper web, using one or more printing units, depending on the number of colors required. The dyeing agent is, in most cases, a solvent that evaporates from the paper. (In some cases, it is necessary to hasten evaporation by feeding in warm air.) Clear varnish is sometimes added to the printed surface. The printed matter is processed off-press, where it is cut, jointed, folded, sewn, bound, packaged, and so on. Printing may also be a step in another manufacturing process-for example, laminating at package printing works, in which layers of paper, plastic and metal foil are joined. Plastic surfaces are treated to facilitate printing using electrical discharges from an electrode system, the "corona treatment."

Pollution Prevention Practices and Opportunities

Emissions into the air mainly consist of organic solvents and other organic compounds. Some substances may cause unpleasant odors or affect health and the environment. Discharges to water bodies mainly consist of silver, copper, chromium, organic solvents, and other toxic organic compounds. Wastes consist of environmentally hazardous wastes such as photographic and residual chemicals, metal hydroxide sludge, dyestuff and solvent residues, wiping material containing dyes and solvents, and oil spills. There are also bulky wastes such as paper. Table 15 provides an action list of pollution prevention opportunities.

Table 15. *P2 Practices and Recommendations for the Printing Industry.*

Treat metal-containing effluents from the manufacture of gravure cylinders and printing blocks by applying the established methods of chemical precipitation, sedimentation, and filtration. Collect fixing baths for recovery or destruction. Evaporate solvents from regeneration of active carbon filters. Perform closed-screen chase washing; recirculate Solvents and separate sludge. Fit developing machines with counterflow fixing or connect them to an organic ion exchanger. Collect film developing agents for destruction. Carry out high-pressure water jet cleaning. Use ultrafiltration to treat washing water.
Where possible, replace chemicals used for form preparation and cleaning with more environmentally friendly alternatives. Maintain a record of chemicals and environmentally hazardous waste. Do not use halogenated solvents and degreasing agents in new plants. Replace them with nonhalogenated substances in existing facilities.
Estimate the quantity of developing bath and fixing bath used per year and maintain these at acceptable levels.
Store chemicals and environmentally hazardous waste such as dyes, inks, and solvents so that the risk of spillage into the wastewater system is minimized. Examples of measures that should be considered are retaining dikes or areas with no outlet, as a means of absorbing spillage. Minimize noise disturbance from fans and presses.
Return toxic materials packaging to the supplier for reuse.
Label and store toxic and hazardous materials in secure, bunded areas.
Recover plates by remelting.
Recover energy from combustion systems, when they are used.
Use equipment washdown waters as make-up solutions for subsequent batches. Use counter-current rinsing.
Use counter-current flow fixing processes. Aim for a closed washing system.
Minimize the rinse water flow in the developing machines by, for example, use of "stand-by" Collect fixing bath, developer, used film, photographic paper, and blackened ends of photosetting paper and manage them properly.
Enclose presses and ovens to avoid diffuse evaporation of organic substances entering the general ventilation system, where feasible. Use suction hoods to collect vapors and other fugitive emissions.
Engrave, rather than etch, gravure cylinders to reduce the quantity of heavy metals used.
Give preference to the use of radiation-setting dyes.
Replace solvent-based dyes and glues with solvent-free or water-based dyes and glues, where feasible. Water-based dyes are preferred for flexographic printing on paper and plastic and for screen printing and rotogravure.
Control emissions of gases from web offset with heat-setting thermic or catalytic incineration. Recover toluene from rotogravure by absorption, using active carbon. Carry out adsorption of solvents, using zeolites, and recover organic solvents.
Evacuate air from printing presses and drying ovens into a ventilation system.
Estimate and control, typically on an annual basis, the quantities of volatile organic solvents used, including the amount used in dyes, inks, glues, and damping water. Estimate and control the proportion that is made up of chlorinated organic solvents.

Because of the relatively small volumes of solid wastes, it is difficult to find acceptable and affordable methods of disposal. Ideally, solid wastes should be sent for incineration in a facility where combustion conditions ensure effective destruction of toxics are maintained.

TEXTILES

Industry Description and Practices

The textile industry uses vegetable fibers such as cotton; animal fibers such as wool and silk; and a wide range of synthetic materials such as nylon, polyester, and acrylics. The production of natural fibers is approximately equal in amount to the production of synthetic fibers. Polyester accounts for about 50% of synthetics. Chemical production of the polymers used to make synthetic fiber is not covered in this section.

The stages of textile production are fiber production, fiber processing and spinning, yarn preparation, fabric production, bleaching, dyeing and printing, and finishing. Each stage generates wastes that require proper management and there are pollution prevention opportunities in each. This section focuses on the wet processes (including wool washing, bleaching, dyeing, printing, and finishing) used in textile processing.

Pollution Prevention Practices and Opportunities

Textile production involves a number of wet processes that may use solvents. Emissions of volatile organic compounds (VOCs) mainly arise from textile finishing, drying processes, and solvent use. VOC concentrations vary from 10 mg/m^3 for the thermosol process to 350 mg carbon/m^3 for the drying and condensation process. Process wastewater is a major source of pollutants. It is typically alkaline and has high BOD-from 700 to 2,000 milligrams per liter (mg/l)-and high chemical oxygen demand (COD), at approximately 2 to 5 times the BOD level. Wastewater also contains solids, oil, and possibly toxic organics, including phenols from dyeing and finishing and halogenated organics from processes such as bleaching. Dye wastewaters are frequently highly colored and may contain heavy metals such as copper and chromium. Wool processing may release bacteria and other pathogens. Pesticides are sometimes used for the preservation of natural fibers, and these are transferred to wastewaters during washing and scouring operations. Pesticides are used for moth-proofing, brominated. flame retardants are used for synthetic fabrics, and isocyanates are used for lamination The use of pesticides and other chemicals that are banned in some countries. Wastewaters should be checked for pesticides such as DDT and PCP and for metals such as mercury, arsenic, and copper. Air emissions include dust, oil mists, acid vapors, odors, and boiler exhausts. Cleaning and production changes result in sludges from tanks and spent process chemicals, which may contain toxic organics and metals. Figures 1 through 5 provide some examples of operations and sources of wastes typically found in a garment manufacturing operation. Pollution prevention programs should focus on reduction of water use and on more efficient use of process chemicals. Process changes might include those listed in Table 16.

Always try to seek simple solutions to waste problems. For example, the author audited a textile manufacturing plant in Macedonia (Tetex Corp.). The management were convinced that they needed a 6 million dollar modular incinerator to eliminate a solid waste problem (scrap fabric from cutting and weaving operations [see Figure 5]). The solution was a broker located in nearby Turkey that purchased 100% of the waste (more than 1,300 tons per year) at 30 cents per pound. Not only did Tetex not have to invest several million dollars in an operation that had tradeoffs between solid waste (nonhazardous) and hazardous air pollution from incineration), but they eliminated about $2,500 per year in transport and disposal costs (wastes were normally sent to Macedonia's only landfill more than 150 km away). A few simple phone calls eliminated the costs associated with this waste stream, and created an incremental revenue stream that more than paid for on-site management of the wastes.

Table 16. *P2 Practices and Recommendations for the Textile Industry.*

Match process variables to type and weight of fabric (reduces wastes by 10-20%).
Manage batches to minimize waste at the end of cycles.
Avoid nondegradable or less degradable surfactants (for washing and scouring) and spinning oils.
Avoid the use, or at least the discharge, of alkyl-phenol ethoxylates. Ozone-depleting substances should not be used, and the use of organic solvents should be minimized.
Use transfer printing for synthetics (reduces water consumption from 250 l/kg to 2 l/kg of material and also reduces dye consumption). Use water-based printing pastes, when feasible.
Use pad batch dyeing (saves up to 80% of energy requirements and 90% of water consumption and reduces dye and salt usage). For knitted goods, exhaust dyeing is preferred.
Use jet rivers, with a liquid-to-fabric ratio of 4:1 to 8:1, instead of winch rivers, with a ratio of 15:1, where feasible.
Avoid benzidine-based azo dyes and dyes containing cadmium and other heavy metals. Do not use chlorine-based dyes.

Use less toxic dye carriers and finishing agents. Avoid carriers containing chlorine, such as chlorinated aromatics.
Replace dichromate oxidation of vat dyes and sulfur dyes with peroxide oxidation.
Reuse dye solution from dye baths.
Use peroxide-based bleaches instead of sulfur and chlorine-based bleaches, where feasible.
Use biodegradable textile preservation chemicals. Do not use polybrominated diphenylethers, dieldrin, arsenic, mercury, or pentachlorophenol in mothproofing, carpet backing, and other finishing processes. Where feasible, use permethrin for mothproofing instead.
Control makeup chemicals.
Reuse and recover process chemicals such as caustic (reduces chemical costs by 30%) and size (up to 50% recovery is feasible).
Replace nondegradable spin finish and size with degradable alternatives.
Use countercurrent rinsing.
Control the quantity and temperature of water used.
Improve cleaning and housekeeping measures (which may reduce water usage to less than 150 m^3/t of textiles produced).
Recover heat from wash water (reduces steam consumption).

Figure 1. *Typical wool feedstock.*

Figure 2. *Wool washing operation.*

Figure 3. *Fabric washing machine.*

Figure 4. *Fabric batch dying machines.*

Figure 5. *Fabric wastes from garment manufacturing.*

For standard pollution control technologies - VOC abatement measures include using scrubbers, employing activated carbon adsorbers, and routing the vapors through a combustion system. A common approach to wastewater treatment consists of screening, flow equalization, and settling to remove suspended solids, followed by biological treatment. Physical-chemical treatment is also practiced: careful control of pH, followed by the addition of a coagulant such as alum before settling, can achieve good first-stage treatment. Further treatment to reduce BOD, if required, can be carried out using oxidation ponds (if space permits) or another aerobic process; up to 95% removal of BOD can be achieved.

Average effluent levels of 30-50 mg/l BOD will be obtained. Anaerobic treatment systems are not widely used for textile wastes. Carbon adsorption is sometimes used to enhance removal. In some cases, precipitation and filtration may also be required. Up to 90% recovery of size is feasible by partial recycling of prewash and additional ultrafiltration of diluted wash water. Disinfection of wastewaters from wool processing may be required to reduce coliform levels.

Residues and sludges often contain toxic organic chemicals and metals. These should be properly managed, with final disposal in an approved, secure landfill. Sludges containing halogenated organics and other toxic organics should be effectively treated by, for example, incineration before disposal of the residue in a secure landfill.

COAL MINING AND PRODUCTION

Industry Description and Practices

Coal is one of the world's most abundant energy resources, and its use is likely to quadruple over the next two decades. Coal occurs in a wide range of forms and qualities; but there are two broad categories: (a) hard coal, which includes coking coal, used to produce steel, and other bituminous and anthracite coals used for steam and power generation, and (b) brown coal (subbituminous and lignite), which is used mostly as on-site fuel. Coal has a wide range of moisture content (2-40%), sulfur content (0.2-8%), and ash content (5-40%). These can affect the value of the coal as a fuel and cause environmental problems in its use.

The depth, thickness, and configuration of the coal seams determine the mode of extraction. Shallow, flat coal deposits are mined by surface processes, which are generally less costly per ton of coal mined than underground mines of similar capacity. Strip mining is one of the most economical surface processes. Here removal of overburden and coal extraction proceed in parallel strips along the face of the coal deposit, with the spoil being deposited behind the operation in the previously mined areas. In open pit mining, thick seams (tens of meters) are mined by traditional quarrying techniques. Underground mining is used for deep seams. Underground mining methods vary according to the site conditions, but all involve the removal of seams followed by more or less controlled subsidence of the overlying strata. Raw coal may be sold as mined or may be processed in a beneficiation/washing plant to remove noncombustible materials (up to 45% reduction in ash content) and inorganic sulfur (up to 25% reduction). Coal beneficiation is based on wet physical processes such as gravity separation and flotation. Beneficiation produces two waste streams: fine materials that are discharged as a slurry to a tailings impoundment, and coarse material (typically greater than 0.5 millimeters) that is hauled away as a solid waste.

Pollution Prevention Practices and Opportunities

The main impacts of surface mining are, in general, massive disturbances of large areas of land and possible disruption of surface and groundwater patterns. In some surface mines, the generation of acid mine drainage (AMD) is a major problem. Other significant impacts include fugitive dust and disposal of overburden and waste rock. In underground mines, the surface disturbance is less obvious, but the extent of subsidence can be very large. Methane generation and release can also be a problem under certain geological conditions.

If groundwater systems are disturbed, the possibility of serious pollution from highly saline or highly acidic water exists. Impacts may continue long after mining ceases. In addition, underground mining operations are dangerous - in addition to cave-ins, underground explosions occur far to often in the industry. Beneficiation plants produce large volumes of tailings and solid wastes. Storage and handling of coal generates dust at rates of as much as 3 kg/t of coal mined, with the ambient dust concentration ranging from 10 to 300 micrograms per cubic meter above the background level at the mine site.

Table 17 provides an action list of pollution prevention practices applicable to the industry. The P2 practices essentially constitute a development plan which defines the sequence and nature of extraction operations and describes in detail the methods to be used in closure and restoration. At a minimum, the plan must address the issues listed in the table. Pollution prevention practices in this industry are often practiced in an irregular fashion. Although in the U.S. mining operations are considered quite advanced compared to standards only a few decades ago, coal mining in countries like Russia, Ukraine, and parts of Central and Eastern Europe not only lack simple environmental management practices, but do not meet universal safety standards.

Table 17. *P2 Practices and Recommendations for the Textile Industry.*

AREA OF OPPORTUNITY	POLLUTION PREVENTION RECOMMENDATION
Mining Operations	1. Removal and proper storage of topsoil. 2. Early restoration of worked-out areas and of spoil heaps to minimize the extent of open areas. 3. Diversion and management of surface and groundwater to minimize water pollution problems. Simple treatment to reduce the discharge of suspended solids may also be necessary. (Treatment of saline groundwater may be difficult.) 4. Identification and management of areas with high potential for AMD generation. 5. Minimization of AMD generation by reducing disturbed areas and isolating drainage streams from contact with sulfur-bearing materials. 6. Preparation of a water management plan for operations and postclosure that includes minimization of liquid wastes by methods such as recycling water from the tailings wash plant.
	7. Minimization of spillage losses by proper design and operation of coal transport and transfer facilities. 8. Reduction of dust by early revegetation and by good maintenance of roads and work areas. Specific dust suppression measures, such as minimizing drop distances, covering equipment, and wetting storage piles, may be required for coal handling and loading facilities. Release of dust from crushing and other coal processing and beneficiation operations should be controlled. 9. Control of the release of chemicals (including floatation chemicals) used in beneficiation processes. 10. Minimization of the effects of subsidence by careful extraction methods in relation to surface uses. 11. Control of methane, a greenhouse gas, to less than 1% by volume, to minimize the risk of explosion in closed mines; recovery of methane where feasible. When methane content is above 25% by volume, it normally should be recovered. Co-generation or use in natural gas vehicles are possibilities (*see Case Studies section in this chapter*). 12. Development of restoration and revegetation methods appropriate to the specific site conditions. 13. Proper storage and handling of fuel and chemicals used on site, to avoid spills.
Mine Closure and Restoration	14. Return of the land to conditions capable of supporting prior land use, equivalent uses, or other environmentally acceptable uses. 15. Use of overburden for backfill and of topsoil (or other plant growth medium) for reclamation. 16. Contouring of slopes to minimize erosion and runoff.

	17. Planting of native vegetation to prevent erosion and encourage self-sustaining development of a productive ecosystem on the reclaimed land. 18. Management of postclosure AMD and beneficiation tailings. 19. Budgeting and scheduling of pre- and post abandonment reclamation activities. 20. Upon mine closure, all shaft openings and mine adits should be sealed or secured.

CASE STUDIES

U.S. Auto Industry Pollution Prevention Practices

The U.S. pollution prevention auto program is a voluntary effort led by Chrysler, Ford, and General Motors. The objective of this dedicated program is to promote pollution prevention throughout business operations, products and practices, concentrating on reductions in the use, generation and release of toxic, persistent toxic substances. This program has been in the works for more than 12 years, and can boast the following accomplishments:

- 63% reduction in releases of Great Lakes Persistent Toxic Substances (GLPTS) on a vehicle produced basis (excluding foundry zinc releases);
- 46% reduction of EPA toxic release inventory (TRI) reportable releases on a vehicle produced basis;
- Millions of dollars in production costs and energy savings;
- Improved dialogue and relationships among government and industry;
- Recognition as a model of government and industry cooperation;
- Development of a forum of combined public accountability and reporting of pollution reduction efforts.

Table 18 provides a summary of aggregate data of pollution reductions achieved by pollution prevention measures since program inception in 1988. The data includes direct releases to air, land, and water, but does not include off-site transfers of GLPT and TRI reportable substances which were recycled or used as alternative fuels for energy recovery. Some interesting observations can be made from a review of the data reported in the table. Most notably, although reductions in pollution levels are huge, it took a significant time period for them to occur over. Critics of pollution prevention programs inevitably point to this drawback, and it is a fair criticism. But the same can be said for any other type of investment. If you invest your money in the stock market, or long-term cash deposits - then the analogy is virtually the same. Over time one is rewarded from the investment. Pollution prevention programs can indeed take years for significant savings to be achieved, but the savings are continuous, which is a point often overlooked. If we eliminate the cost of off-site disposal of a waste from a P2 project, then that savings continues year after year, for as long as we continue the manufacturing operation. Hence the P2 economic benefit is not a one time windfall.

Table 19 provides a P2 cost savings matrix. The chart provides examples of the types of P2 practices applied throughout the U.S. auto industry, along with the associated economic benefits. Table 19 only provides a few examples. A key point to remember is that the successes in this industry sector were achieved by many small to mid level investments. That is basically the best way that pollution prevention practices work. By implementing many small projects over a period of time, significant reductions in operating costs can be achieved. Pollution prevention just doesn't eliminate an emission or waste discharge. It has an economic benefit associated with it - namely the cost normally incurred in the waste or pollution management is eliminated or reduced. Additionally, there can be other value-added benefits such as improved yields, raw materials savings, energy savings, product quality improvements, and others. In many ways, the U.S. auto industry has been a leader in pollution prevention, and has developed unique approaches to sophisticated life cycle analysis methods.

Table 18. *Aggregate Data on Pollution Reductions Achieved by P2 Projects.*

Yr.	No. Vehicles Produced	GLPTs Released (kg)	Total TRI Released (kg)	GLPTs (kg) per Vehicle	% Reduct. GLPT per Vehicle Since 1988	TRI (kgs) per Vehicle	% Reduct. TRI per Vehicle Since 1988
'88	4,659,973	12,377,281	72,813,542	1.21		7.10	
'89	4,383,892	13,948,865	56,305,719	1.44	-19.0	5.83	17.9
'90	3,770,781	10,876,382	40,712,968	1.31	-8.3	4.90	31.0
'91	3,651,323	9,214,587	37,751,768	1.28	-5.8	5.24	26.2
'92	3,651,323	7,431,708	31,035,470	0.93	23.1	3.86	45.6
'93	4,014,858	7,556,456	31,651,643	0.85	29.8	3.58	49.6
'94	4,483,711	9,578,591	32,703,974	0.97	19.8	3.31	53.4
'95	4,290,037	10,996,754	31,394,933	1.16	4.1	3.32	53.2
'96	4,203,387	9,695,881	27,517,203	1.05	13.2	2.97	58.2
'97	4,312,009	9,061,561	26,858,280	0.95	21.5	2.83	60.1

Table 19. *Examples of U.S. Auto Pollution Prevention Cases Studies That Improved Bottom Line Performance.*

Target Substance(s)	Project Description	Results	Savings $
General Motors, Inland Fisher Guide (Livonia, Michigan)			
Pollution Prevention Project: *Solvent-Free Spray Adhesives for Interior Trim*			
VOCs: methylene chloride, methyl ethyl ketone, hexane, and toluene	Substitution with a solvent free adhesive for interior trim.	1. Substituted water-based adhesive eliminating 20 tons of VOC emissions; 2. Conversion of solid waste stream from hazardous to nonhazardous; 3. Annual disposal cost savings	26,000
General Motors Hamtramck Assembly			
Pollution Prevention Project: *Copper and Nickel Reclamation from Plating Waste*			
Copper and Nickel	Off-site reclamation of nickel and copper from nitric acid, made possible by a small tanker used to ferry the acid from an otherwise inaccessible storage tank to the reclamation facility's large tanker truck	Annual reclamation of 68 tons of copper and 40 tons of nickel, which would have been disposed of as hazardous waste sludge. Resulted in annual savings in wastewater treatment costs	23,000
General Motors Hamtramck Assembly			
Pollution Prevention Project: *Adjusting Paint Equipment Reduces Emissions and Solid Waste*			
VOCs and paint sludge (toluene, xylene, methanol, and butyl cellosolve acetate)	Paint spray equipment timing was fine-tuned, so that excess paint is no longer sprayed after the target body moves out of range	1. Source reduction of 5.5 tons of VOCs; 2. Reduction of 4 tons of paint sludge from paint overspray	85,000
Chrysler Newark Assembly Plant			
Pollution Prevention Project: *Glycol Ether Reduction in Surface Preparation Materials*			
Glycol ethers	Reformulation of a surfactant to remove glycol ether content.	1. 91 % source reduction of 71,920 kg of glycol ether 2. Reduced employee exposure.	500,000
Ford Utica Plant			
Pollution Prevention Project: *Molded Fiberglass Headliner Offal Reduction Project*			

Target Substance(s)	Project Description	Results	Savings $
Chalet cloth offal	Changed the specs for chalet cloth, fiberglass matte, and foam/glue composite used to manufacture headliners so as to produce less offal.	15 % (45 ton) reduction of offal generated per year.	430,000
Ford Assembly Plant Pollution Prevention Project: *Basecoat and Solvent Reduction Project*			
Paint and solvent	Fine-tuned the timing of a paint spray nozzle to improve its transfer efficiency.	1. Reduction of 12,862 kg of paint and solvent use per year; 2. Reduction in usage of paint sludge treatment chemical.	63,224
Chrysler Warren Stamping Plant Pollution Prevention Project: *Adhesive Waste Reduction*			
Waste adhesives	Inefficiencies in the adhesive pumping process resulted in the disposal of usable adhesive material. Increasing cost led to the identification and reduction of these inefficiencies.	1. Elimination of 50 m^3 of waste adhesive per year; 2. Improved quality.	500,000
Ford Norfolk Assembly Plant Pollution Prevention Project: *Paint Shop VOC Reduction Program*			
VOC Emissions from paint solvents	Paint area management asked suppliers to review and improve, where possible, the current method of solvent usage and to also review the spray booth cleaning procedure.	1. VOC solvent reduction of approximately 4 m^3 per week; 2. VOC reduction of 155,000 kg/year.	95,355
GM Fairfax Assembly Plant Pollution Prevention Project: *Turning off the Water Saves Millions of Gallons*			
Wastewater	Installation of a photoelectric cell connected to a timer to control the flow of rinse water.	Annual reduction of 25,000 m^3 of water.	33,000
GM Technical Center Pollution Prevention Project: *Reducing the Volume and Cost of Plating Wastes*			

Target Substance(s)	Project Description	Results	Savings $
Nickel, copper, cyanide, and zinc	Treated rinse water from the plating operations was reduced due to extending the retention time in the settling tank which improved the separation of water from heavy metal sludge.	Reduction of wastewater pump-out by 62.4 tons during the first year of implementation.	8,282

Miscellaneous P2 Practices by Ukrainian Companies

The author was involved with a number of programs throughout Central and Eastern Europe for the United States Agency for International Development (USAID), and World Bank missions which focused on developing in-country expertise in pollution prevention, and in assessing potential investment opportunities. Some examples of P2 projects in Ukraine are summarized in Table 20, along with the accompanying photographs referred to in the table. Table 20 simply provides some examples of the diversity of applications and range of potential savings associated with various projects among industry.

Among the Newly Independent States (NIS) of the former Soviet Union, it is relatively easy to identify many pollution prevention opportunities simply because there are few modern facilities and practices. Most technologies in this part of the world were generally copied from Western companies and often implemented in haphazard fashion during Soviet times. NIS enterprises, left with the remnants of old technologies and little in the way of modern controllers and real-time monitoring methods are forced to operate at low efficiencies, and very often with industry practices that died in the 1970s. Although some operations might be justified by this simply because of a large, cheap labor force, there are enormous losses in almost every industrial operation one visits.

A great deal of the inefficiencies among industrial operations are due to government's refusal to rationalize many industry sectors, and to raise utility prices to levels that are competitive. Russia has now had 10 years to implement its privatization program, and to a very large extent it has been a failure. Many companies still rely almost entirely upon government subsidies, or they don't pay their bills in many cases (or employees for that matter). If the price of raw water, energy, coal, heavy fuel oil are maintained at artificially low levels, and environmental enforcement is not practiced or environmental statutes imposed nonuniformly, then industries in these regions have few incentives to identify savings, let alone reduce pollution levels. If there are incentives, then they are largely external - such as attracting foreign investors, loans, or partners where the lending institutions or third part Western investors impose production efficiency measures as contingencies to partnering or loans.

Even when senior management understands the need for pollution prevention practices and that there are incremental cost savings for their operations, it is difficult to get management buy-ins into dedicated P2 projects. Management simply is focused only on large-scale infrastructure investments that they feel are needed in order to make the enterprise competitive in international markets. These types of investments are highly questionable for many reasons. Two major ones are - first, investments into old, gigantic industrial plants that are at best dinosaurs from a by-gone age doesn't make sense. Heavy polluting, old plants designed for 3 to 5 times the capacity that the market will bear today are destined to die. The reader may recall the vodka bottling plant example in an earlier chapter. This facility (in Russia), despite being shown the savings from water recycling, continued to press for finances to upgrade automated bottle washing machines - seeking 2 or more million dollars for larger capacity, more efficient machines. If this were an expanding business operation, that might be the proper decision - but unfortunately, the plant was only operating at around 30% of its nameplate capacity. Having the larger, more efficient machines will not make this facility more competitive - but reducing its manufacturing costs, especially through P2 projects, could help it to sustain operations for better times.

Second - the corporate mentality at many enterprises is still greatly different than that found in Western corporations. By this, we mean that quality principles (even in many companies that have gained off-shore ISO 9001 Certification) are not practiced across the board in a company. If the quality management principles are not present in an organization, then it is difficult to establish environmental management systems - and it is even more difficult to develop dedicated P2 programs.

Despite these criticisms, there are indeed some interesting and potentially profitable projects. The project financing for the projects listed in Table 20 are not provided as some of this information is of a proprietary nature. However, the reader can obtain some further details from USAID or by contacting the World Bank Organization. Not all the projects listed in Table 20 were actually implemented. In this example, approximately 12 million dollars is needed for the investment. The payback period is therefore on the order of 5 to maybe 7 years, when inflation, taxes and loan interest rates are taken into consideration. A problem however in investing in large scale projects like this is obtaining guarantees. Sovereign guarantee is the most frequent method of securing foreign loans from international lending institutions - but if a company has been privatized, as in the case of many of the examples in Table 20, then it must work with regional lending institutions that charge rates far above Western level commercial rates. This of course makes a potential large-scale P2 project even less attractive.

A major lesson that companies in this part of the world need to learn is that incremental savings derived from small-scale investments will collectively add up to sizable savings for an operation. In addition, many pollution prevention projects often have side benefits such as improved yield, improved product quality, energy savings associated with them. In many Western corporations it is still very difficult to sell P2 projects on the basis of improved worker productivity and reduced health risks. A semi-sophisticated total cost analysis needs to be done. But in the NIS, these types of less tangible incentives are virtually impossible to sell to senior management.

Table 20. *Examples of Miscellaneous Pollution Prevention Projects.*

Target Substance(s)	Project Description	Results	Yearly Savings $
Example I **Avdeyevka Coke Chemical Plant, Avdeyevka** Pollution Prevention Project: *Coke Oven Gas Reutilization Project*			
Coke Oven Gas (COG), carbon monoxide, NO_x, SO_x, and H_2S.	Replace practice of COG flaring by cogeneration by the installation of a gas turbine, which will enable electricity and steam generation for on-site consumption.	1. Elimination of flaring > 30,000 cmph of COG; 2. Enable the generation of 87,000 MW-hr of electricity per year; 3. Enable 108,000 Giga-Calories of Heat per year; 4. Reduces on-site SO_x emissions by 333 tons/yr and NO_x emissions by 3.1 tons/yr; 5. Eliminates dependency of off-site power purchase from a coal fired power plant. Off-site yearly emissions reductions are 1305 tons SO_x and 261 tons of NO_x; 6.Increased chemicals manufacturing of C_6 and H_2SO_4.	3,585,568
Example II **Azovstal Iron and Steel Works, Mariupol** Pollution Prevention Project: Graphite Recovery Project			

Target Substance(s)	Project Description	Results	Yearly Savings $
Graphite and molten slag vapors.	High quality graphite from sintering and slag operations were captured using baghouses and cyclone separators.	Approximately 200 tons per month of graphite recovered and sold into a local markets Worker exposure reduced.	108,000
Example III **Donetsk Iron and Steel Works, Donetsk** Pollution Prevention Project: *Waste Pickling Project*			
Spent sulfuric acid and sludge.	Modifications in the form of baffles and compartments increased residence time and mixing of sulfuric acid baths used in pickling operations.	A 25% reduction in wastewater to be treated Improved product quality. This increased the life of the baths by 25%.	12,000
Example IV **Donetsk Iron and Steel Works, Donetsk** Pollution Prevention Project: *Flue Gas Monitoring and Control*			
CO and Fuel Economy.	By use of a CO/O_2 analyzer, the plant open hearth furnace operations increase fuel economy by 15%.	Reduction in natural gas consumption by nearly 15% Reductions in CO by 5%.	18,000
Example V **Donugledegasatsia Coal Mines, Pavlovskaya** Pollution Prevention Project: *Coal Bed Methane Utilization*			
Natural Gas (methane).	By installing a compressor station, high purity methane gas is used to fuel natural gas vehicles (NGVs). A fleet of buses have been retrofitted as NGVs to transport coal miners to mining operations. *Refer to Figures 6 and 7.*	Replaces nearly 2,200 tons per year of diesel fuel used for motor transport, resulting in reductions in vehicle emissions of SO_x, NO_x, CO, CO_2, HCs, and PM. Reduces 20 to 30% of methane emissions into mine workings, thus creating safer operations. Eliminates direct discharge of over 2 million m^3 of methane (a greenhouse gas) to atmosphere from mine degassing practices.	201,000

Target Substance(s)	Project Description	Results	Yearly Savings $
Example VI **Druzhkovka Hardware Plant, Druzhkovka** Pollution Prevention Project: *Scrap Metal Recycling*			
Steel.	Plant-wide P2 program aimed at identifying recycling options. Scrap steel wire from bolt making operations stockpiled for long periods of time, and segregated from other wastes and sludge. *Refer to Figures 8-10.*	Reduced on-site landfill by 1100 tons per year by recycling steel to local steel works.	71,500
Example VII **Druzhkovka Hardware Plant, Druzhkovka** Pollution Prevention Project: *Waste Oil Recycling*			
Cutting oils.	Spent cutting oils from machine operations collected and sent to an on-site waste-water treatment facility. Installation of baffles on the oil-water separator - 35% increase in spent oil recovery achieved.	Approximately 500 barrels of spent oils were recovered and recycled into a local market.	6,000

Older plant operations are often-most the ideal candidates for pollution prevention programs, largely because obvious or simple solutions to waste and pollution problems have been overlooked. Quite often, these are good candidates for best practices, which tend to be low-cost to no-cost type P2 investments. Also, as seen by a number of examples here and elsewhere in our discussions, the use of simple instrumentation can help to achieve better control of a process.

The sampling of small to medium scale projects shown in the accompanying photographs help to illustrate the interest in P2 in this region of the world. But unfortunately, these tend to represent exceptions rather than normal practice at many enterprises.

Figure 6. *A modular compressor station that takes degassed methane from a local mine and uses it for fueling vehicles. Units like this can be situated in a number of locations. The major limitation to implementing the technology is a pipeline distribution system.*

Figure 7. *Gas cylinders being filled with degassed mine gas. The cylinders or bottles are the fuel tanks for buses. It is a relatively inexpensive conversion to retrofit a diesel engine bus to a NGV.*

Figure 8. *Shows hot steel rod being drawn from a furnace to be made into railway pins and bolts at the Druzhkovka Hardware Plant in Ukraine.*

Figure 9. *Shows railway pin production from the plant.*

Figure 10. *Steel rod feedstock to the plant.*

Pollution Prevention in The Food Processing Industry

This case study provides an analysis of a pollution prevention project conducted at a meat processing plant in Estonia. The pollution prevention demonstration included several small scale investments that collectively added up to sizable savings and pollution reductions. The program was conducted as a part of USAID's waste minimization programs in the CEE region. The company that participated in this project was the Rakvere Meat Processing Plant, located in Rakvere, Estonia. The plant operations were started in 1990 and is a full line red meat processing operation for beef, pork and some lamb. The plant operations include:

- slaughtering with separate pig and cattle processing lines
- manufacturing of edible by-products including fresh cut packaging, cooked hams, and sausages;
- a technical department for by-products processing, water treatment, and wastewater treatment.

The plant has an average daily production capacity of 42 tons of sausage products and 17 tons of semi-manufactured products.

To illustrate the benefits of low-cost pollution prevention measures, a rapid in-plant assessment was made. The in-plant assessment consisted of a team of engineers performing a walk-through of the plant operations, with the intent of identifying one or more opportunities for reducing wastes, pollution, saving raw materials and energy, and improving yields and efficiencies. These opportunities are then quantified by performing material and energy balances, from which dollar savings and returns on investments can be calculated for a proposed technical solution.

In this case study, four pollution prevention opportunities were identified in various stages of the manufacturing operation, all aimed at wastewater reductions. In each case, preliminary estimates were made as to estimated pollution reductions and estimated cost savings associated with eliminating the waste or pollution problem. It was recognized at the start that no one opportunity would be a significant savings or pollution reduction, but that collectively, the emissions reductions and possible savings could be significant.

Individual opportunities to reduce wastewaters and pollution, and achieve savings were identified from the pollution prevention audit. The individual opportunities are briefly described below.

Reduction of BOD in wastewaters by recycling waste products to farmers - At Rakvere inedible solids from the slaughtering process are rendered in the technical department. The wastewater discharge from the fat separation step known as "glue water", has a fat content ranging from 8 to 30 percent. In the past this wastewater was discharged to the wastewater treatment facility contributing to high biological oxygen demand (BOD) loadings. During the waste minimization project an evaluation was performed to identify opportunities to reuse wastewater streams with high BOD. Using a chemical oxygen demand (COD) analyzer it was determined that the glue water stream had a high COD content and was a good candidate for recovery and reuse. A decision was made to install a pump to divert and recover the 'glue water" for use by local farmers as feed. As a result of the project the facility reduced COD levels in its wastewater by 54 tons of oxygen/year, and thereby capturing savings from reduced pollution fees.

Recovery of organic solids and reduction in water consumption with low cost process modification - At Rakvere processing of cattle stomachs to recover salable products such as tripe requires extensive handling and cleaning of stomachs which contain manure. In the past the cattle stomachs underwent two washes after removal of manure. To increase the value of the product a third wash step was added. The wastewater from all washing steps were sent to a manure press to concentrate solids before discharge to the wastewater treatment plant. The stomach cleaning and washing procedure was performed on a flat surface which increased the amount of handling and water required.

During the waste minimization project an evaluation was performed to identify opportunities to reduce the volume of wastewater streams with high biological oxygen demand (BOD). Using a chemical oxygen demand (COD) analyzer it was determined that the wash water from cattle stomach cleaning operations had a high COD content and was a good candidate for reduction. A decision was made to construct a "birdcage" on which cattle stomachs could be stretched during washing. This device facilitated handling by exposing more stomach area for cleaning, and it reduced water use. In addition, screens were added to the wash drums to recover organic solids from the final two wash steps. As a result of the project the facility reduced COD levels in its wastewater by 9 tons of oxygen/year and reduced fresh water consumption significantly.

Implementation of a water conservation program and good housekeeping practices conserves raw water supplies - Cleanup of the slaughtering lines involved extensive use of water to hose solids into sewers leading to the wastewater treatment plant. Many hoses were not equipped with shutoff nozzles and were left running when not in use. These practices resulted in excessive water use and high operating costs. During the waste minimization project an evaluation was performed to identify opportunities to reduce the volume of wastewater streams with high biological oxygen demand (BOD). Using a chemical oxygen demand (COD) analyzer it was determined that the wastewater generated from cleanup of slaughter lines had a high COD content and was a good candidate for reduction.

To accomplish this a training program for supervisors and workers was implemented to emphasize the importance of water conservation and good operating practices. In addition, dry cleanup procedures using squeegees and clean as you work methods were introduced. Finally a program was undertaken to ensure timely repair of leaking nozzles and replacement of missing nozzles. As a result of the project the facility reduced water usage by 316 800 m^3/year.

All of the P2 activities conducted at the enterprise can be described as low cost investments. The principle investment made was in a spectrophotometer, that served as the key analytical instrument for water quality testing. Table 21 provides a pollution prevention matrix, which summarizes the reductions of wastewaters and the savings that were achieved based on dollars. In total, $13,411 was invested, for which $4,400 was spent on the spectrophotometer alone. The collective results were 84 ton/yr reductions in BOD, 373,009 tons/yr reduction in raw water consumption, and a yearly dollar savings of $149,200.

This case study illustrates how one instrument investment can be applied in a versatile way to capture a number of pollution reductions and dollar savings. The collective financial benefits derived and pollution reductions achieved from the P2 activities performed at the meat processing plant are significant. They help to illustrate that by applying pollution prevention practices to a number of small scale problems, sizable savings and emissions reductions can be achieved over time, and with almost immediate payback periods.

Table 21. *Pollution Prevention Matrix for Wastewater Reductions.*

Pollution Prevention Measure	Reductions in tins per year		Inves t-ment $	Savings $	Payback Period (months)
	COD	Waste-water			
Reduction of BOD in wastewaters by recycling waste products to farmers	54		2,447	11,600	<3
Recovery of organic solids and reduction in water consumption with low cost process modification	9	24,221	2,847	9,500	<4
Implementation of a water conservation program and good housekeeping practices conserves raw water supplies		348,78 8	5,117	114,00 0	<1
Changes in operating practices and employee training help to reduce BOD and solid waste	21		3,000	14,100	<3
Totals	**84**	**373,00 9**	**13,41 1**	**149,20 0**	**1**

Many CEE companies believe that the only way they can become more profitable, or to sustain their operations during difficult market conditions is by making large scale investments that modernize their facilities. This case study illustrates that pollution prevention based on low levels of investments, including good housekeeping practices can result in sizable savings that can redirect operating revenues to the modernization projects needed.

Chapter 9
Epilogue

INTRODUCTION

This final chapter provides some general comments and guidelines on applying the principles and tools provided in the handbook. Its purpose is to provide the reader with some additional areas of consideration in applying pollution prevention principles on a company-wide basis.

THE AUDIT

Remember that the audit is a tool. The twenty steps outlined in Chapter 4 are general, and the applicator should decide which steps really apply to their specific case or exercise. Forward planning is essential in order to apply this tool effectively to the total cost assessment. The areas that you should give consideration to include:

Developing a Set of Clear Objectives for The Audit

It's not enough to say, let's go out into the plant and find some savings or pollution prevention opportunities (*although the author has been given such nebulous assignments from time to time*). The right approach is to prioritize the environmental issues first, and then tailor the audit so that the most effective and useful information can be gathered. A meeting with production and environmental people within an organization should quickly identify general areas of concern from an environmental management or pollution standpoint, and areas where potential cost savings might be achievable. For example, if we are manufacturing HDPE in a batchwise solution process, we are faced with the disposal of unused and spent catalysts (most notable EADC [ethyl aluminum dichloride]). This material can be pyrophoric and if disposal is off-site, then we have a transport and disposal cost issue, as well as a compliance issue (under RCRA it's a regulated hazardous waste). A possible objective of the audit would be to see if there are operational changes to reduce the volume of off-spec or unused EADC. The meeting should focus on whether or not its worth the effort to conduct an audit. It may be that the volume of waste generated is infrequent and that the cost for T/D are small. On the other hand, even if current disposal costs are small, there may be a liability issue. As an example, a third party responsible for disposing of this waste has an explosion and fire, and a personal injury results. Even if the generator complies totally with the RCRA rules on proper labeling, waste characterization, and dealing with licensed transporters and disposal facilities, it will ultimately be dragged into a legal dispute. Does your company need this degree of exposure and can it deal with it? That in itself may be a justification for implementing an audit. We can quickly come to an understanding as to what the "ballpark" economic incentives are for better managing a pollution or waste problem, and further we can certainly develop a list of priorities in terms of multiple waste/pollution streams.

This forward planning exercise accomplishes two things. First and foremost, it provides the basis for senior management buy-in, and secondly - it will enable us to tailor the steps and spreadsheets needed for data collection and post analysis.

Getting Management Buy-In

It is essential from the start to have senior management support or buy-in. Without management support and a clear understanding of the potential savings and priorities, the greatest, most cost-effective project in the world goes nowhere but into a final company report. A meeting with senior management (section head, division manager, general manager, vice president, CEO - whoever makes the decision or can help promote the program at a senior level) must be given a clear understanding of what the status quo is and what may be the potential benefits derived if a suitable P2 alternative is identified. Under no circumstances should one oversell or overstate incentives on what might possibly be achieved. At this stage, we are only asking at best to commit resources to implement a study (audit and total cost assessment) to review possible alternatives. A possible outcome may very well be to maintain status quo in operations, and this should be stated. On the other hand, if we argue for a limited, rapid assessment to obtain better than an order of magnitude estimate of savings/incentives, we have a good chance of being given priority and resources to conduct the audit and analysis. For management to make the decision to commit, we must also be prepared to define what the level of effort is. Is it simply a review of records and one or more walkthroughs and interviews, or do we need to commit to some limited plant trials. If there are plant trials needed to get at certain data or information, how do they impact on operations? Are there downtimes associated with trials? Is manpower/operator time diverted away from normal production activities? Can impacts be minimized by performing trials during normally scheduled turnaround times? The more up front thinking and logical layout of the potential benefits and the resources needed to identify ways to capture the benefits, the better chance we have of getting the resources needed to implement the audit effectively.

The Exit Meeting

From a protocol standpoint, there really are two distinct meetings with management. One is at the start, where we get management support. The second we can call an exit meeting. The exit meeting is sort of a judgement call as to when it should be conducted, and in fact it may be a series of meetings. The first opportunity for the exit meeting is at the conclusion of the audit itself, at which point preliminary conclusions and recommendations have been arrived at. This keeps management in the loop, and further defines how much effort remains to arrive at a final recommendation. Once the project cost or total cost accounting analysis has been completed, we certainly want to present the results to management and go the next step. The next step being - if a pollution prevention project can be economically justified using the principles and criteria outlined in Chapter 3, we argue for the funds to implement the project.

The Audit Team Mix

An effective P2 audit is a team approach. Although consultants like to sell themselves as the experts, the facts of life are that no-one knows it all! Again, look at the objectives of the audit and the specifics of the part of the plant where the audit will be conducted. Although it is dangerous to generalize, many P2 opportunities fall into the category of low cost/no cost, which often can be addressed by simple operational changes, or investments into instrumentation or even monitoring equipment. Therefore, a process engineer with a strong background in instrumentation is a good candidate for the team. In the author's opinion, it is a big mistake to rely heavily on the environmental engineer. He or she is an important resource to consult, but more often this individual is more versed in compliance and control technologies and practices. The team needs to really focus on resources that understand what the "cause and effect" parameters are in the process. That is quite often a process or operations engineer.

If the focus of the audit is on raw materials handling, then of course we need an inventory or materials handling specialist. If the focus is on identifying a less toxic or hazardous substitute for a raw material, then we may need a chemist or a chemical engineer as a part of the team. And if, through the course of the audit we identify that a major or high level of investment may be needed, then either an experienced cost estimator and/or business analyst may be called in for consultation - but certainly not early on in the audit.

The proper mix of specialists will result in some interesting brainstorming sessions during analysis and post-analysis sessions of the audit, and quite often, very simple opportunities that can result in significant savings are identified.

Having the proper mix specialists will help to ensure that a wide range of alternatives are identified.

DEVELOPING A TRACK RECORD

There are very few examples where pollution prevention has made a single contribution to the bottom line performance of a business, let alone industry, on the basis of a single project or green technology. The U.S. auto industry (the Big Three - Ford, General Motors, Chrysler) boast of enormous savings achieved over the last 15 to 20 years from pollution prevention practices. Their successes however are from a string of projects - many of which are quite small. Collectively, over a substantial period of time, P2 programs have shown strong returns on investments in this industry sector.

Tracking the performance of pollution prevention projects that are implemented is essential. Without formal tracking, documentation and reporting the achievements - no matter how small or incremental, there really is no program. A *Pollution Prevention Program* means having dedicated resources in place to identify continual improvements in savings through reductions in all forms of waste (not just pollution). The only way that continual improvements can be made is by monitoring the results of a project(s). From these monitoring results we can identify such needs as to whether we should increase resources for more frequent audits, to expand audits to other operating divisions or environmental priorities, to identify a next generation of improvements or modifications to a particular project or process that can provide even greater savings, and even to justify that status quo is perhaps the best option and should not be changed unless future environmental regulations and or business plans make it necessary. Tracking and reporting or publicizing the results helps to gain further management commitment as well as support by the staff. The use of the pollution prevention matrix, for which a number of examples have been offered in case studies throughout the book, is a simple way to keep track of raw materials savings, energy savings, water consumption reductions, reduced emissions, wastewater discharges and solid waste disposals, and actual dollar savings. The matrix approach enables us to quickly identify the ROIs and payback periods for projects.

DEVELOPING CORPORATE PHILOSOPHY

P2 should not be the occasional practice of a corporation. It must be interwoven into the corporate environmental policy, the company's environmental statement, into the actual environmental management plans and the management system itself, which means it must be part of the overall business practices of an organization. This can only happen when senior management understand the benefits and commits dedicated manpower and other resources to continual improvement methods. To develop this corporate philosophy and to apply P2 practices, two things must happen. First, corporate mentality needs to be forward thinking or pro-active in addressing environmental issues. Management must believe, or at the very least, commit resources and efforts to ascertain the specific benefits that can be achieved. Hence, the importance of tracking the performance of P2 projects cannot be overstated. Collective achievements of such projects provide over the long run, provide further momentum for expanding and improving on approaches to implementing P2 programs.

The second important element is employee buy-in. This can be a difficult sell in many organizations. Training programs on the value of P2 projects/programs can certainly help spread the need for the commitment. A lessons-learned approach is most effective - i.e., providing actual case studies or practical examples is a good way to convince employees of the value of practicing P2 and in making voluntary suggestions that could feed into technical solutions for cost savings and a waste or pollution reduction. Publicizing the achievements of P2 projects in company newsletters and posting P2 matrixes that summarize savings helps to raise employee awareness, and to get personnel thinking of innovative approaches to improving the efficiencies of their areas of responsibility within the plant operations.

Management also needs to consider employee incentives as well. Peer recognition, external publications of project and even individual employee achievements in the area of pollution prevention can be small but important rewards/incentives. The approaches of a bonus to an individual or group of individuals, even in the form of stock shares, cash, payed time-off are even higher levels of incentives that are worthy of consideration. If one good suggestion saves a company 200,000 dollars a years for the next ten years - why not reward the inventor(s)?

A very good exercise to convince an organization of the need for P2 practices is a *Gap Analysis*. The gap analysis is most often applied to assess either the due diligence of a company, or among EMS specialists - to determine what is needed in order to achieve ISO 14000 certification or voluntary adoption. However, in this case, we suggest applying the principles of a gap analysis to assessing how well your company is performing compared to the rest of the industry sector in terms of environmental, waste, raw materials, and energy costs. There is a wealth of well publicized information on industry profiles - a lot of which is on the Internet. Industry averages on units costs of production and simple emission factors are readily available. A simple table of key areas that consider unit cost factors such as energy expenditure per unit production, yearly average quantities of priority pollutants per unit production (one type of emission factor), water usage per unit production, raw materials consumption per unit production, employee health insurance costs per unit production, and other factors you may think of, can give one a sense for how well your organization is doing compared to the rest of the industry. If your organization is far below the industry averages, then perhaps the current practices and your EMS need only some refinement, but if indeed there are areas that stand out, such as high costs for raw materials, excessive water consumption, high costs for solid waste management or other forms of pollution control, etc. - then these are areas where pollution prevention activities are likely to benefit.

FINAL REMARKS

Pollution prevention works well if it is applied uniformly and in a dedicated fashion throughout a company or organization. The identical approach to developing quality management principles within a company apply to EMSs, of which P2 is an integral part of. Unfortunately, there are far too many examples of organizations that claim to have an EMS, and even have achieved ISO 14000 certification or self declaration, that simply do not practice P2. Pollution prevention is far from being practiced on a global basis, and indeed, in many countries that are undergoing a transition in their economies, including privatization of industries, simply do not have a fundamental understanding of what P2 truly means and how it can be applied. To many, P2 is recycling, product recovery - and yes, these could be viable projects that fall within the realm of P2 practices. But the subject goes far beyond the occasional cost savings measure, as we have tired to illustrate in this volume. The tools presented in this handbook are meant to be illustrative, and should be tailored to the needs of a particular organization. The case studies and industry profiles should provide some ideas that may be applicable to your own organization.

INDEX